4th International Conference on Artificial Intelligence and Computer Science (AICS 2022)

Journal of Physics: Conference Series Volume 2363

Beijing, China
30-31 July 2022

ISBN: 978-1-7138-6471-4
ISSN: 1742-6588

Printed from e-media with permission by:

Curran Associates, Inc.
57 Morehouse Lane
Red Hook, NY 12571

Some format issues inherent in the e-media version may also appear in this print version.

Institute of Physics
Dirac House, Temple Back
Bristol BS1 6BE UK

Phone: 44 1 17 929 7481
Fax: 44 1 17 920 0979

techtracking@iop.org

Additional copies of this publication are available from:

Curran Associates, Inc.
57 Morehouse Lane
Red Hook, NY 12571 USA
Phone: 845-758-0400
Fax: 845-758-2633
Email: curran@proceedings.com
Web: www.proceedings.com

TABLE OF CONTENTS

Preface

2022 4th International Conference on Artificial Intelligence and Computer Science (AICS2022) has been held successfully on July 30-31, 2022. Due to the COVID-19 pandemic around the world, and with the strict travelling rules in China, it is still difficult to take international travel for our attendees abroad. Therefore, AICS2022 was held both in physical (Beijing, China) and online (Zoom).

About 80 attendees have participated in the conference. The conference was divided into three parts: five keynote speeches in the first day, each with 45min, and 18 oral presentations followed in the first afternoon and the second day, each with 15min. Besides, 20 e-posters have been presented in the conference website. The session chairs and keynote speakers have worked together to selected the best oral presentation and poster presentation.

This year, AICS2022 has attracted about 90 submissions, and 38 of them are accepted to be included in this collection. The process of evaluation and peer-review has last for about 6 months, with over 50 TPC members and reviewers involved in. We would like to express our sincere gratitude to all experts for their professional and detailed review comments, which have contributed to the high standard of this collection.

This collection includes three major parts: (1) Algorithms and Programming, (2) Machine Learning and Computer Vision; (3) Artificial Intelligence and Innovations, with most up-to-date findings and technologies being discussed. We believe participants of AICS2022 have gained knowledge as well as met new friends in the same field. And we hope this collection can benefit the readers in the same way.

General Chair
Habib Zaidi, Geneva University, Switzerland
Program Chair
Huiyu Zhou, University of Leicester, UK
Publication Chair
Hamid Reza Karimi, Politecnico di Milano, Italy
Yuriy S. Shmaliy, Universidad de Guanajuato, Mexico
Organizing Chair
Wenbin Zhang, Carnegie Mellon University, USA

Technical Program Committee
Alessandro Sebastian Podda, Università degli Studi di Cagliari, Italy
Ali Al-Sharadqah, California State University, Northridge, USA
Anabela Moreira Bernardino, Polytechnic Institute of Leiria, Portugal
Anand Nayyar, Duy Tan University, Vietnam
Arun Agarwal, Siksha 'O' Anusandhan University, India
Aswani Poosapati, University of Maryland Baltimore County, USA
Basel A. Mahafzah, The University of Jordan, Jordan
Biju Theruvil Sayed, Dhofar University, Sultanate of Oman
Bin Xue, National University of Defense Technology, China
Cheng Siong Chin, Newcastle University in Singapore, Singapore
Chenyu Wang, Beijing University of Posts and Telecommunications, China
Cong Pu, Marshall University, USA
Dragana S. Krstic, University of Nis, Serbia
Fan Yang, Walmart Global Tech, USA
Gen Chen, Guangdong University of Finance & Economics, China
Hong Lin, University of Houston-Downtown, USA
Hongyang Chen, Zhejiang Lab, China
Houbing Song, Embry-Riddle Aeronautical University, USA
Houda HARKAT, Instituto de Telecomunicações (IT) – Lisboa, Portugal
Hyunsung Kim, Kyungil University, Korea
James Lu, Genentech, USA
Jayachander Surbiryala, University of Stavanger, Norway
Ji Zhang, University of Southern Queensland, Australia
Jianping Luo, Shenzhen University, China
Jun Liu, Carnegie Mellon University, USA
Liang Luo, University of Electronic Science and Technology, China
Loc Nguyen, Loc Nguyen's Academic Network, Vietnam
M. Hassaballah, South Valley University, Egypt
Marran Aldossari, Shaqra University, Saudi Arabia
Masoud Rahiminezhad Galankashi, Imam Khomeini International University, Iran
Mengyu Wang, Harvard University, USA
Min Shi, Harvard University, USA

Peer Review Statement

All papers published in this volume have been reviewed through processes administered by the Editors. Reviews were conducted by expert referees to the professional and scientific standards expected of a proceedings journal published by IOP Publishing Publishing.

- **Type of peer review:** Double Anonymous
- **Conference submission management system:** Morressier
- **Number of submissions received:** 89
- **Number of submissions sent for review:** 80
- **Number of submissions accepted:** 38
- **Acceptance Rate (Submissions Accepted / Submissions Received × 100):** 42.7
- **Average number of reviews per paper:** 2
- **Total number of reviewers involved:** 27
- **Contact person for queries:**
  Name: Amanda Chen
  Email: amandachen@easyace.cn
  Affiliation: Hubei Zhongke Institute of Geology and Environment Technology

APInf: Adaptive Policy Inference Based on Hierarchical Framework

Hongjie Zhang[1,2], Shuo Wang[2] and Jing Li[2]

[1] School of Computer Science, Sichuan Normal University, Chengdu, Sichuan, China
[2] School of Computer Science and Technology, University of Science and Technology of China, Hefei, Anhui, China
Email: zhanghongjie@sicnu.edu.cn; shuowang@ustc.edu.cn; lj@ustc.edu.cn

Abstract. Deep reinforcement learning has made adequate progress in addressing complex visual tasks. However, policy inference in relatively large neural networks is costly. Inspired by the fact that some states in a specific task can easily make decisions, we propose an APInf framework that can achieve low policy inference costs with quality guarantees. Two-stage training is proposed: First, to accelerate inference at easy states, sub-policy networks ordered by increasing capacity are generated; Second, to keep the quality of policy, master-policy is designed to choose a suitable sub-policy network subjected to state difficulty dynamically. We also improve the efficiency of master-policy inference by sharing the convolution layer between sub-policy networks and master-policy and then train master-policy under an extended MDP. Extensive experiments conducted in gym show that the adaptive inference framework reduces the FLOPs by 41.2% while maintaining similar quality. Furthermore, APInf outperforms the existing dynamic network structure in terms of policy quality and inference cost.
Keywords. Deep reinforcement learning; dynamic neural network; neural network compression; adaptive policy inference.

1. Introduction

Deep Reinforcement Learning (DRL) combines a deep neural network (DNN) with reinforcement learning to make decisions in various tasks such as video games [1], auto-driving [2], financial investment [3], and so on. Specifically, the agent makes an action according to the current state of the environment, and the environment changes its state and returns a reward. The goal of DRL is to learn the state-action function, which maximizes the cumulative reward. So the task completion time is composed of agent inference time and environment progress time [4]. However, using complex neural networks to solve challenging tasks will increase policy inference costs, resulting in long task completion time in most cases. Thus, saving inference time without sacrificing policy quality is vital.

To save the prediction time of DNN, many acceleration techniques are proposed. They can be classified into two categories: One is making architectural changes by neural network pruning [5-9] or parameters quantization [10-13]; The other is the dynamic neural networks method to reduce the average inference time. Orthogonal to architectural changes, it utilizes more computation for "harder" inputs and fewer budgets for "easier" inputs [14-16]. Our work is based on the dynamic neural network, which aims to reduce the inference cost of the policy in DRL tasks. Recent dynamic neural network technologies are mainly used in image processing (classification, image segmentation, etc.), which try to distinguish the difficulty of each input based on its content and vision details.

(a) Image classification in the real world. The left image is "hard" and the right image is "easy".

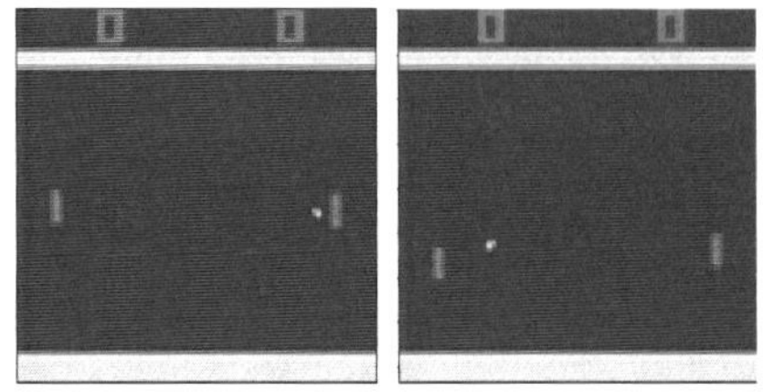

(b) Different states in game pong. The left image is "hard" and the right image is "easy".

Figure 1. The difficulty level between image classification and atari game pong.

For example, the left image in figure 1a is more difficult than the right one because of its complex texture and features, and the dynamic neural network will use more computations (larger neural network) for the left image [14]. However, the dynamic method cannot be applied easily to reinforcement learning tasks. Three challenges exist in DRL tasks. First, the difficulty level is hard to distinguish since the visual content per state is similar. For example, the states in figure 1b have similar content. But the left state in figure 1b is difficult because it shows that the "ball" is approaching the agent, in which case the agent should choose a smarter policy to fight back against the "ball". On the contrary, the right state in figure 1b is easy because it shows that the "ball" is approaching the enemy, in which case the agent can do any actions without influencing the final score. Second, there is no ground truth and confidence level in each state to adopt, which are necessary for dynamic networks. Third, future influence should be taken into consideration when different inference process flow is selected. Therefore, it is vital to model the dynamic neural network as a reinforcement learning problem, which simultaneously considers both the task's goal and the image content.

To overcome the above challenges, we propose a novel Adaptive Policy Inference (APInf) framework based on the hierarchical framework in this work. The APInf likes the hierarchical DRL framework and contains two levels of policies, including the lower-level sub-policies and the high-level master-polices. First, the sub-policies are designed to accelerate inference at easy states. Given a pre-trained "teacher" policy, we train multiple sub-policy networks of increasing capacity using policy distillation [17]. Second, the master-policy is designed to keep the quality of APInf. We train the master-policy to learn how to choose a sub-policy based on the current state. Specifically, we use the policy gradient to train the master-policy, which minimizes the inference time under the constraint of cumulative rewards. Furthermore, we share the feature extraction layer (e.g., the convolutional layer) with the sub-policy to reduce the inference cost of the master-policy. However, the master-policy must choose the sub-policy before the sub-policy executes. To overcome this challenge, we propose the extended Markov Decision Process (eMDP) that utilizes the previous state and action to predict which sub-policy is used in the current state effectively.

The contributions in this paper can be summarized as follows:

(1) We design a two-stage training scheme on DRL tasks using the dynamic neural network method, which can accelerate the policy inference while maintaining similar quality. As far as we know, our paper is the first work that introduces the dynamic neural network method to DRL acceleration.

(2) The general framework APInf is proposed to implement the two-stage training scheme. It can generalize well to neural network compression methods and DRL algorithms.

(3) We evaluate APInf's performance on atari tasks, which shows that APInf saves about 41.2% floating-point of operations (FLOPs) and 22% inference running time while maintaining similar quality.

2. Methodology

2.1. Overview

Figure 2 shows an overview of the proposed APInf. It contains two main modules. The first one is the sub-policy generator, which produces the sub-policies by using policy distillation. The second one is the master-policy, which selects a proper sub-policy for inference. Intuitively, an independent master-policy network can be used to select sub-policies according to the current state. However, the inference cost of the master-policy network becomes the bottleneck. A variety of master-policy network models are designed and tested, including the classic convolutional neural network (CNN), CNN + attention, and Binary CNN. The result shows that the inference cost is about 31% higher than that of APInf. To save the inference cost of master-policy, we share the feature extraction layer with sub-policy (as shown in figure 2). Since the sub-policy networks are separated from each other and have different structures, multiple master-policy networks are required, which is corresponding to the sub-policy networks. However, the master-policy must choose which sub-policy k_t to use in the current state s_t before that sub-policy runs. To overcome this challenge, the master-policy predicts k_t based on the embedding e_{t-1} of state s_{t-1} and action a_{t-1} on the previous step. The next sections will explain these two modules in detail.

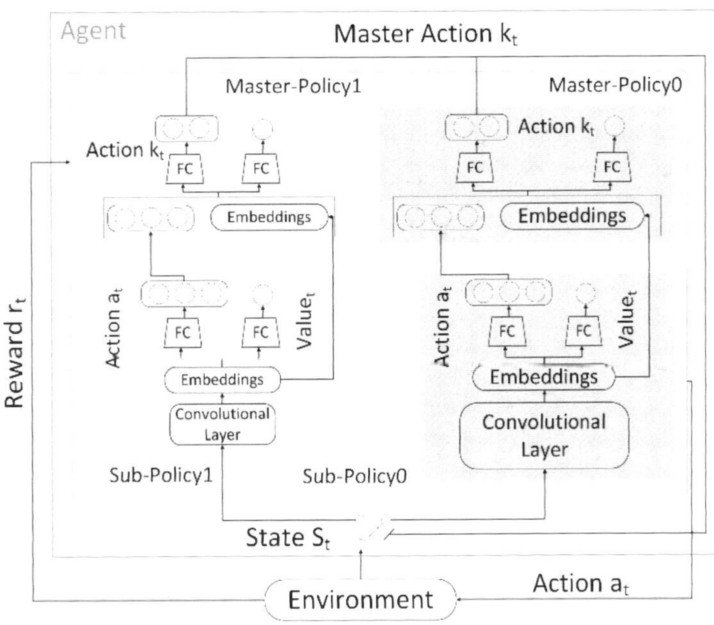

Figure 2. An overview of the APInf framework.

2.2. Sub-policy Generator

Our first training stage is the sub-policy generation process based on policy distillation. As aforementioned, we already have a pre-trained optimal policy called "teacher". All sub-policies are compressed by channel pruning and layer deletion from the teacher network. The sub-policy mimics the outputs of the teacher network. And the sub-policy is represented by π_θ with parameter θ. Specifically, we use superscript π_θ^k to identify different sub-policy. $Loss_{PG}$ is the loss function of policy gradient. α_0 and α_1 are hyperparameters that control the influence of the teacher. The loss function of sub-policy generation is defined as equation (1).

$$Loss(\theta) = Loss_{PG} + \alpha_0 KL(\pi_0(\cdot \mid s_t) \,||\, \pi_\theta(\cdot \mid s_t)) + \alpha_1 MSE(V_0(s_t), V_{\pi_\theta}(s_t)) \quad (1)$$

Using policy distillation brings an advantage that the sub-policies have a similar action distribution. In this way, a smaller sub-policy can be selected to predict action at most states in APInf framework.

2.3. Master Policy

The details of the master policies are shown in figure 2. All policy networks are Convolutional Neural Networks (CNN). The master-policy network shares the convolutional layers with sub-policies. In detail, each sub-policy has its master-policy branch, which is trained separately during the second stage.

Formal problem definition. We define master-policy learning as a constrained policy optimization problem in equation (2). C_{master_t} is the inference cost of the master-network, and C_t is the inference cost of the sub-network selected by master-policy at time step t. V_{mix} represents the state-value of mixed policy π_{mix}, which is defined in equation (3). V_0 is a constant value that represents the state-value of the teacher policy. π_{master_t} is the optimized policy and π_k represents the sub-policy k. We assume that the capacity of the sub-network is decreasing with the increment of the index number k. The formal definition means that the goal is to minimize the total inference cost while keeping a similar final state value as the teacher policy.

$$\begin{aligned} \min \quad & \mathrm{E}_{\tau \sim p(\tau)}\left[\sum_{t=0}^{\infty} \gamma^t (C_{master_t} + C_t)\right] \\ s.t. \quad & |V_{mix} - V_0| \leq \varepsilon \end{aligned} \quad (2)$$

$$\pi_{mix}(\cdot \mid s_t) = \sum_{k=0}^{n} \pi_{master_t}(k \mid s_t) \pi_k(\cdot \mid s_t) \quad (3)$$

We assume that $V_{mix} = \mathrm{E}_{\tau \sim p(\tau)}[\sum_{t=0}^{\infty} \gamma^t r_t]$ is always less than V_0. The unconstrained optimization problem is derived using the Lagrange multiplier in equation (4). r_t is the reward of the environment at time step t, and α is a hyperparameter that controls the strength of inference cost. When α equals 0, the master-policy only considers maximizing the cumulative rewards (e.g., always choose π_0). As α gets bigger, the master-policy tends to minimize the inference cost while decreasing the cumulative rewards. Users set different α at various practical tasks and devices. For example, when we deploy policy on the embedded device, it tends to set a larger α to improve speed and sacrifices a little policy quality.

$$\max \quad \mathrm{E}_{\tau \sim p(\tau)}\left[\sum_{t=0}^{\infty} \gamma^t (r_t - \alpha(C_{master_t} + C_t))\right] \quad (4)$$

Extended MDP. The action space of master-policy is growing with the number of sub-policy. The reward function becomes complex, including the original reward and inference cost. To reduce the inference cost of master-network, we share the feature extract layer with sub-networks, as shown in figure 2. We define the master-policy as $\pi_\phi(\cdot \mid s_{t-1}, a_{t-1})$, where ϕ is the trainable parameter. So, the mixed policy is rewritten as equation (5). Specifically, we use superscript π_ϕ^k to identify different master-policy.

$$\pi_{mix}(\cdot \mid s_t) = \sum_{k=0}^{n} \pi_\phi(k \mid s_{t-1}, a_{t-1}) \pi_k(\cdot \mid s_t) \tag{5}$$

To train the master-policy, we extend the classic MDP and derive the policy gradient for parameter ϕ. Figure 3 shows the difference between the classic MDP and eMDP. In eMDP, the new state X_t is a 3-tuple (s_{t-1}, a_{t-1}, s_t) rather than s_t shown in figure 3b (blue shade).

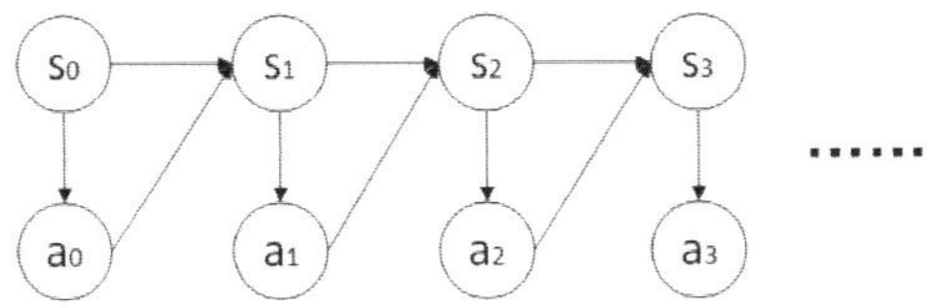

(a) The classic MDP

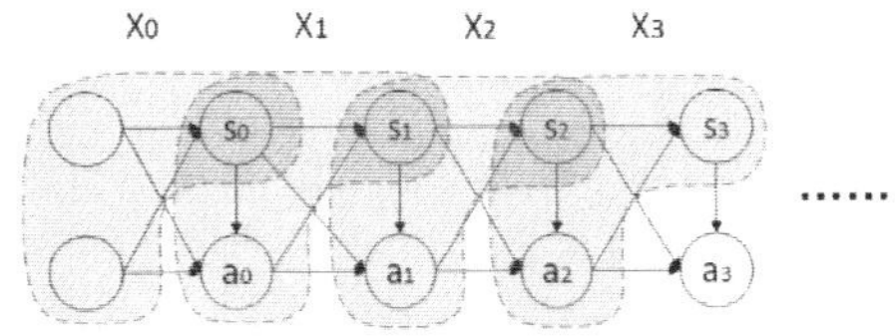

(b) The extended MDP

Figure 3. The difference between MDP and eMDP.

The new action space is the same as the classic MDP. The reward function is $R(X_t, a_t) = R((s_{t-1}, a_{t-1}, s_t), a_t)$. In theory, the reward $R((s_{t-1}, a_{t-1}, s_t), a_t) = R(s_{t-1}, a_{t-1}) + R(s_t, a_t)$ contains a two-step reward, which is calculated repeatedly when estimating the cumulative reward. We only count once for each reward $R(X_t, a_t) = R(s_t, a_t)$ in eMDP without influencing the optimal policy. $M(X_t \mid X_{t-1}, a_{t-1})$ is the transition function. From the definition of eMDP, it has the same transition function as the classic MDP (as shown in equation (6)). Additionally, the initial distribution of state $\rho_0(X_0)$ equals $\rho_0(s_0)$. In summary, the eMDP has the same basic rule as the classic one, except for the state space.

$$\begin{aligned} & M(X_t \mid X_{t-1}, a_{t-1}) \\ &= P((s_{t-1}, a_{t-1}, s_t) \mid (s_{t-2}, a_{t-2}, s_{t-1}), a_{t-1}) \\ &= P(s_{t-1}, a_{t-1} \mid s_{t-2}, a_{t-2}, s_{t-1}, a_{t-1}) P(s_t \mid s_{t-2}, a_{t-2}, s_{t-1}, a_{t-1}) \\ &= P(s_t \mid s_{t-1}, a_{t-1}) \end{aligned} \tag{6}$$

PPO on eMDP. According to the eMDP, we derive the policy gradient for the parameter ϕ in master-policy π_ϕ. Unlike the classic MDP, there is a branch action k_t. The probability of trajectory τ is expressed by equation (7).

$$\begin{aligned} p(\tau) &= p(X_0, k_0, a_0, X_1, k_1, a_1, \cdots) \\ &= \rho_0(X_0) \prod_{t\geq 0} \pi_\phi(k_t \mid X_t) \prod_{t\geq 0} \pi(a_t \mid X_t; k_t) \prod_{t\geq 0} M(X_{t+1} \mid X_t, a_t) \\ &= \rho_0(s_0) \prod_{t\geq 0} \pi_\phi(k_t \mid s_{t-1}, a_{t-1}) \prod_{t\geq 0} \pi(a_t \mid s_t; k_t) \prod_{t\geq 0} P(s_{t+1} \mid s_t, a_t) \end{aligned} \tag{7}$$

Based on the definition of $J(\pi)$, the policy gradient is derived with $p(\tau)$ shown in equation (8).

$$\begin{aligned}
&\nabla_\phi \mathrm{E}_{\tau\sim p(\tau)}\left[\sum_{t\geq 0}\gamma^t R(X_t,a_t)\right]\\
&=\mathrm{E}_{\tau\sim p(\tau)}\left[\nabla_\phi \log p(\tau)\sum_{t\geq 0}\gamma^t R(X_t,a_t)\right]\\
&=\mathrm{E}_{\tau\sim p(\tau)}\left[\sum_{t\geq 0}\nabla_\phi \log \pi_\phi(k_t \mid s_{t-1},a_{t-1})\sum_{t'\geq t}\gamma^{t'-t} R(X_{t'},a_{t'})\right]
\end{aligned} \tag{8}$$

Clearly, the parameter ϕ only appears in master-policy π_ϕ. We derive the equation by using the fact that $\nabla_\phi \log p(\tau)=\nabla_\phi \sum_{t\geq 0}\log \pi_\phi(k_t \mid s_{t-1},a_{t-1})$. To reduce the variance of equation (8), we replace the cumulative reward with the advantage function $A_{\pi_\phi}(X_t,a_t)=[R(X_t,a_t)+\gamma V_{\pi_\phi}(X_{t+1})-V_{\pi_\phi}(X_t)]$.

Specifically, we remove the state s_t from X_t in the value function, which means $V_{\pi_\phi}(X_t)=V_{\pi_\phi}(s_{t-1},a_{t-1})$. We let the value function share the network with policy π_ϕ (as shown in figure 2). We derive the PPO algorithm based on eMDP, which is shown in equation (9).

$$\begin{aligned}
&\nabla_\phi \mathrm{E}_{\tau\sim p(\tau)}\left[\sum_{t\geq 0}\frac{\pi_\phi(k_t \mid s_{t-1},a_{t-1})}{\pi_{\phi_{old}}(k_t \mid s_{t-1},a_{t-1})}A_{\pi_\phi}(X_t,a_t)\right]\\
&s.t.\quad KL(\pi_\phi(k_t \mid s_{t-1},a_{t-1}) \,\|\, \pi_{\phi_{old}}(k_t \mid s_{t-1},a_{t-1}))<\delta
\end{aligned} \tag{9}$$

3. Experiments

Training setups. The agents are developed in python using PyTorch 1.1.0. All networks are trained on GPU with GeForce GTX-1080 Ti 12GB GPU and tested on CPU with Intel(R) Xeon(R) Gold 5118 CPU @ 2.30GHz processor. In the first stage, the training process uses a parallel framework PAAC while using a parallel framework Distributed PPO (DPPO) in the second stage.

Networks setups. Table 1 shows the details of each network for atari games. In this work, we only use two sub-policies to verify our contribution, including teacher policy π_θ^0 and student policy π_θ^1. π_ϕ^0 is the master branch of π_θ^0 and π_ϕ^1 is the master branch of π_θ^1. π_{sep} represents a separate master policy (not share backbone with sub-policy). The "Conv" columns show the filer shape of the convolution. Specifically, the hyperparameter α is 0.001.

Table 1. Network architectures and FLOPs of each policy for atari games.

Agents	Input	Conv.1	Conv.2	Conv.3	F.C.1	F.C.2	Output	FLOPs	Time(ms)
π_θ^0	4x84x84	32(8)	64(4)	64(3)	512	-	[1-18]	7764992	2.00
π_θ^1	4x84x84	4(8)	4(4)	-	32	-	[1-18]	433088	0.85
π_ϕ^0	512	-	-	-	128	[128+[1-18]]	2	65802	0.08
π_ϕ^1	32	-	-	-	256	[256+[1-18]]	2	9497	0.06
π_{sep}	4x84x84	4(8)	4(4)	-	32	-	2	432992	0.71

3.1. Final Score Compare

Table 2 shows the final score of each network on 11 atari games. From the "scores" column, the relative score decreases about 3% in atari games, which saves about 41.2% of FLOPs and 22% of inference time. We also notice that scores of the sub-policy π_θ^1 decrease significantly because of its small capacity. Specifically, the APInf on "alien" decreases more than in other games. The reason is that the sub-policy π_θ^1 is too bad to make any contribution to our framework APInf. On the contrary, the game "amidar" keeps the same score as π_θ^0 because the sub-policy π_θ^1 is good enough to make more contribution to APInf (about 30% speedup). Furthermore, we can control the strength of inference cost α in equation (4) to balance the score and inference time. In real-world applications, we pre-define how much score decrease is acceptable and control the hyperparameter α to keep score during the master-policy training.

Table 2. The final score reached by each network.

Tasks	π_θ^0	π_θ^1	π_ϕ^0	π_ϕ^1	APInf	scores	FLOPs	time
Pong	**21.0**	15.0	20.6	20.4	20.6	1.9% ↓	46.7% ↓	25.5% ↓
Breakout	**422.4**	288.4	416.4	407.4	410.3	2.8% ↓	46.7% ↓	25.5% ↓
Spaceinvader	**2975.8**	1130.1	2889.8	2704.6	2800.2	5.9% ↓	37.2% ↓	19.5% ↓
H.E.R.O.	**28622.2**	16450.1	28490.5	28368.6	28479.1	0.5% ↓	36.2% ↓	19.0% ↓
Qbert	**15570.7**	15075.2	15250.7	15200.5	15175.2	2.5% ↓	51.5% ↓	28.5% ↓
Seaquest	1706.6	1560.1	1695.2	**1727.5**	1704.1	0.1% ↓	43.8% ↓	23.5% ↓
Alien	**2200.0**	1288.0	416.4	407.4	1965.0	2.8% ↓	46.7% ↓	25.5% ↓
Amidar	**253.0**	247.6	253.0	252.0	253.0	0.0% ↓	53.4% ↓	29.5% ↓
Bankheist	**949.7**	505.8	941.9	917.6	947.5	0.2% ↓	20.1% ↓	9.0% ↓
Beamrider	**1990.2**	1116.5	1949.1	1880.2	1932.5	2.9% ↓	56.2% ↓	31.5% ↓
Riverraid	**6929.5**	5376.1	6163.4	6364.5	6462.5	6.7% ↓	47.7% ↓	26.0% ↓
Average	-	-	-	-	-	3.0% ↓	41.2% ↓	22.0% ↓

Table 3. The influence of hyperparameter α on "breakout" and "spaceInvader".

Tasks	α	Scores	FLOPs
Breakout	0.01	377.6	1550816
Breakout	0.001	410.3	4136689
Breakout	0.0001	424.0	5614331
Spaceinvader	0.01	2610	3767279
Spaceinvader	0.001	2800	4875510
Spaceinvader	0.0001	3020	5614331

We also make a sensitivity analysis of α on score and FLOPs. As a hyperparameter in the reward function α balances the reward and the policy inference cost. When the value of α is large, we pay more attention to the reward and weaken the cost of policy inference. Otherwise, we preference we prefer to reduce the cost of inference. We verify the influence of α on "breakout" and "spaceInvader". The result is shown in table 3. We notice that the score and FLOPs become larger as the alpha increases. Furthermore, when α is lower than 0.0001, both tasks outperform the teacher policy π_0 on score and FLOPs. Therefore, in many states, it doesn't matter to choose any action.

3.2. Visualization on Toy Example

Figure 4 shows the states of breakout (left column), master action k_t (middle column) and sub-policy action a_t (right column). The "teacher" and "student" represent the sub-policy π_θ^0 and π_θ^1 respectively. We notice that the outputs of sub-policies are similar in the "easy" state. In detail, the ball in the white box is moving to the wall. The agent takes any actions that can not affect the reward (figure 4a). So the APInf chooses π_θ^1 to save inference cost. In the bottom, the ball is closing to our racket, and the agent should take the best action to get a higher score (figure 4b). So the APInf will choose the π_θ^0 to take that action.

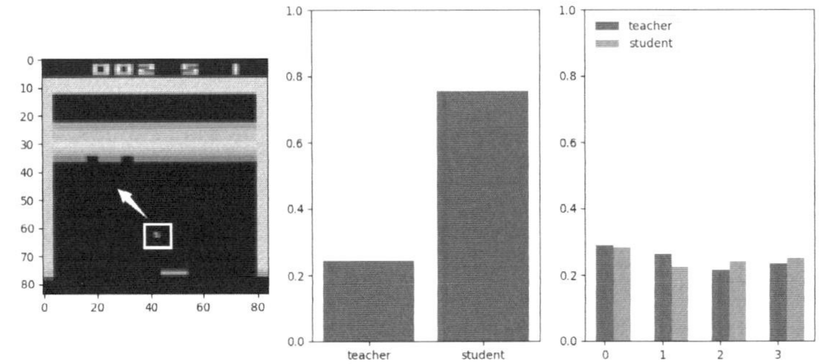

(a) The "easy" state in game which uses π_θ^1

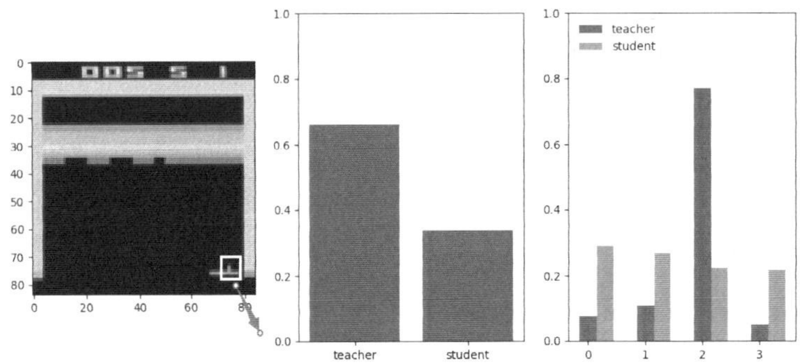

(b) The "hard" state in game which uses π_θ^0

Figure 4. Visualization results of the APInf on breakout.

3.3. Ablation Study

To verify the usefulness of master-policy and eMDP, we conduct four ablation study cases. The ablation study results are summarized in table 4. APInf has the best score in most tasks.

(1) random master-policy π_{random} : the sub-policy is random selected.

(2) learning from KL π_{KL} : the master-policy is trained to classify the data with labels based on the similarity among sub-policies. For example, if the similarity $KL(\pi_0 \| \pi_1) < 0.1$, the label is 1, which means choose sub-policy π_1 .

(3) separate master-policy π_{sep} (the similar architecture as IF-CNN [17]): an independent master-policy trained based on the classic MDP.

(4) MSDNet-liked policy π_{msdnet} [14]: the π_1 shares the convolution layer with π_0. Specifically, π_0 is the same as teacher network in APInf, and π_1 is a branch of the first convolution layer in the teacher network. The master policy also shares the first convolution layer with teacher network.

Table 4. The ablation study results of difference master-policy.

Tasks	APInf	π_{random}	π_{KL}	π_{sep} [17]	π_{msdnet} [14]
Pong	**20.6**	18.3	19.5	**20.6**	19.2
Breakout	**410.3**	330.4	373.8	407.3	406.8
spaceInvader	**2800.2**	2423.2	2207.3	2597.3	2175
H.E.R.O.	**28479.1**	20875.0	25798.1	27133.3	28582
Qbert	**15175.2**	15091.1	13503.3	13870.3	15029
Seaquest	1704.1	1668.0	1651.1	**1707.0**	1633
Alien	**1965.0**	863.0	1556.0	1690.0	1789
Amidar	**253.0**	186.2	240.3	241.1	243
Bankheist	**947.5**	620.0	760.8	899.0	890
Beamrider	**1932.5**	1273.5	1764.3	1806.0	1934
Riverraid	**6462.5**	5075.8	5457.5	5537.1	5782
Average	**3.0% ↓**	24.6% ↓	14.3% ↓	8.8% ↓	6% ↓

4. Conclusion and Future Work

In this paper, we designed a novel framework APInf, which speeds up the inference time while keeping a similar final score as the best policy. Specifically, APInf tries to select the efficient sub-policy from the policy pool according to the current state's difficulty. The master policy is optimized by using PPO based on an eMDP problem. Extensive experiments on atari games show that APInf can significantly speed up the inference at test time. In this work, the strength of inference cost α is tuned by hand. Our next work will design an automatic algorithm to adjust α.

References

[1] Vinyals O, Babuschkin I, Czarnecki W M, Mathieu M, Dudzik A, Chung J Y, Choi D H, Powell R, Ewalds T and Georgiev P 2019 Grandmaster level in StarCraft II using multi-agent reinforcement learning *Nature* **575** (7782) 350-354.

[2] Zhang Y, Sun P, Yin Y H, Lin L and Wang X S 2018 Human-like autonomous vehicle speed control by deep reinforcement learning with double Q-learning *Proc. of the 2018 IEEE Intelligent Vehicles Symp. (IV'8)* pp 1251-1256.

[3] Yang X, Liu W Q, Zhou D, Bian J and Liu T Y 2020 Qlib: An ai-oriented quantitative investment platform *arXiv:2009.11189.* https://arxiv.org/abs/2009.11189

[4] Zhang H J, He Z C and Li J 2019 Accelerating the deep reinforcement learning with neural network compression *Proc. of 2019 Inter. Joint Conf. on Neural Networks (IJCNN'19)* pp 1-8.

[5] Han S, Mao H Z and Dally W J 2015 Deep compression: Compressing deep neural networks with pruning, trained quantization and huffman coding *arXiv:1510.00149.* https://arxiv.org/abs/1510.00149

[6] Han S, Pool J, Tran J and Dally W 2015 Learning both weights and connections for efficient neural network *Advances in Neural Information Processing Systems 28(NIPS'15)* pp 1135-1143.

[7] See A, Luong M T and Manning C D 2016 Compression of neural machine translation models via pruning *Proc. of the 20th SIGNLL Conf. on Computational Natural Language Learning (CoNLL'16)* pp 291-301.
[8] Molchanov P, Mallya A, Tyree S, Frosio I and Kautz J 2019 Importance estimation for neural network pruning *Proc. of the IEEE/CVF Conf. on Computer Vision and Pattern Recognition (CVPR'19)* pp 11264-11272.
[9] Srinivas S and Babu R V 2015 Data-free parameter pruning for deep neural networks. *arXiv:1507.06149*. https://arxiv.org/abs/1507.06149
[10] Polino A, Pascanu R and Alistarh D 2018 Model compression via distillation and quantization *Proc. of the 6th Inter. Conf. on Learning Representations (ICLR'18)*.
[11] Lin D, Talathi S and Annapureddy S 2016 Fixed point quantization of deep convolutional networks *Proc. of the 33rd Inter. Conf. on Machine Learning (ICML'16)* pp 2849-2858.
[12] Gupta S, Agrawal A, Gopalakrishnan K and Narayanan P 2015 Deep learning with limited numerical precision *Proc. of the 33rd Inter. Conf. on Machine Learning (ICML'15)* pp 1737-1746.
[13] Lane N D and Georgiev P 2015 Can deep learning revolutionize mobile sensing? *Proc. of the 16th Inter. Workshop on Mobile Computing Systems and Applications (HotMobile'15)* pp 117-122.
[14] Huang G, Chen D L, Li T H, Wu Felix, van der Maaten Laurens and Weinberger K 2018 Multi-scale dense networks for resource efficient image classification *Pro. of the 6th Inter. Conf. on Learning Representations (ICLR'18)*.
[15] Xie Z D, Zhang Z, Zhu X Z, Huang G and Lin S 2020 Spatially adaptive inference with stochastic feature sampling and interpolation *European Conf. on Computer Vision* pp 531-548.
[16] Wang Y L, Lv K, Huang R, Song S J, Yang L and Huang G 2020 Glance and focus: A dynamic approach to reducing spatial redundancy in image classification *Advances in Neural Information Processing Systems*.
[17] Shu G S, Liu W Q, Zheng X J and Li J 2018 IF-CNN: Image-aware inference framework for CNN with the collaboration of mobile devices and cloud *IEEE Access* **6** 68621-68633.

Pipe Thread Parameter Detection System Based on Machine Vision

Yingying Li, Yuxiu Wu and Donghong Zhang

School of Electrical and Information Engineering, Anhui University of Technology, Maanshan, Anhui, China

Email: 2950754147@qq.com; yuxiuwu@ahut.edu.cn; 1792943248@qq.com

Abstract. In order to achieve non-contact pipe thread detection with high-precision, high-efficiency and make the use of pipe threads safer, using machine vision and image processing technology to measure and analyze the geometric parameters of pipe threads. The projection matrix and distortion coefficient of the camera are obtained by Zhang Zhengyou's calibration method, and the distortion coefficient is used to correct the distortion of the collected image. After that, grayscale processing is performed on the RGB (Red, Green, Blue) image obtained by the collection and correction, fusion filtering and denoising, threshold segmentation to obtain the binary image, and then the Canny operator is used to detect the thread edge information on the binary image. Then the pipe thread parameter measurement algorithm is proposed based on the extracted pipe thread image characteristics. Finally, the feasibility of the measurement algorithm is verified by using the measured data of 60 sets of pipe threads. The test results show that the measured values of tooth height and crest height obtained by this algorithm are concentrated near the standard value, with high accuracy and precision, and the overall data fluctuation is small.
Keywords. Thread detection; machine vision; image processing; canny operator; measurement algorithm.

1. Introduction

Tubing is widely used to transport oil in petroleum exploitation, when the depth of an oil field increases, so does the length of the pipe, which requires more pipe threads to be used to connect the pipe. In addition, oil pipelines are subject to harsh operating conditions, so the quality of the pipe threads used to connect them must be guaranteed [1]. Due to the complexity of the connection between the oil pipe and the tubing, it is easy for the tubing to fail, which can lead to significant economic losses and, in some cases, serious injuries [2]. The common failures of tubing are: tubing disengagement, tubing leakage and tubing fracture, but the direct cause of tubing failure is the failure of the threads. It is therefore vital that pipe thread parameters are tested.

Machine vision is a comprehensive technology that gathers multidisciplinary content and has applications in many fields [3]. Research on machine vision in thread detection includes Guo, who proposed a thread parameter detection method based on the problem of low precision and difficulty in thread detection [4]. Fan optimized the algorithm in traditional image processing to achieve accurate image edge contour extraction. The geometrical parameters of the pipe thread were obtained by calculation [5]. Currently, thread parameter detection is mainly focused on the study of small and common threads, with relatively few real applications in oil pipeline threads. Moreover, most of the research only stays in the theoretical analysis stage.

Therefore, based on the above reasons, this paper designs a machine vision based on pipe thread inspection system. A sliding platform is designed and built, and a combination of machine vision and image processing techniques is used to achieve multi-parameter detection.

2. Platform Construction and Image Processing

2.1. Platform Building and Image Acquisition

The experimental platform built for the pipe thread parameter detection system is shown in figure 1. Considering the size of the oil pipe threads, a motor and slide unit was chosen to control the camera to capture the images. The image acquisition uses the ME2P-1230-23U3 colour surface array camera with a lens of WP-10M7528-C, the lighting source is backlit with a white LED strip light source and the motor control uses the TSMD5710B-485-A integrated torque servo motor. The test samples were pipe threads of the same model in the same batch in different areas. Figure 2 shows the physical image of the pipe thread and figure 3 shows the RGB image of the pipe thread.

Figure 1. Experiment platform.

Figure 2. Physical map.

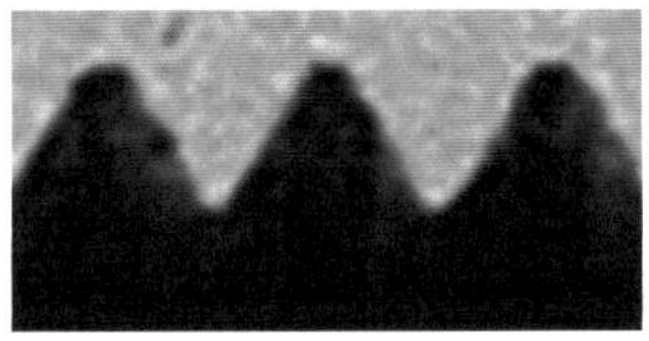

Figure 3. Original RGB image.

2.2. Image Analysis and Processing

2.2.1. Grayscale Processing. The RGB image collected by color camera in this paper contains a large amount of data and is complicated to calculate. In order to simplify the matrix and increase the speed of calculation, transform the color image collected by the system into gray image [6]. The effect of the threaded image before and after conversion is shown in figure 4.

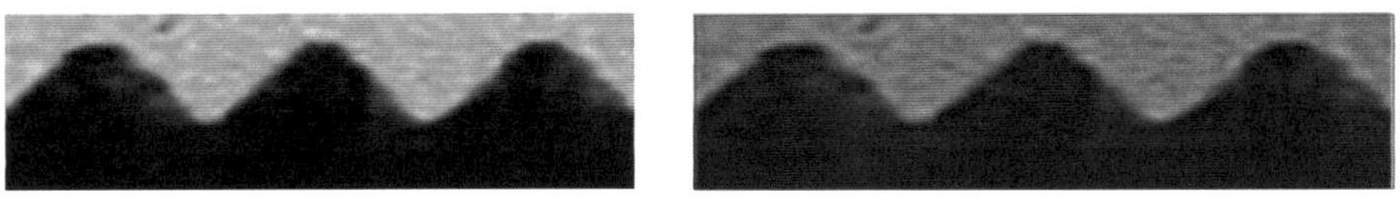

(a) Original image (b) Grayscale images

Figure 4. Gray processing effect chart.

2.2.2. Image Filtering. The effectiveness and reliability of the subsequent image processing is directly affected by the image filtering results [7, 8]. After several experiments to compare the effects, the 9×9 convolution kernel is finally used to do median filtering on the image, and $N=19$, $\delta_d=39$, $\delta_r=21$ is used to do bilateral filtering [9-11]. It can be found that the outline of the image is clearer, the edge information is well preserved and the smoothness is better after using this fusion filter. Figure 5 shows the effect of fusing the median filter and bilateral filter on the threaded image.

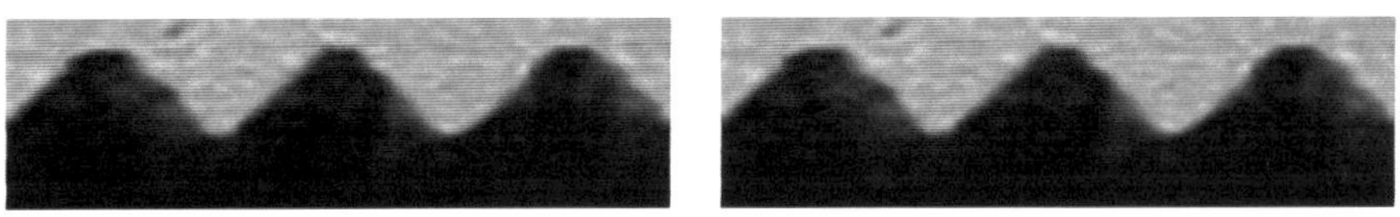

(a) Original image (b) Fusion filter map

Figure 5. Fusion filtering effect map.

2.2.3. Image Segmentation. In this paper, a fixed segmentation algorithm based on histogram thresholding is used to segment the threaded image. The processing process is mainly to determine the segmentation threshold through the histogram [12], and then compare the threshold with the grey value of the image, and finally convert it to a binary image. The image after segmentation using the histogram threshold segmentation method is shown in figure 6.

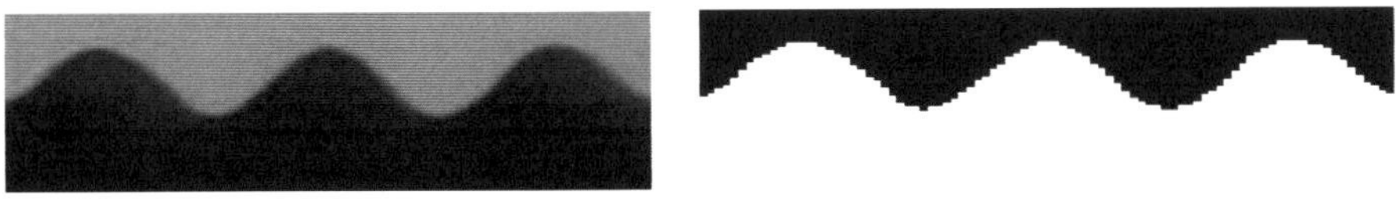

(a) Fusion filter map (b) Binary images

Figure 6. Image binarization.

2.2.4. Threaded Edge Detection. The Canny operator based on bilinear interpolation was chosen for pipe thread edge detection in this paper. Edge detection is first performed on the binary image. Then, the detected image is refined. Finally, the refined image is repaired for breakpoints, which is shown in figure 7.

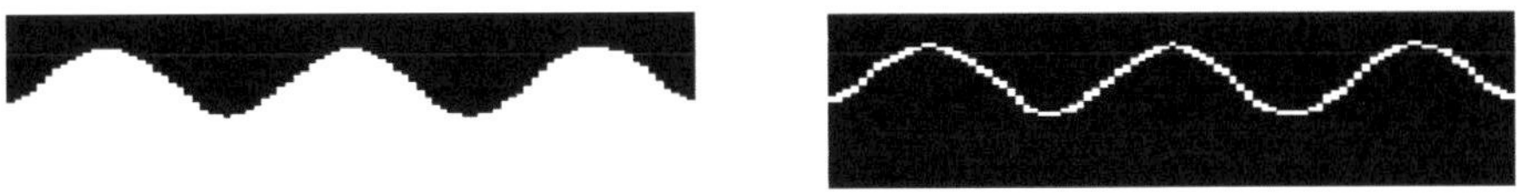

(a) Binary images (b) Canny operator edge detection image

(c) Refine image (d) Breakpoint defect repair image

Figure 7. Thread edge detection image.

3. Pipe Thread Parameter Measurement

This paper focuses on measuring parameters such as tooth height and top height of pipe threads. The distribution characteristics of the pipe thread are shown in figure 8, where E, G and I are the top thread diameter points, F and H are the bottom thread diameter points and km is the ideal pipe thread centre diameter line.

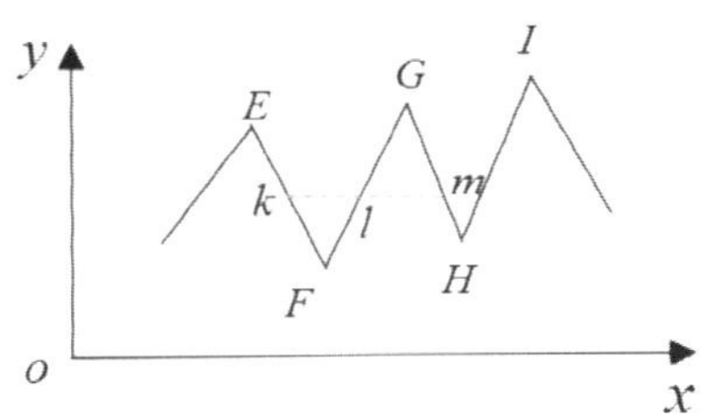

Figure 8. Pipe thread distribution characteristic diagram.

3.1. Tooth Height Measurement

The tooth height of a pipe thread refers to the distance between the top diameter point and the bottom diameter point in the direction perpendicular to the thread axis. This paper proposes a straight-line fitting algorithm based on the pixels of the edge feature points, and the flow chart of the algorithm is shown in figure 9.

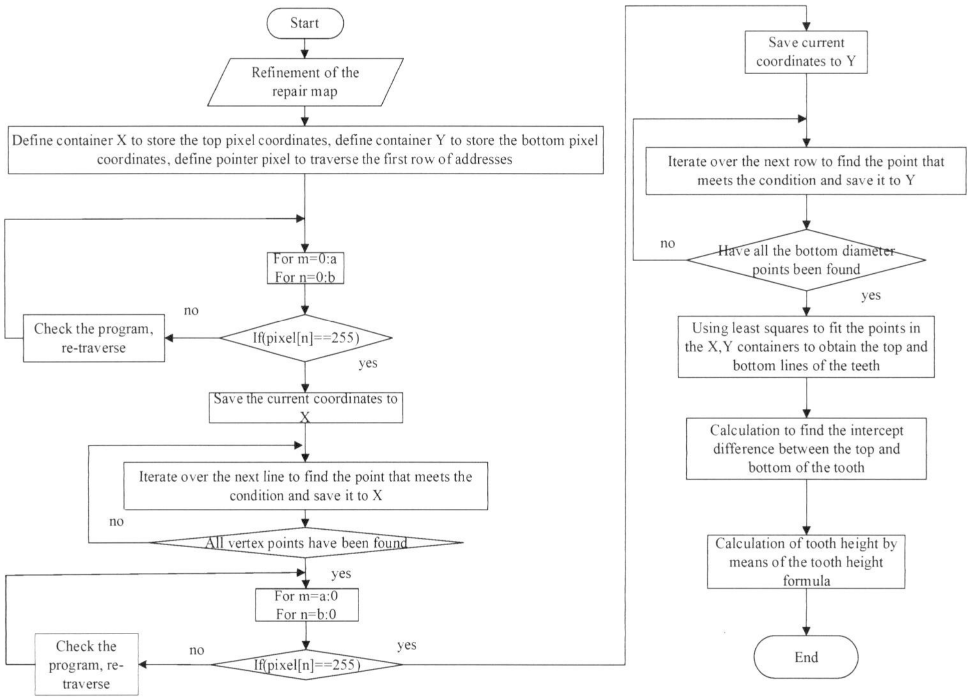

Figure 9. Tooth height algorithm flow chart.

As shown in figure 10, assuming that the top diameter point fits a straight line of $y = ax + b$, the bottom diameter point fits a straight line of $y = cx + d$, a point $M(x_1, y_1)$ on the top diameter point straight line, and a point $N(x_2, y_2)$ on the bottom diameter point straight line, then.

$$ML = y_1 - y_2 \tag{1}$$

$$LO = (x_2 - x_1)\tan\theta \tag{2}$$

In this formula, $\theta = \arctan(a_2)$, so the tooth height h is

$$h = ML + LO = (y_1 - y_2) + (x_2 - x_1)\tan\theta \tag{3}$$

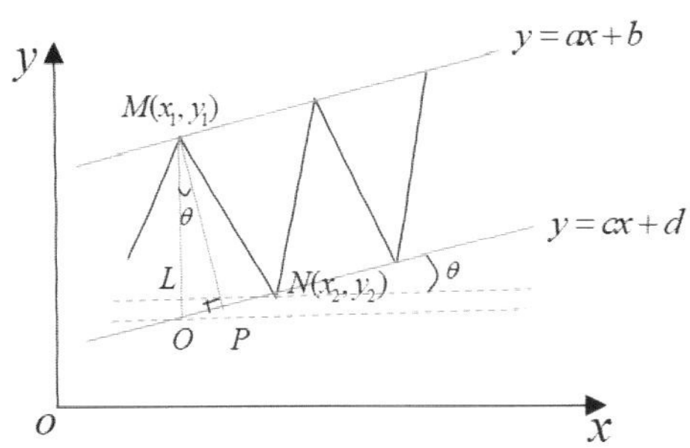

Figure 10. Thread tooth height calculation.

3.2. Top of Tooth Height Measurement

The tooth top height is defined as the radial distance from the top of the tooth to the centre diameter line on a threaded tooth profile. This paper proposes a verification algorithm for mid-diameter extraction based on Floyd's progressive pixel points. The core of the algorithm is to first determine the location of the bottom diameter in the contour and intercept the ROI region of the contour based on this location. Then all pixel points are traversed within this region. Finally, the Floyd algorithm is used to reject and validate the traversed points to obtain the exact mid-diameter position.

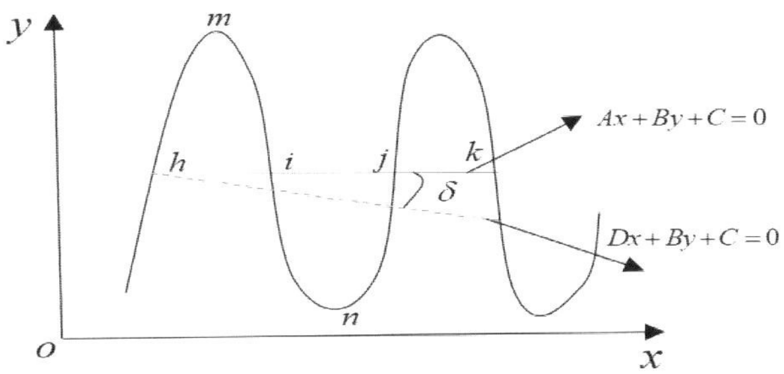

Figure 11. Thread pitch lin.

As shown in figure 11, the Floyd algorithm idea is to assume that $h(a_0,b_0)$, $i(a_1,b_1)$, $j(a_2,b_2)$, $k(a_3,b_3)$, $m(x_0,y_0)$, $n(x_1,y_1)$. The distance between m point and n point is calculated, denoted $dist\_\min$:

$$dist\_\min = \sqrt{(x_1 - x_0)^2 + (y_1 - y_0)^2} \tag{4}$$

The sum of the distances of mi and ni is calculated, denoted as $dist$:

$$dist = \sqrt{(x_0 - a_1)^2 + (y_0 - b_1)^2} + \sqrt{(x_1 - a_1)^2 + (y_1 - b_1)^2} \tag{5}$$

At this time, if the result of the calculation yields $dist \le \text{dist\_min}$, then the point is a point on the median. Then proceed to find the other points. Assuming that the ideal centre diameter equation is $Ax + By + C = 0$ and the true tapered centre diameter equation is $Dx + By + C = 0$, dh is the ideal top height intercept, dh_z is the true top height intercept, Z is the true top height axial value, m is the top coordinate (x_0, y_0) and δ is the taper of the thread, then.

$$dh = \left| \frac{Ax_0 + By_0 + C}{\sqrt{A^2 + B^2}} \right| \tag{6}$$

$$dh\_z = \left| \frac{Dx_0 + By_0 + C}{\sqrt{D^2 + B^2}} \right| \tag{7}$$

$$Z = -B \frac{B\tan(\delta) - A}{A\tan(\delta) + B} \tag{8}$$

4. Experimental Results and Analysis

4.1. Tooth Height Data Processing

Firstly, the tooth height of the same model in different areas is measured. Then, the tooth height of the pipe thread is extracted and calculated by using the tooth height measurement algorithm and the motor control system. Table 1 shows some of the measured and calculated tooth height values.

Table 1. Partial tooth height measured and calculated values.

Number of measurements	Tooth height (mm)	Absolute deviation (mm)	Relative error (%)
1	1.8454	0.0259	1.42
2	1.8216	0.0021	0.12
3	1.8217	0.0022	0.12
4	1.8455	0.0260	1.43
5	1.8012	-0.0183	-1.01
6	1.7969	-0.0226	-1.24
...	...	...	...
57	1.8022	-0.0173	-0.95
58	1.8160	-0.0035	-0.19
59	1.8453	0.0258	1.42
60	1.8262	0.0067	0.37

4.2. Top of Tooth Height Data Processing

Measuring the top height of the thread for the same model in different areas. Using the top height measurement algorithm and the motor control system, the top height of the pipe thread is extracted and calculated. Table 2 shows some of the measured and calculated values.

Table 2. Partial top of tooth height measured and calculated values.

Number of measurements	Top of tooth height (mm)	Absolute deviation (mm)	Relative error (%)
1	1.3261	0.0054	0.41
2	1.3260	0.0053	0.40
3	1.3152	-0.0055	-0.42
4	1.3322	0.0115	0.87
5	1.3150	-0.0057	-0.43
6	1.3260	0.0053	0.40
...	...	...	...
57	1.3384	0.0177	1.34
58	1.2998	-0.0209	-1.58
59	1.2999	-0.0208	-1.57
60	1.3400	0.0193	1.46

4.3. Analysis of Data Results

In this paper, 60 sets of data were measured and analysed for the same batch of pipe threads of the same type in different areas, with table 3 showing the standard values for the geometric parameters of the pipe threads and table 4 showing the measured values for the geometric parameters of the pipe threads in different areas.

Table 3. Standard value of geometric parameters of pipe thread.

Manufacturer's specified value	Tooth height	Top of tooth height
Standard values (mm)	1.82	1.3260
Standard error (mm)	± 0.0300	± 0.0380
Positive relative error (%)	1.65	2.87
Negative relative error (%)	-1.65	-2.87

Table 4. Measured values of geometric parameters of pipe threads in different areas.

Measured values	Tooth height	Top of tooth height
Average (mm)	1.8195	1.3207
Maximum value (mm)	1.8457	1.3472
Minimum value (mm)	1.7965	1.2982
Average deviation (mm)	0.0135	0.0112
Relative mean deviation (%)	0.74	0.85
Standard deviation of a single measurement (mm)	0.0169	0.0134
Standard deviation (mm)	0.0021	0.0022
Limit deviations (mm)	±0.0037	±0.0029
Maximum measurement error (mm)	0.0041	0.0080

In combination with tables 3 and 4, it can be seen that the maximum, minimum and mean values of the 60 measured pipe thread tooth heights are all within the standard values and standard errors specified by the manufacturer. Besides the sum of the average and limit errors is always around the standard values and the relative average deviation is much less than the specified relative error. Therefore, the algorithm proposed in this paper for measuring pipe thread parameters is feasible.

5. Conclusion

In this paper, a non-contact inspection algorithm based on machine vision is used for the inspection of oil pipe threads. A high-resolution camera + slide mechanism + motor control platform is built according to actual needs. The camera and the system are calibrated to correct the image and to obtain the scale factor of the object to the image. Then, the image is collected, processed and the parameters are measured in combination with the pipe thread image feature extraction and measurement algorithm. Finally, the measurement algorithm is validated with error analysis and data processing. At the same time, it also provides an idea for other non-contact pipe thread measurements.

References

[1] Ding J F, Kang Y H and Wu X J 2006 Tubing thread inspection by magnetic flux leakage *Ndt. & E. International* **39** (1) 53-56.
[2] Idachaba F and Rabiei M 2021 Current technologies and the applications of data analytics for crude oil leak detection in surface pipelines *Journal of Pipeline Science and Engineering* **1** (4) 436-451.
[3] Jones R E 1991 Machine vision applications *Mechatronics* **1** (4) 439-446.
[4] Guo Y, Ma L T and Su C 2019 The method of thread defect detection based on machine vision *2019 2nd Inter. Conf. on Information Systems and Computer Aided Education (ICISCAE)* pp 579-582.
[5] Fan J P, Li Y L, Fan H, et al. 2018 Pipe tread parameters detection based on machine vision *Sixth Inter. Conf. on Optical and Photonic Engineering.*
[6] Sui X B, Chen Q and Gu G H 2013 Adaptive grayscale adjustment-based stripe noise removal method of single image *Infrared Physics & Technology* **60** (5) 121-128.
[7] Deng T Q, Dong T Z, Xie W, et al. 2013 Median filtering method based on adaptive central weighting for color images *Control and Decision* **28** (9) 1372-1376.
[8] Dang M Y, Meng Q K, et al. 2020 Rapid recognition of potato late blight based on machine vision *Transactions of the Chinese Society of Agricultural Engineering* **36** (2) 193-200.
[9] Gavaskar R G and Chaudhury K N 2019 Fast adaptive bilateral filtering *IEEE Transactions on Image Processing* **28** (2) 779-790.
[10] Asokan A and Anitha J 2020 Adaptive cuckoo search based optimal bilateral filtering for denoising of satellite images *ISA Transactions* **100** (5) 308-321.

[11] Yu H C, Zhao L and Wang H X. 2011 An efficient edge-based bilateral filter for restoring real noisy image *IEEE Transactions on Consumer Electronics* **57** (2) 682-686.
[12] Lu Y Z and Lu R F 2017 Histogram-based automatic thresholding for bruise detection of apples by structured-illumination reference imaging *Biosystems Engineering* **160** (8) 30-41.

Session Recommendation Based on Edge Information Clustering

Manfu Ma, Dongliang Yang and Yong Li

School of Computer Science and Engineering, Northwest Normal University, Lanzhou 730000, Gansu, China
Email: 1242726467@qq.com; mamanfu@nwnu.edu.cn; facingworld@126.com

Abstract. Traditional session recommendation mainly uses the time sequence of users clicking items to construct a user session graph, which often ignores the similarity and differences between user groups. To improve the effect of recommendation, an E-SGNN (E-SGNN, Edge-Session Graph Neural Network) method combining edge information clustering and session recommendation is proposed. Firstly, similar users are clustered by edge information and divided into different session user groups. After extracting the data features of the user site relationship graph in the session, it is reset and updated through the gated graph neural network (GGNN); Secondly, a self-attention mechanism is introduced to adjust the proportion of users' current preference and historical preference; Finally, the ranking score is obtained through linear transformation and softmax classifier. The higher the score, the more obvious the user's preference for the item. Experiments show that compared with session-based graph neural network and cross-session information recommendation, the E-SGNN algorithm proposed in this paper has a significant improvement in recall rate and average reciprocal ranking. When the three edge parameters are combined, the recall rate reaches 98.97% and the average reciprocal ranking reaches 45.77%.
Keywords. Edge Information; clustering; groups; session recommendation; gated graph neural network.

1. Introduction

Facing the situation of information overload and unclear user needs, it is extremely important to filter and extract effective information, and recommendation systems are born. Recommendation systems do not require users to provide explicit needs and model their interests by analyzing their historical behaviors, to proactively recommend information to users that can meet their interests and needs [1]. Currently, recommendation systems are heavily used in many fields, such as movies, songs, travel, medical care, e-commerce, short videos, and other various fields are very widely used [2]. In the recommendation system, it is difficult to accurately judge users' interests from a large number of users' historical data, which is the challenge that the recommendation system is currently facing. Facing this challenge, the recommendation system effectively prunes a large amount of data through sequence-based recommendation, session recommendation, group recommendation, graph neural network recommendation, self-attention recommendation, and other methods, to make more accurate recommendations [3].

2. Related Work

2.1. Traditional Recommendation Method

Traditional recommendation systems are divided into content-based recommendations and sequence-based recommendations [4]. The general content-based recommendation regards the relationship between users and sites as static, and the accuracy of this method is not high. Therefore, most recommendation systems currently use sequence-based recommendations. The traditional sequence-based recommendation is not perfect and accurate for processing large amounts of data. Taking Taobao and JD.com as examples, a large number of users accumulate historical behavior data for several years or even ten years on shopping websites [5]. Facing such a huge amount of data, it is impossible to predict directly by traditional methods, because traditional methods, such as collaborative filtering, Markov chains, and other sequence modeling techniques are generally applicable to the case of sequence data length within 100, and when the sequence length increases by an order of magnitude, there will be difficulties for data processing. In addition, although the matrix decomposition-based approach can handle a large amount of data, it loses a large number of features when processing user relationship graph data and does not take into account changes in users' historical preferences and users' current preferences [6].

2.2. Graph Neural Network

The graph neural network approach can be used to process graph structure data and obtain the potential vectors of each node to make recommendations [7]. The new graph embedding representation learning method GSE has further development for the embedding of node feature vectors [8]. Although recurrent neural networks provide a large improvement in the data space, they do not distinguish the historical behavior of users from their current behavior [9]. The gated graph neural network approach based on feature extraction only takes into account some of the characteristics of the user, so the recall rate is not high [10].

Many recommendation algorithms convert the user-site sequence into a graph structure for the recommendation, and such methods have achieved good results in some recommendation fields [11]. Compared with the traditional sequence-based recommendation, the graph-based recommendation is not only more complete in structure but also more accurate in information. The most important thing is that it can effectively process a large amount of user historical behavior data, so the final recommendation effect is also more accurate [12, 13].

2.3. Session Recommendation Based on Edge Information

Session-based recommendation refers to predicting user behavior in the next session from the previous session, and the sessions in a recommendation system have a well-bounded list of interactions [14]. Where boundary means the beginning and the end of an event [15]. Session-based recommendations abstract users into individual sessions, but do not take into account the differences in users within sessions [16]. The recommended method for cross-session information does not identify and classify this information into user groups, although it takes into account edge information beyond the user [17]. Traditional session recommendation does not consider the similarities and differences between user groups, but simply builds session graphs in chronological order, so it is necessary to classify user groups [18]. In addition, group recommendation is also a common recommendation method, where group users usually have the same or similar preferences [19]. Therefore, edge information is generally used to cluster users in group recommendations [20]. Group recommendations can balance the needs of multiple users, or an appropriate model can be built for a group of users to improve the quality of recommendations [21].

The previous session recommendation only considers the current session and does not consider the similarities and differences between similar session groups, but only relies on the information of the session itself to make recommendations, resulting in low recommendation performance. The group recommendation only groups users and does not use the current graph neural network session recommendation which has high recommendation performance. Therefore, the performance of a single group recommendation is not high. Considering some similarities and differences between sessions, it is an attempt to use cross-session information for the recommendation, and to make cross-session recommendations for similar sessions it is necessary to take into account the edge information of the users themselves. Therefore, we propose E-SGNN by considering the three ideas of edge information of users, similarity and difference of user groups, and cross-session graph neural network. The method clusters users for edge information such as age, gender, and income, and uses a hierarchical partitioning method to divide users into different user groups and divide users with the same or similar attributes into the same sessions. Therefore, after hierarchical clustering, the similarity between users who are divided into the same conversation group will be relatively high, and their interests and preferences are relatively similar. Secondly, we introduce a self-attention mechanism, which adjusts the ratio of the global feature vector and the current feature vector of the user when extracting the feature vector, which can significantly improve the recommendation performance.

3. Method

3.1. Definition of the Problem
Let $U=\{U_1, U_2,\dots, U_m\}$ denotes the set of m users, $S=\{S_1, S_2,\dots, S_n\}$ denotes the set of n site, where m users are divided into t groups, and $G=\{g_1, g_2,\dots, g_t\}$ represents a set of user groups.

3.2. User Grouping under Edge Information
This is sample data, from which we can see that U_1, U_2 and U_4 are the same set, U_3 and U_5 are the same set. In the experiment, U_1 and U_2 in the first group are the training set, and U_4 is the test set. Similarly, U_3 in the second group is the training set and U_5 is the test set.

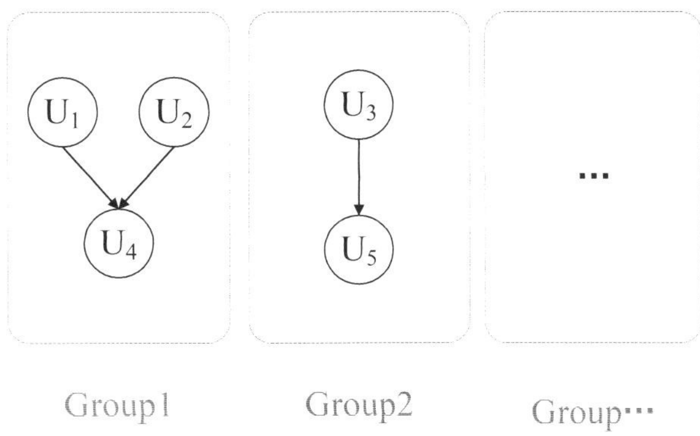

Figure 1. Group-user diagram.

Figure 1 shows the grouping of users with similar or identical characteristics using edge information such as gender, age, and income to construct a group-user relationship graph.

The grouping strategy is to group these users according to factors such as age, gender, education level, income, household registration, social class, etc., to set edge parameters, and the number of edge parameters is the number of groups. For example, all users can be divided into two groups according to gender, one for men and one for women, with a default edge parameter of 2. Table 1 is an example of hierarchical classification.

In addition, feature extraction methods can be used to group users. If there are many attributes and a large number of users, similarity methods or attention feature extraction methods can be used to group users [22].

Journal of Physics: Conference Series **2363** (2022) 012003 doi:10.1088/1742-6596/2363/1/012003

Table 1. User attribute information.

Group	User	Sex	Age	Educational level	Income
G1	U1	Male	34	Undergraduate	5500-6000
G1	U2	Male	29	Undergraduate	4500-5500
G1	U4	Male	33	Undergraduate	5000-6000
G2	U3	Female	35	Undergraduate	3000-4000
G2	U5	Female	30	Undergraduate	3000-4500

3.3. User-Site Relationship Diagram

From figure 2, we can see that the click order of user U_1 is $\{S_1, S_2, S_3\}$, the click order of user U_2 is $\{S_2, S_4, S_5\}$, and so on. Only a small part of the sample data is drawn in the figure, and the rest is omitted. Thus, the input in this paper is a list of a set of users and the sites they clicked on.

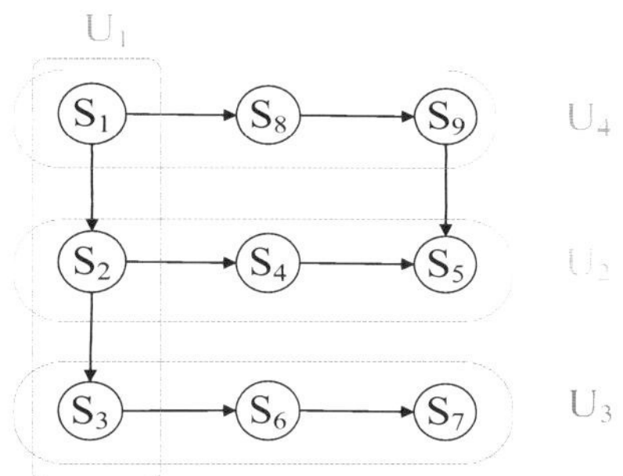

Figure 2. User-site diagram.

Figure 2 shows that after grouping user groups, a user-site relationship graph is constructed for the users in each group and the sites they click on.

3.4. Preferred Modeling

In the process of user-site modeling, a user-session-site relationship diagram is first constructed for the *U*1 user. For example, the site {*S*1, *S*2, *S*3} clicked by the user U1 is taken as a session in the diagram. Secondly, the potential vector *Si* of nodes is obtained by the method of graph neural network for the sites in the session, and then the feature vector *Sp* of the user's current behavior and the feature vector Sh of the user's historical behavior are extracted by the gated neural network, after which the *SL* is obtained by the method of matrix switching and linear transformation, and finally, the ranking score is obtained by the softmax classifier.

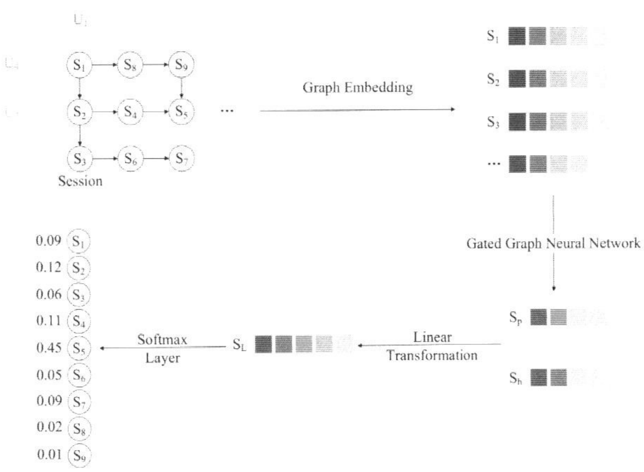

Figure 3. The GGNN modeling graph.

Figure 3 represents modeling the user-site relationship in each group with GNN feature extraction and self-attention mechanism.

3.4.1. Feature VECTOR EMBEdding on the RELATIONSHIP GRAph. For each set of user-site relationship graphs, the sequence g can construct a directed graph $G_g=(S_g, E_g)$. In this relational graph, each node can be regarded as a site S_i, and each edge $V=\{S_{g,i-1}, S_{g,i}\}$ can be regarded as a user in a certain group g clicking from site S_{i-1} to site S_i. For the sake of brevity of the formula, $S_{g,i}$ is denoted by S_i in the following formula. We quantify the learning process of graph embedding representation of node feature vectors in the user-site relationship graph by some mathematical formulas to obtain the potential vectors S_i of nodes.

$$a_{g,i}^{t} = A_{v,i:}\left[S_1^{t-1},\dots,S_n^{t-1}\right]^{\mathrm{T}} H + b \quad (1)$$

where a is the node vector representation of the embedding, H and b are the weights and parameters, $A_{v,i}$ is the out-degree and in-degree splicing matrix, and $[s^{t-1}_1,\dots, s^{t-1}_n]$ is the list of node vectors in the relationship graph.

The matrix $A_{v,i}$ is stitched together by an out-degree matrix and an in-degree matrix. For example, let a user's click order be $\{S_1, S_2, S_3, S_2\}$, then its adjacency matrix is shown in figure 4, where the value of the out (in) degree matrix is the degree of the out (in) degree divided by the degree of the out-degree added to the degree of the in-degree, and the rest is filled with zeros.

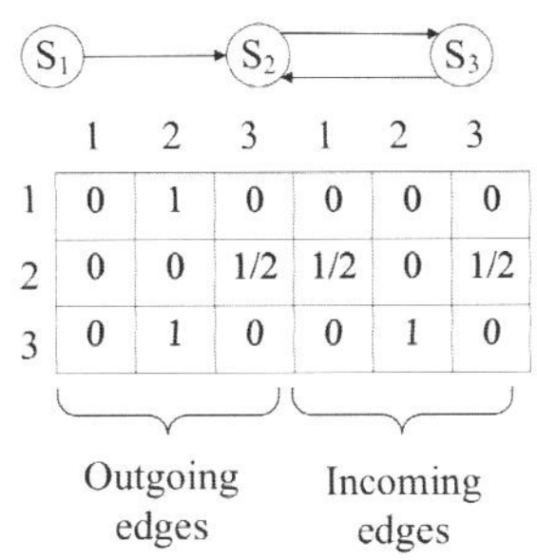

Figure 4. An example of matrix splicing.

3.4.2. Gated Graph Neural Network. To obtain the feature vectors of the user's historical behavior and the feature vectors of the current behavior, the node vectors after the graph processing is completed are trained by a GGNN.

$$Z_{g,i}^{t} = \sigma\left(W_z a_{g,i}^{t} + U_z S_i^{t-1}\right) \tag{2}$$

$$r_{g,i}^{t} = \sigma\left(W_r a_{g,i}^{t} + U_r S_i^{t-1}\right) \tag{3}$$

$$\overline{S_i^{t}} = \tanh\left(W_o a_{g,i}^{t} + U_o\left(r_{g,i}^{t} \Theta S_i^{t-1}\right)\right) \tag{4}$$

$$S_i^{t} = \left(1 - Z_{g,i}^{t}\right)\Theta S_i^{t-1} + Z_{g,i}^{t}\Theta \overline{S_i^{t}} \tag{5}$$

where $z_{g,i}$ and $r_{g,i}$ are the reset gate and update gates respectively, $\sigma(.)$ is the Sigmoid function, W and U are the parameters, and Θ is the operation of multiplying and adding the matrix elements.

For each relational graph, the graph neural network extracts the potential vectors of the matrix neighborhoods and feeds them into the graph neural network as inputs. $z_{g,i}$ and $r_{g,i}$ determine which information is retained and discarded respectively. After that, the candidate vectors are constructed by equation (4), where tanh is the activation function and the final vector S_i^t is the combination of the previously hidden vectors and the candidate vectors. When all the nodes in the relation graph converge, all the node vectors are obtained.

3.4.3. Attention Preference Integration. The previous model did not distinguish the current and historical preferences of group users, we integrate them by combining the current and historical preferences of group users.

$$\partial_i = q^T \sigma\left(\mathrm{W}_1 S_n + \mathrm{W}_2 S_i + c\right) \tag{6}$$

$$S_p = \sum_{i=1}^{n} \partial_i \mathrm{S}_i \tag{7}$$

The feature vector S_h of all nodes is obtained after inputting all relationship maps into the gated graph neural network. To consider the current preferences of the user, we embed the local vector into S_p. For example, for the vector $[S_1, S_2, \ldots, S_n]$, the local embedding vector S_p is defined as the last site S_n clicked by the user.

$$S_L = W_3\left[S_h : S_p\right] \tag{8}$$

where S_h and S_p denote user historical preferences and user current preferences respectively.

Finally, the final feature vector S_L is calculated by linearly transforming the connection of the historical and current feature vectors.

3.4.4. Model Training. After obtaining the feature vectors for each group of users, we calculate the score $\hat{z}_i$ for each candidate site $S_{g,i}$ by multiplying them together.

$$\hat{Z}_i = \mathrm{S}_L^{\mathrm{T}} S_{g,i} \tag{9}$$

Then, a softmax function is used to obtain the output vector of the model $\hat{y}$.

$$\hat{y} = \mathrm{softmax}\left(\hat{z}\right) \tag{10}$$

where $\hat{z}_i$ denotes the recommendation score for all candidates and ŷ denotes the probability of the next click of a node in cluster g. For each relationship graph, the loss function is defined in terms of cross-entropy.

$$L(\hat{y}) = -\sum_{i=1}^{m} y_i \log(\hat{y}_i) + (1 - y_i)\log(1 - \hat{y}_i) \tag{11}$$

Finally, the proposed E-SGNN model is trained using a back propagation algorithm [23].

So, the output of the model is a list of the next click sites for the recommended users.

4. Experiment

4.1. Datasets

The data we use is online user behavior log data collected from the China Internet Center, which is collected from a data collection client installed on a personal computer [24]. There are now tens of thousands of volunteers across the country who have installed the client.

Up to now, the datasets have terabytes of data. To facilitate the study, we randomly select 1000 users' online behavior browsing records in one month for the experiment. Assuming that the online behavior of each user in a month has stability, this paper extends the obtained conclusions to other times for verification, to ensure the real validity of the data.

Table 2. Datasets statistics.

Statistics	User	Site
Quantity	905	20636
Number of groups	2	-
Average user Clickstream length	442	
Total number of hits	794497	

Table 2 shows the statistical information of the data. To ensure the accuracy of the results, 95 users with a clickstream length of less than 5 were deleted. Considering the privacy of users, the user number was used instead of the user name during the experiment. The data recorded the user's Internet time, browser address, and user demographic attributes.

Table 3. User online log data sample.

User	Browser Address	Timestamp	Date
DD5	http://www.2345.com/	1343794023	2012/8/1
DD5	http://236jj.com/	1343794039	2012/8/1
DD5	http://www.youku.com/	1343794074	2012/8/1
DD5	http://www.6vdy.com/	1343794089	2012/8/1

Table 4. User attribute information.

User	Sex	Age	Education	Job	Income
DD5	Male	40	Specialty	Professional and technical staff	2001-3000 RMB
D77	Male	49	Undergraduate	Business managers	5001-8000 RMB
7D8	Male	45	Undergraduate	General staff	2001-3000 RMB
2A3	Male	39	Undergraduate	Career workers	5001-8000 RMB
5D9	Male	23	Middle school	Students	No

To better predict the online user's preferences and recommend websites of interest to him, the consecutive repeated records were censored to obtain the sample data shown in table 3.

In addition, table 4 shows the demographic attribute information of users such as age, gender, education, job, and income.

4.2. Baseline Algorithms

To analyze the performance of our proposed algorithm, the following five representative benchmark algorithms are compared in this paper. (1) GRU4Rec is a method for making recommendations using a modified version of GRU with recurrent neural networks [25]; (2) NARM uses an RNN with an attention mechanism to determine the user's preferences [26]; (3) STAMP uses the attention network to replace the RNN encoder, and finally, the short-term memory to make the recommendation model more accurate [27]; (4) SR-GNN is an approach to modeling using a session-based GNN [16]; (5) GNN-SR uses cross-session information to build a new graph neural network FGNN for recommendation [17].

4.3. Evaluation Metrics

The following two metrics are used to evaluate compared methods.

(1) Recall Rate (recall@N): It refers to the ratio of the number of successful recommendations to the total number of recommendations. Specifically, it means that the recommendation system recommends N sites to the user, and the success is indicated when the user is interested in one or more of the N sites, and the recall does not indicate the ranking order.

$$recall = \frac{TP}{TP + FN} \tag{12}$$

(2) Mean Reciprocal Rank (MRR@N): The best case is that the site recommended to the user is the more forward the better, once the site is backward then it will give the user a bad experience and affect the recall rate, so the larger mean reciprocal rank the better.

$$\mathrm{MRR} = \frac{1}{|\mathrm{N}|}\sum_{i=1}^{|N|}\frac{1}{rank_i} \tag{13}$$

where rank_i denotes the recommended rank and N denotes the number of recommended sites.

4.4. Experimental Results

To demonstrate the performance of our proposed model, we compare it with five other existing benchmark algorithms and split the datasets into two parts, a training set, and a test set. Of which, the training set accounts for 90% of the entire data and the test set is the remaining 10%. In addition, the edge parameter is set to 2 and all other parameters are initialized using a Gaussian distribution with a mean of 0 and a standard deviation of 0.1, where the initial learning rate is set to 0.001 and will decay by 0.1 after every 3 cycles [28].

Table 5. Comparison with other algorithms.

Method	Recall@20	MRR@20
GRU4REC	67.64	24.89
NARM	68.20	28.93
STAMP	69.32	30.76
SR-GNN	70.54	31.94
FGNN-SG	71.21	33.32
E-SGNN	**74.34**	**34.56**

Table 5 shows the results of comparing E-SGNN with the other five benchmark algorithms, and it can be seen that the E-SGNN algorithm outperforms the other algorithms in terms of recall and mean reciprocal rank.

The E-SGNN algorithm proposed in this paper consumes more resources than previous algorithms in both time complexity and space complexity due to its high computational complexity.

In the E-SGNN algorithm, three categories of edge parameters are set, which are gender, age, and social class. For example, the edge parameter of gender can be subdivided into male and female, and the default edge parameter is 2. The edge parameter of age can be divided into a teenager, youth, and middle-aged, and the default edge parameter is 3. The social class can be divided into high, middle, and low-income people, and can be further divided according to their specific income. We find that the results are different after the classification according to different edge parameters in the E-SGNN method.

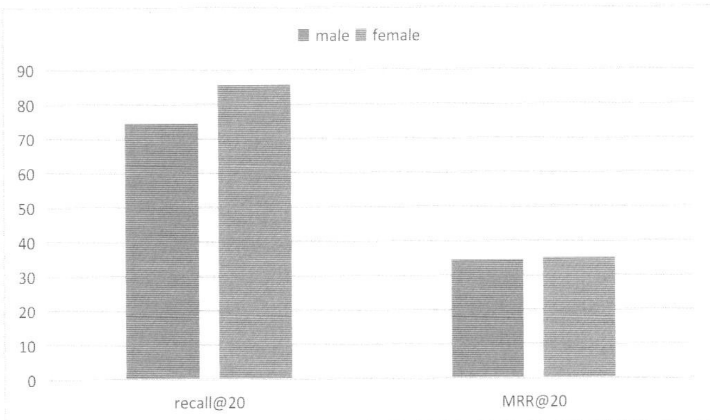

Figure 5. Grouped by gender.

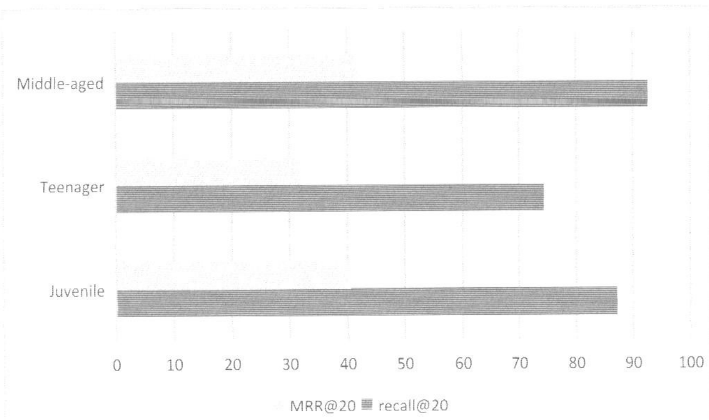

Figure 6. Grouped by gender.

As can be seen in figure 5, both male and female recall rates and mean reciprocal rank are the same, and the difference between male and female recall rates is likely caused by the number of male to female ratios, which is in line with the current number of male to female ratios in China.

As seen in figure 6, the E-SGNN algorithm has superior performance in terms of recall rate and mean inverse rating for adolescents aged 7-17 years. This is because the interests of young people are mostly similar, and the period of this age is not as large as the latter two periods of youth and middle age, so the performance will be better. In the youth group of 18-40 years old, it can be seen that the algorithm of this paper slightly outperforms the others, which is due to the consideration of group grouping and putting people of the same age together. In the last group, the middle-aged group of 41-65 years old, the recall rate ranks in the middle of the three groups because the elderly user group is slightly more than the teenagers but less than the youth. The differences exhibited by the three groups are also consistent with the division of the overall Chinese population groups.

4.5. Ablation Experiments

We performed ablation experiments according to the different edge parameters.

(1) E-SGNN-1: According to the gender grouping of users, all users are divided into male and female groups for session recommendation.

(2) E-SGNN-2: According to the age grouping of users, all users are divided into three groups teenagers, young people, and middle-aged people for session recommendation.

(3) E-SGNN-3: According to the income grouping of users, all users are divided into three groups low, medium, and high income for session recommendation.

(4) E-SGNN-4: The users' gender and age are combined, and they are divided into six groups, which are teenage male, teenage female, young male, young female, and other six groups for session recommendation.

(5) E-SGNN-5: Users are grouped by combining gender, age, and income, and divided into eighteen groups for session recommendation.

Table 6. Performance comparison of ablation experiment results of E-SGNN.

Method	Recall@20	MRR@20	Edge parameters
E-SGNN-1	74.34	34.56	2
E-SGNN-2	84.76	38.07	3
E-SGNN-3	83.41	37.05	3
E-SGNN-4	93.68	41.75	6
E-SGNN-5	98.97	45.77	18

It is obvious from table 6 that the recommended results are more similar for the same edge parameters. The larger the edge parameters are set, the finer the grouping for users and the more accurate the recommendation results for users, but this requires higher hardware facilities such as servers.

Through the above experimental analysis, it is found that the same groups have universal preferences, so this paper considers the similarity between session groups, performs group identification of users in modeling, and then adjusts the edge parameters on users' feedback. The experiments show that the proposed method in this paper is much better than previous algorithms in terms of average recall and mean reciprocal rank.

5. Conclusions

Through the mining and analysis of website data, it was found that the existing methods did not take into account the similarities and differences between sessions. We designed the E-SGNN model through the idea of group recommendation and using edge information. The model takes advantage of the existence of gender, age, income, and other differences among users to effectively screen out different types of group users when making recommendations for them, and make session recommendations for different groups, thus improving the accuracy of the experiment. Finally, we used gated neural networks and attention mechanisms to distinguish the historical preferences of users from their current preferences to further improve the accuracy of recommendations. Theoretically, the larger the edge parameter is set, the more accurately it can recommend websites of interest to users. The E-SGNN model can be applied not only to website recommendations for online users, but also to a wide range of Internet applications such as shopping, travel, and hotel accommodation. In addition, we conducted the study on a site-level basis and did not cover a smaller level of sites, from which more in-depth research can be conducted in the future.

References

[1] Wu S W, Sun F and Zhang W T 2021 Graph neural networks in recommender systems: A survey.*ACM Computing Survers* **3** (7) 34.
[2] Kanwal S, Nawaz S and Malik M 2021 A review of text-based recommendation systems *IEEE Access* **9** 31638-31661.
[3] Huang L W, Jiang B T and Lv S Y 2018 Review on recommendation system based on deep learning *Chinese Journal of Computers* **41** (07) 1619-1647.
[4] Fayyaz Z, Ebrahimiann M and Nawara D 2020 Recommendation systems: Algorithms, challenges, netrics, and business opportunities *Applied Sciences* **10** (21) 7725-7748.
[5] Pi Q, Zhou G R and Zhang Y J 2020 Search-based user interest modeling with lifelong sequential behavior data for click-through rate prediction *Proc. of the 29th ACM Inter. Conf. on Information & Knowledge Management* pp 2685-2692.
[6] Liu H F, Jing L P and Yu J 2018 Review on matrix decomposition recommendation methods integrating social information *Journal of Software* **29** (02) 340-362.
[7] Bacciu D, Errica F and Micheli A 2020 A gentle introduction to deep learning for graphs *Neural Networks* **129** 203-221.
[8] Yang J Y, You J G and Wan X R 2021 Graph embedding via graph summarization *IEEE Access* **9** 45163-45174.
[9] Saab S, Fu Y W, Ray A and Hauser M 2021 A dynamically stabilized recurrent neural network *Neural Processing Letters* **54** 1195-1209.
[10] Liang Z Y, Du J P, Shao Y X and Ji H Y 2021 Gated graph neural attention networks for abstractive summarization. *Neurocomputing* **431** 128-136.
[11] Velammal B 2019 Typicality-based collaborative filtering for book recommendation *Expert Systems* **36** (3) 8-25.
[12] Li Y J, Daniel T and Marc B 2015 Gated graph sequence neural networks *International Conference on Learning Representations* pp 11-17.
[13] Zeng Y F, Mou Q L and Zhou L 2020 Perceptual graph based learning representation model *Journal of Computer Research and Development* **57** (03) 590-603.
[14] Wang H W and Guo M Y 2020 Recurrent memory networks: Modeling long short-term user preferences for session-based recommendation *Scientia Sinica (Informationis)* **50** (12) 1867-1881.
[15] Wang S J, Cao L B and Wang Y 2021 A survey on session-based recommender systems *ACM Computing Surveys* **54** (7) 38-59.
[16] Wu S, Tang Y Y and Zhu Y Q 2019 Session-based recommendation with graph neural networks *Proc. of the AAAI Conf. on Artificial Intelligence* pp 346-353.
[17] Qiu R H, Huang Z and Li J J 2020 Exploiting cross-session Information for session-based recommendation with graph neural networks *ACM Transactions on Information Systems* **38** (3) 1-23.
[18] Phuong T M, Thanh T C and Bach N X 2018 Combining user-based and session-based recommendations with recurrent neural networks *Neural Information Processing Systems* 487-498.
[19] Yang Q, Zhou S J and Li H Y 2020 Embedding implicit user importance for group recommendation *Cmc-computers Materials & Continua* **64** (3) 1691-1704.
[20] Rapecka A and Dzemyda G 2015 A new recommendation model for the user clustering-based recommendation system *Proc. of the ACM on Measurement and Analysis of Computing Systems* **44** (1) 54-63.
[21] Wang P P, Li L and Wang R 2021 Socially-driven multi-interaction attentive group representation learning for group recommendation *Pattern Recognition Letters* **145** 74-80.

[22] Wang X N, Tan Q M and Goh M 2020 Attention-based deep neural network for Internet platform group users' dynamic identification and recommendation *Expert Systems with Applications* **160** (5) 113728.
[23] Che Z G, Chiang T A and Che Z H 2011 Feed-forward neural networks training: A comparison between genetic algorithm and back-propagation learning algorithm *International Journal of Innovative Computing Information and Control* **7** (10) 5839-5850.
[24] Li Y, Meng X F and Zhang Q 2017 Common patterns of online collective attention flow *Science China Information Sciences* **60** (5) 59102.
[25] Hidasi B, Karazoglou A and Baltrunas L 2015 Session-based recommendations with recurrent neural networks *Inter. Conf. on Learning Representations* pp 1-15.
[26] Li J, Ren P J and Chen Z M 2017 Neural attentive session-based recommendation *Proc. of the 2017 ACM Conf. on Information and Knowledge Management* pp 1419-1428.
[27] Liu Q, Zeng Y F and Mokhosi R 2018 STAMP: Short-term attention memory priority model for session-based recommendation *Proc. of the 24th ACM SIGKDD Inter. Conf. on Knowledge Discovery & Data Mining* pp 1831-1839.
[28] Glorot X and Bengio Y 2010 Understanding the difficulty of training deep feed forward neural networks *Proc. of the Thirteenth Inter. Conf. on Artificial Intelligence and Statistics* pp 249-256.

Flow Field Measurement and Display Technology Based on Hand-Held Probe

Jin Song, Sheng Xu, Weiguo Zhang, Yi Gu and Jun Ma

Low Speed Aerodynamics Institute, China Aerodynamics Research and Development Center, Mianyang, Sichuan, China
Email: songjin830131@sina.com, 281896582@qq.com, 289458950@qq.com, guyi@cardc.cn, majun@cardc.cn

Abstract. Measuring flow field information in wind tunnel is an indispensable basic condition for aerodynamic research. Traditional measuring equipments include phase Doppler particle analyzer, hot wire anemometer, particle image velocimeter, etc. Particle image velocimeter can visualize the flow field, but its installation and debugging are complicated. To obtain flow field information quickly and realize visualization, a kind of technology based on hand-held probe is designed. The velocity of the flow field is calculated by the probe, and the position and attitude information of the probe is obtained by a kind of visual measurement system (Optotrak system), The spatial velocity distribution is fused with the position and attitude information of the probe, and then the flow field in the test area is graphically displayed by interpolation with mathematical tools. By using this technique to carry out test in wind tunnels, taking segment of the airfoil as the test object, and the dynamic measurement of the probe at the rear of the airfoil, the complex vortical flow generated by segment of the airfoil is successfully plotted. The test result shows the device used in this technology is simple in structure and flexible in use and can quickly acquire flow field information, which provides a new method and idea for flow field measurement.
Keywords. Flow field; seven-hole probe; Optotrak system.

1. Introduction
Flow measurement and display is an important way to study aerodynamic performance of aircraft. Although fluid flow measurement has entered a complex and expensive measurement era of optical and electrical equipment, for example, phase Doppler particle analyzer, particle image velocimeter, hot wire anemometer, etc., but in some measuring situations, especially the average velocity of measurement, the porous pressure probe is still one of the main choices [1, 2]. In 3D measurement, a carefully designed, well calibrated seven-hole probe has better accuracy than the hot wire anemometer, therefore, the seven-hole probe is often used as a technical means to measure the velocity and pressure, and is widely used in the quantitative measurement of complex flow around the wind tunnel test model and its rear wake [3-5], especially in the measurement of complex flow field with large flow angle, such as complex flow measurement of bluff body wake. It also has a good application background in the quantitative measurement of forebody vortices [6-8], edge vortices, leading edge vortices, tip vortices and rear body vortices. But in the process of operation, it is generally necessary to install multiple probes on the moving measuring frame, and adjust the probes to horizontal state in the initial installation. Aiming at the inconvenience in use, an experimental technique based on seven-hole probe is studied. The rapid measurement and display of local flow field can be realized by using only one

probe and hand-held method. The method can be used for rapid diagnosis of flow field when abnormal aerodynamic load occurs in the wind tunnel test. After the initial problem is found, PIV and other equipment can be used for fine measurement of flow field according to the specific situation. Therefore, it has the better flexibility to improve test efficiency.

2. Composition of the System

The system hardware of this technology is mainly composed of seven-hole probe and Optotrak system. Optotrak system is an optoelectronic device based on stereo vision measurement principle. It can be used to measure the 3D coordinates of special markers. Its sensor consists of three linear array CCD cameras, as shown in figure 1.

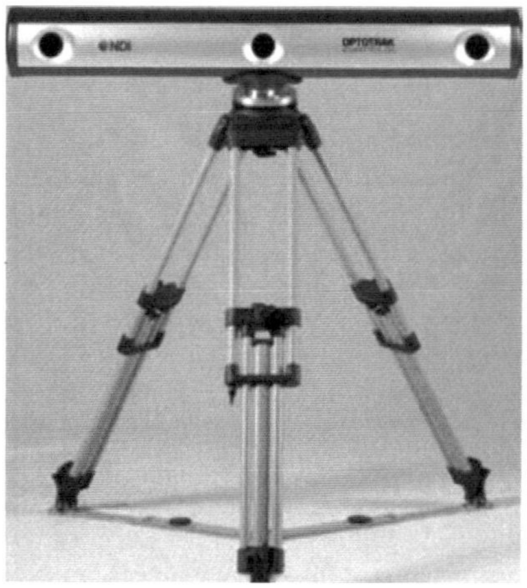

Figure 1. Optotrak system.

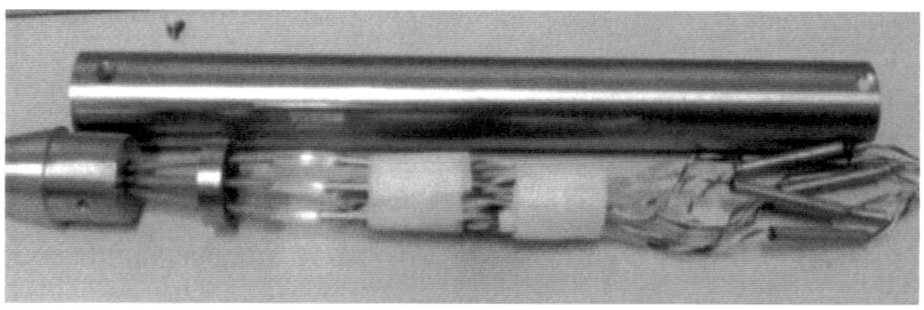

Figure 2. Internal structure of seven-hole Probe.

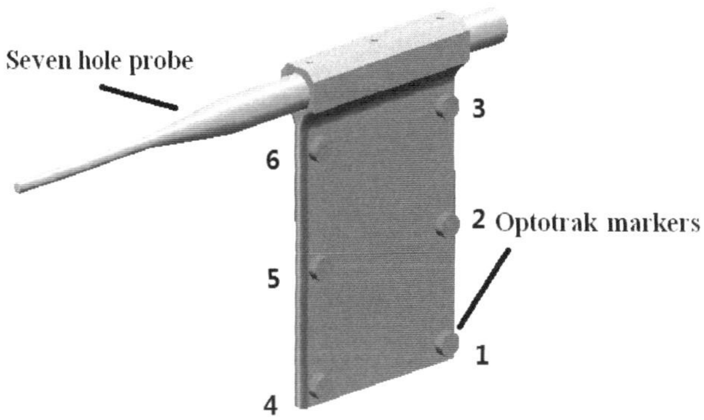

Figure 3. Probe with Optotrak markers.

After the precise calibration, two independent 2D rectangular coordinate systems are formed in orthogonal ways, and then the 3D rectangular coordinate system is synthesized. The intersection field of view of three CCD cameras is the effective measuring area of the system. The Optotrak system markers send out infrared light of specific frequency, and the CCD cameras capture the infrared light emitted by the markers, and the spatial three-dimensional coordinate of the mark point can be calculated [9-11]. Seven-hole probe is equipped with seven dynamic pressure sensors and corresponding mechanical and electronic connectors inside to convert pressure information into analog voltage signals, which are collected directly by the data acquisition system, as shown in figure 2.

A target plate is connected to a seven-hole probe. Six Optotrak markers are installed on the target plate. The structure is shown in figure 3. By the specific layout and mathematical calculation of these markers, the probe attitude and 3D coordinates of contact can be obtained. In this system, seven-hole probe is used to collect flow field pressure and convert it into velocity information. Optotrak system is used to acquire probe attitude and position information.

3. Measurement Method

By calculating the pressure information of the seven-hole probe, the velocity information of the measured points in the flow field can be obtained. The probe can be placed in the wind tunnel in any attitude, so the velocity calculated by pressure information is relative to the probe rather than relative to the wind tunnel. Therefore, it is necessary to acquire the probe attitude at this time to convert the velocity to relative to the wind tunnel coordinate system. If the flow field information of a region is to be obtained, besides pressure and probe attitude data, the coordinates of probe contact in the wind tunnel coordinate system should also be obtained. The pressure information can be directly acquired by seven-hole probe, and the attitude and contact coordinates of the probe need to be tracked by the Optotrak system markers on the target plate, so the key problem of measurement is how to get the attitude and contact coordinates of the probe.

3.1. Method for Measuring Coordinates of Probe Contact

The location relation between the markers on the target plate is known. Make marker 1 on the target plate as the origin, and the marker 1 to marker 4 is the positive direction of the X axis. The marker 1 to marker 3 is the positive direction of the Y axis, and the direction of the Z axis is determined by the right-hand rule. Thus, the probe body coordinate system $O_bX_bY_bZ_b$ is established, and the coordinates of each marker are known and recorded as $(x_i, y_i, z_i)(i=1,2,\ldots,6)$.

Using Optotrak system, the coordinates of six markers on the target plate can be obtained in the Optotrak default coordinate system, recorded as $(X_i, Y_i, Z_i)(i=1,2,\ldots,6)$. The relation between probe body coordinate system and Optotrak default coordinate system (rotation translation matrix) is C_{bc}, as follows:

$$C_{bc} = \begin{bmatrix} R_{bc} & T_{bc} \\ 0 & 1 \end{bmatrix} \tag{1}$$

Then:

$$\begin{bmatrix} X_i \\ Y_i \\ Z_i \\ 1 \end{bmatrix} = \begin{bmatrix} R_{bc} & T_{bc} \\ 0 & 1 \end{bmatrix} \begin{bmatrix} x_i \\ y_i \\ z_i \\ 1 \end{bmatrix} \tag{2}$$

In the calibration process, the probe contact is fixed at a certain position, and the probe rotates around its contact to get the coordinates of the six markers at different positions.

When the probe system is at any position, the equation (2) is true. Since the rotation translation matrix has 12 unknowns, each marker can provide 3 equations. In general, at least 4 markers are needed to get the rotation translation matrix completely. Because the markers on the target plate are on the same plane, for $z_i = 0(i = 1,2,...,6)$, there are only 9 unknowns in the rotation translation matrix, as follows:

$$C_{bc} = \begin{bmatrix} r_{11} & r_{12} & r_{13} \\ r_{21} & r_{22} & r_{23} \\ r_{31} & r_{32} & r_{33} \end{bmatrix} \tag{3}$$

Complete determination can be achieved by using at least three markers. In this paper, six markers are adopted, in order to optimize the processing and improve the accuracy of the results.

The coordinates of probe contact in the probe body coordinate system are set as (x_0, y_0, z_0). The rotation translation matrix of each position calculated by the above method is $C_{bc}^{(j)}(j\, is\, position count)$. When the probe rotates around the contact, the coordinates of the probe contact in the Optotrak default coordinate system remains unchanged, getting the following equation.

$$C_{bc}^{(1)} \begin{bmatrix} x_0 \\ y_0 \\ z_0 \\ 1 \end{bmatrix} = C_{bc}^{(2)} \begin{bmatrix} x_0 \\ y_0 \\ z_0 \\ 1 \end{bmatrix} = \ldots = C_{bc}^{(j)} \begin{bmatrix} x_0 \\ y_0 \\ z_0 \\ 1 \end{bmatrix} \tag{4}$$

The overdetermined linear equations are solved by the least square method, the coordinates (x_0, y_0, z_0) of the probe contact in the probe body coordinate system can be obtained.

Since the coordinates of the probe contact in the probe body coordinate system is known, as long as the transformation matrix from the probe body coordinate system to the wind tunnel coordinate system is obtained, the coordinates of the probe contact in the wind tunnel coordinate system can be obtained.

The method of setting up the wind tunnel coordinate system is as follows: the probe is placed in the center of the wind tunnel test section. With the aid of horizontal ruler and plumb line, the direction from marker 1 to marker 4 on the target plate is parallel to the direction of wind axis, representing positive direction of X axis. The direction from marker 1 to marker 3 is parallel to the vertical direction, representing positive direction of Y axis, and he Z axis is determined by the right-hand rule. The location of marker 1 is set as the origin of the wind tunnel coordinate system.

The coordinates of markers on the target plate are known in probe body coordinate system, and the coordinates of markers in the wind tunnel coordinate system can be directly obtained by the Optotrak system. Thus, the transformation relation of probe body coordinate to wind tunnel coordinate system can be established. The transformation matrix of Optotrak coordinate system to wind tunnel coordinate system is recorded as C_{cw}, then:

$$C_{cw} = \begin{bmatrix} R_{cw} & T_{cw} \\ 0 & 1 \end{bmatrix} \tag{5}$$

$$\begin{bmatrix} X_i \\ Y_i \\ Z_i \\ 1 \end{bmatrix} = C_{cw} \begin{bmatrix} x_i \\ y_i \\ z_i \\ 1 \end{bmatrix} \tag{6}$$

The coordinates of six markers in probe body coordinate system and wind tunnel coordinate system can be brought into equation (6) to solve C_{cw}, and the coordinates (X_0,Y_0,Z_0) of probe contact in wind tunnel coordinate system can be calculated as follows:

$$\begin{bmatrix} X_0 \\ Y_0 \\ Z_0 \\ 1 \end{bmatrix} = C_{cw} \begin{bmatrix} x_0 \\ y_0 \\ z_0 \\ 1 \end{bmatrix} \tag{7}$$

3.2. Method for Measuring Probe Attitude

One point $P(x_p,0,0)$ in the positive direction of X axis of probe body coordinate system is taken, and the coordinates in the wind tunnel coordinate system are:

$$\begin{bmatrix} x'_p \\ y'_p \\ z'_p \end{bmatrix} = R_{cw} \bullet \begin{bmatrix} x_p \\ 0 \\ 0 \end{bmatrix} \tag{8}$$

The angle between the X axis of probe body coordinate system and the horizontal plane is the pitch θ [12]:

$$\theta = \arcsin({y'_p}^2 / |OP|) \tag{9}$$

$$|OP| = \sqrt{{x'_p}^2 + {y'_p}^2 + {z'_p}^2} \tag{10}$$

The projection of the point P on the horizontal plane is defined as $Q(x'_p,0,z'_p)$. OQ is the projection of OP on the horizontal plane. The angle between OQ and X axis of the wind tunnel coordinate system is the yaw ψ.

$$\psi = \arccos(x'_p / |OQ|) \tag{11}$$

$$|OQ| = \sqrt{{x'_p}^2 + {z'_p}^2} \tag{12}$$

One point $M(0,y_m,0)$ in the positive direction of Y axis of probe body coordinate system is taken, and the coordinates in the wind tunnel coordinate system are:

$$\begin{bmatrix} x'_m \\ y'_m \\ z'_m \end{bmatrix} = R_{cw} \begin{bmatrix} 0 \\ y_m \\ 0 \end{bmatrix} \tag{13}$$

The equation of OM is

$$\frac{x}{x'_m} = \frac{y}{y'_m} = \frac{z}{z'_m} \tag{14}$$

The equation of the longitudinal symmetry plane in the probe body coordinate system is

$$-z'_p x + x'_p z = 0 \tag{15}$$

According to the method of calculating the angle between straight line and plane, the roll angle γ between straight line OM and longitudinal symmetrical plane is obtained as follows:

$$\gamma = \arcsin\left(\frac{\left|-z_p' x_m' + x_p' z_m'\right|}{\sqrt{x_p'^2 + z_p'^2}\sqrt{x_m'^2 + y_m'^2 + z_m'^2}}\right) \tag{16}$$

So far, the coordinates of the probe contact (X_0, Y_0, Z_0) and the attitude angle (θ, ψ, λ) of the probe in the wind tunnel coordinate system can be obtained.

4. Data Processing Method

According to the data processing method of the seven-hole probe, the total pressure P_{oL} and static pressure $P_{\infty L}$, as well as the angle of attack α and sideslip angle β of the airflow relative to the probe are obtained. Thus wind speed can be obtained.

$$V = \sqrt{2(P_{oL} - P_{\infty L}) / \rho} \tag{17}$$

ρ is the air density, the three velocity components of the airflow relative to the axis probe body coordinate system are:

$$\begin{aligned} u &= V\cos(\alpha)\cos(\beta) \\ v &= V\cos(\beta)\sin(\beta) \\ w &= V\sin(\beta) \end{aligned} \tag{18}$$

Since the attitude of the probe relative to the wind tunnel coordinate system is described by angle (θ, ψ, λ), it is necessary to convert the velocity components to the wind tunnel coordinate system.

$$V_0 = \begin{bmatrix} u \\ v \\ w \end{bmatrix} \tag{19}$$

$$R_1 = \begin{bmatrix} 1 & 0 & 0 \\ 0 & \cos(\gamma) & \sin(\gamma) \\ 0 & -\sin(\gamma) & \cos(\gamma) \end{bmatrix} \quad R_2 = \begin{bmatrix} \cos(\psi) & 0 & -\sin(\psi) \\ 0 & 1 & 0 \\ \sin(\psi) & 0 & \cos(\psi) \end{bmatrix} \quad R_3 = \begin{bmatrix} \cos(\theta) & \sin(\theta) & 0 \\ -\sin(\theta) & \cos(\theta) & 0 \\ 0 & 0 & 1 \end{bmatrix} \tag{20}$$

then:

$$V_1 = \begin{bmatrix} u_1 \\ v_1 \\ w_1 \end{bmatrix} = R_3 \times R_2 \times R_1 \times V_0 \tag{21}$$

In the formula, the velocity vector V_1 of the airflow is relative to the wind tunnel coordinate system. So far, the velocity vector relative to the wind tunnel coordinate system can be obtained for each point in the space to be measured, since the 3D coordinates of the probe contact in the wind tunnel coordinate system have also been obtained, using the position and the corresponding velocity information, the flow field distribution in the area to be measured can be easily plotted by MATLAB.

5. Test Verification

Flow field measurement test was carried out in FL-14 wind tunnel of China Aerodynamic Research and Development Center. The test wind speed was 20 m/s. The probe is fixed to the target plate and mounted on a long pole. Hold the pole by hand and insert the probe into the area to be measured in the wind tunnel. Slowly scan the local area of a section (YZ plane in the wind tunnel coordinate system) in the direction of incoming airflow in the wind tunnel test section, as shown in figure 4. In order to verify the correctness of this technology, it is necessary to test the flow field with known shape. In the test, a segment of the airfoil is used as the test object, and the tip of the airfoil has the characteristics of the flow field which generates the tip vortices along the direction of incoming airflow, so it can be used to test and verify the technology. Optotrak system and probe pressure data acquisition system are synchronously collected. The acquisition frequency is 2 Hz and the acquisition time is about 3 minutes. The velocity of each point in the region is calculated by pressure of the probe, and the position and attitude information obtained by the Optotrak system are combined. The velocity distribution in the region can be calculated by MATLAB and the flow field shape can be plotted, as shown in figure 5. The figure reflects the distinct vortices, which are consistent with the expected shape.

Figure 4. Wind tunnel test site.

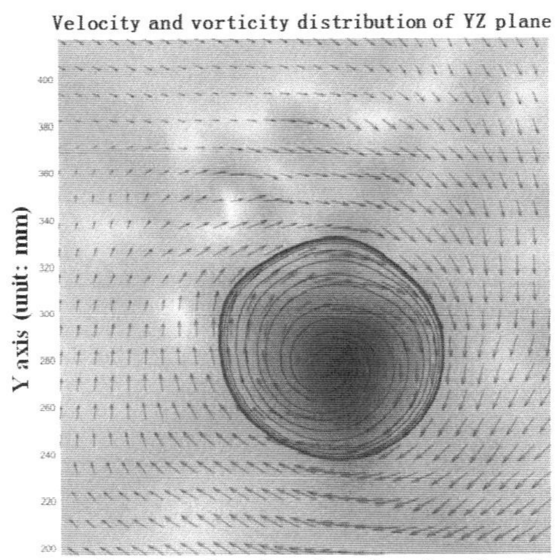

Figure 5. Distribution of velocity and vorticity.

6. Conclusion

The flow field measurement and display technology introduced in the paper obtains pressure information by probe, and obtains probe attitude and contact position by the Optotrak system, then the flow field distribution in the test area is calculated and plotted, and the visualization effect of the flow field is realized. Compared with the traditional flow field measurement equipment, this technology has the characteristics of simple structure and convenient use. Especially in the analysis of unsteady flow field aerodynamic problems, it can be used for the rapid diagnosis of flow field in wind tunnel, and provide a new method and idea for the measurement of flow field.

References

[1] Li S Z. 2015. The application of seven-hole probe in the flight test[D] (Nanjing: Nanjing university of aeronautics and astronautics): 2-5.
[2] Wang L. 2010. Research of wind tunnel experiment measurement for seven-hole probes[D] (Nanjing: Nanjing university of science and technology): 1-3.
[3] Sun M T, Zhang Y Q, LI Peng. 2008. Meterage and study on flow field beneath bottom of a helicopter in flight[J]. Acta aeronautica et astronautica sinica, 29(6): 1467-1471.
[4] Xu C Y, Zheng J, Wang Z, etc. 2021. Shear layer and wake characteristics of square cylinder in transonic flow[J]. Journal of Shanghai Jiao Tong university, 55(4):403-411.
[5] Li Y R, Yang M, Zhang Z. 2021. Flow transition and bifurcation of laminar wake in elliptical tube bundles[J]. Journal of Chinese society of power engineering, 41(4):294-300.
[6] Zhang Y, Zhang W L, Deng X G, etc. 2018. POD and DMD analysis of complex separation flows over a aircraft model at high angle of attack[J]. Physics of gases, 3(5):30-40.
[7] Liu T Z, Wang J J, He H W. 2011. Experimental study of the leading edge vortex-flap of double-delta wing[J]. Journal of experiments in fluid mechanics, 25(3):6-9.
[8] Zhang D, Hu M Q, Wang X, etc. 2016. Flow mechanism of canard vortex over forward and backward-swept wing configurations[J]. Flight dynamics, 34(1):36-39.
[9] NDI. Optotrak Certus User Guide[G]. Canada, 2009.
[10] Song J, Ma J, Jiang M. 2011. Application research of Optotrak system in model deformation and attitude measurement in wind tunnel[J]. Transducer and microsystem technologies, 30(12): 65-67.
[11] Tang L, Yang Z H, Xie Y. 2015. Application of Optotrak Optical tracking measurement instrument in wind tunnel angle measurement[J]. Ordance industry automation, 34(3): 68-70.
[12] Gu Y, Cen F, Wen Y C. 2013. Research about continuous scanning test technique based on non-contact measurement technique in low speed wind tunnel[J]. Journal of experiments in fluid mechnics, 27(5): 98-103.

Noise Suppression of Towed Line Array Sonar Platform Based on Spatial Filtering Technique

Meijiao Yang, Dong Han and Ning Li

Dalian Naval Academy, Dalian, Liaoning, China
Email: zuijiaoyan@163.com; small_hand@163.com; 1076124373@qq.com

Abstract. The problem of interference caused by the radiated noise of the towed line array sonar platform to the hydrophone line array and the formation of detection blind area in the direction of the end head of the line array is addressed. In this paper, the optimal spatial matrix filter is designed using the plane wave array manifold, the platform radiation noise copy vector, and the weighted coefficient matrix. Through iteration of the weighted coefficient matrix, the null matrix filter achieves the effect of equalizing the response of each direction of the far-field plane wave signal under the condition of zero response constraint of the platform radiation noise. The method of generalized singular value simplified decomposition is used to analyze and check the correctness of the optimal solution of the filter in theory. From the simulation results, the filter with iterative calculation is more effective in processing far-field plane waves and the signal has lower distortion.
Keywords. Line array sonar; spatial matrix filter; noise suppression; iterative calculation.

1. Introduction

Target detection in the ocean is primarily based on the target's radiation noise. The towed line array sonar is an important means of detecting low-frequency and low-noise underwater targets, and it can also locate targets. Sonar hydrophone arrays are connected by cables and are located hundreds to thousands of meters behind ships or submarines. Although the distance between ships, submarines, and other platforms and the hydrophone array is great, due to cable length limitations and other factors, the radiation noise from the platform has a significant impact on the hydrophone array, forming a blind spot in the head and tail directions of towed line array sonar and affecting target detection performance in other directions [1-3].

In recent years, the use of spatial matrix filtering technology in ocean target detection has proven to be an effective way to reduce the radiated noise of towed line array sonar platforms. The spatial matrix filter has different response effects to the passband and stopband when the passband and stopband are set to the target detection area. At the moment, spatial matrix filters, such as broadband, adaptive, discrete and continuous, are primarily used in target azimuth estimation and matching field location [4-11]. Before using array data for target azimuth estimation, spatial domain filtering can ensure that the target signal in the passband can pass through with minimal distortion while suppressing strong interference in the stopband.

References [12] and [13] address the issue of how near-field interference caused by towed line array sonar platform radiation noise affects sonar weak target detection. The matching field positioning technology and the far-field plane wave target azimuth estimation technology are combined, and the spatial matrix filter with strong interference suppression function is designed by using the copy vector of platform radiation noise reaching the receiving array and the far-field plane wave target azimuth estimation technology. Different platform radiated noise suppression effects are obtained by designing two optimization problems. The response of the designed filter to the platform noise copy vector is zero in the first optimization problem, so the platform radiation noise can be eliminated when the far-field plane wave is estimated using conventional beamforming. The absolute value of the response of the filter designed to the platform radiation noise copy vector is less than or equal to a set specific value in the second optimization problem, in order to achieve the goal of quantitatively suppressing the platform radiation noise. Despite the fact that this method can completely suppress platform radiation noise, the azimuth of the platform radiation noise will be excessively suppressed in the plane wave beam formation. As a result, the response weighting method is proposed for designing the spatial matrix filter. The filter's response value to the azimuth of the platform radiated noise is increased to reduce over-suppression, while the response error of other azimuths is reduced to ensure that far-field weak target information can be restored without distortion.

2. Weighted Spatial Filter

2.1. Design of Spatial Filter

The calculation result of towed linear array sonar target azimuth estimation is based on the assumption that the received acoustic signal is a far-field plane wave. Because of the short distance between the towed linear array sonar platform and the sonar hydrophone array, radiation noise from the platform should be treated as near-field interference caused by seawater refraction and reflection.

Assume that the incidence azimuth of far-field plane wave signal in spatial domain is Θ_P, and the array manifold matrix formed by the direction vector of far-field plane wave signal is A. Then

$$A=\left[a(\theta_1),...,a(\theta_p),...,a(\theta_P)\right],1\le p\le P,\theta_p\in\Theta_P \tag{1}$$

where $\theta=\left[\theta_1,...,\theta_p,...,\theta_P\right]$ is the plane wave incident direction, $a(\theta_p)$ is the p th direction vector of the far-field plane wave signal incident to the array by space domain discretization, and P corresponds to the total number of direction vectors of the far-field plane wave signal discretization.

Assume that the number of towline array sonar hydrophone array elements is N, and $V=[v_1,v_2,...,v_S]$ is the set of copy vectors of platform radiation noise interference arriving at the receiving array via multiple paths in the ocean, where $v_1,v_2,...,v_S$ is the copy vector of platform radiation noise at multiple locations, respectively. The received array signal including far-field plane wave and near-field platform noise interference is $x(t)$. It can be expressed as

$$x(t)=As_1(t)+Vs_0(t)+n(t) \tag{2}$$

where $s_1(t)$ is the P-dimensional far-field plane wave signal, $s_0(t)$ is the S-dimensional near-field platform noise interference, S denotes the total number of discrete direction vectors of near-field platform radiation noise, and $n(t)$ is the N-dimensional ambient noise.

The spatial matrix filter H of dimension $N\times N$ filters the received array data, and the output signal is given by

$$y(t)=HAs_1(t)+HVs_0(t)+Hn(t) \tag{3}$$

The response error of the spatial matrix filter $\boldsymbol{H}$ to the far-field plane wave signal direction vector is

$$E(\theta_p)=\left\|Ha(\theta_p)-a(\theta_p)\right\|_F^2,p=1,...,P,\theta_p\in\Theta_P \tag{4}$$

Let $w(\theta_p)$ be the weighting coefficient of the direction vector response error of the far-field plane wave signal, and the weighted overall response error of the spatial matrix filter to the far-field plane wave signal is:

$$\sum_{p=1}^{P} w(\theta_p)\left\|Ha(\theta_p)-a(\theta_p)\right\|_F^2 = \left\|(HA-A)\cdot R_{1/2}\right\|_F^2 \tag{5}$$

where $R_{1/2}$ is a diagonal matrix formed by the square root of the weighted value of the direction vector of the far-field plane wave signal.

$$R_{1/2} = \mathrm{diag}\left[\sqrt{w(\theta_1)},\sqrt{w(\theta_2)},\ldots,\sqrt{w(\theta_P)}\right]_{P\times P} \tag{6}$$

The optimization problem is constructed so that the response of the spatial matrix filter to the platform radiation noise is 0. Under this condition, the overall response error weighting of the spatial matrix filter to the far-field plane wave signal is minimized.

$$\begin{aligned}&\min_{H} J(H) = \sum_{p=1}^{P} w(\theta_p)\left\|Ha(\theta_p)-a(\theta_p)\right\|_F^2, \theta_p \in \Theta_P \\ &s.t.\ HV = 0_{N\times S}\end{aligned} \tag{7}$$

The solution process and optimal solution of this optimization problem are given. The constraint conditions of this optimization problem are equivalent to the following formula:

$$\begin{cases}\mathrm{Re}(Hv_n) = 0_{N\times 1}, 1\le n\le S \\ \mathrm{Im}(Hv_n) = 0_{N\times 1}, 1\le n\le S\end{cases} \tag{8}$$

where, $\mathrm{Re}(\cdot)$ and $\mathrm{Im}(\cdot)$ represent the real and imaginary parts of the vector, respectively.

The Lagrangian function is defined as:

$$\begin{aligned}L(H,\lambda_1,\ldots,\lambda_S,\delta_1,\ldots,\delta_S) &= \sum_{p=1}^{P} w(\theta_p)\left\|Ha(\theta_p)-a(\theta_p)\right\|_F^2 - \sum_{n=1}^{S}\lambda_n^T \mathrm{Re}(Hv_n) - \sum_{n=1}^{S}\delta_n^T \mathrm{Im}(Hv_n) \\ &= \left\|(HA-A)\cdot R_{1/2}\right\|_F^2 - \sum_{n=1}^{S}\lambda_n^T \mathrm{Re}(Hv_n) - \sum_{n=1}^{S}\delta_n^T \mathrm{Im}(Hv_n) \\ &= tr\left(HARA^H H^H\right) - tr\left(ARA^H H^H\right) - tr\left(HARA^H\right) + tr\left(ARA^H\right) \\ &\quad - \sum_{n=1}^{S}\left(\frac{\lambda_n^T}{2}+\frac{\delta_n^T}{2j}\right)Hv_n - \sum_{n=1}^{S}\left(\frac{\lambda_n^T}{2}-\frac{\delta_n^T}{2j}\right)H^* v_n^*\end{aligned} \tag{9}$$

$tr(\cdot)$ represents the trace of the matrix, $(\cdot)^*$ represents the conjugate, $(\cdot)^H$ represents the conjugate transpose, and $\lambda = [\lambda_1,\ldots,\lambda_S] \in R^{N\times S}$ and $\delta = [\delta_1,\ldots,\delta_S] \in R^{N\times S}$ are Lagrange multipliers.

The partial derivatives of $L(H,\lambda_1,\ldots,\lambda_S,\delta_1,\ldots,\delta_S)$ with respect to $(H,\lambda_1,\ldots,\lambda_S,\delta_1,\ldots,\delta_S)$ are respectively given by

$$\begin{cases} \dfrac{\partial L(H,\lambda_1,\cdots,\lambda_S,\delta_1,\cdots,\delta_S)}{\partial H} = \left(ARA^H H^H\right)^T - \left(ARA^H\right)^T - \sum_{n=1}^{S}\left(\dfrac{\lambda_n^T}{2}+\dfrac{\delta_n^T}{2j}\right)v_n^T \\ \qquad = H^* A^* RA^T - A^* RA^T - \left(\dfrac{\lambda}{2}+\dfrac{\delta}{2j}\right)V^T \\ \qquad = H^* A^* RA^T - A^* RA^T - \gamma V^T \\ \dfrac{\partial L(H,\lambda_1,\cdots,\lambda_S,\delta_1,\cdots,\delta_S)}{\partial \lambda_n} = \mathrm{Re}\left(Hv_n\right), 1\le n\le S \\ \dfrac{\partial L(H,\lambda_1,\cdots,\lambda_S,\delta_1,\cdots,\delta_S)}{\partial \delta_n} = \mathrm{Im}\left(Hv_n\right), 1\le n\le S \end{cases} \tag{10}$$

where $\gamma = \dfrac{\lambda}{2}+\dfrac{\delta}{2j} \in C^{N\times S}$ is the Lagrange multiplier constructed from λ and δ.

Assume that the stable point of the Lagrange function be $(\hat{H},\hat{\lambda},\hat{\delta})$. $(\hat{\cdot})$ denotes the optimal, namely the optimal null field filter and the optimal Lagrange multiplier satisfy the conditions that

$$\hat{H}^* A^* RA^T - A^* RA^T - \hat{\gamma} V^T = 0_{N\times N} \tag{11}$$

$$\begin{cases} \mathrm{Re}\left(\hat{H}v_n\right) = 0_{N\times 1}, 1\le n\le S \\ \mathrm{Im}\left(\hat{H}v_n\right) = 0_{N\times 1}, 1\le n\le S \end{cases} \tag{12}$$

where $\hat{\gamma} = \dfrac{\hat{\lambda}}{2}+\dfrac{\hat{\delta}}{2j}$. As shown in equation (12), the constraint condition of zero response of platform noise is

$$\hat{H}V_S = 0_{N\times S} \tag{13}$$

It can be obtained from equation (11):

$$\hat{H} = \left(ARA^H + \hat{\gamma}^* V^H\right)\left(ARA^H\right)^{-1} \tag{14}$$

where $(\cdot)^{-1}$ means finding the inverse of the full rank square matrix. From the construction of A, it follows that A is a Vandermonde matrix with full rank rows, so AA^H is a full rank square matrix and it is invertible.

The equation (14) is brought into the equation (13) to find the constructed optimal Lagrange multiplier.

$$\hat{\gamma}^* = -V\left[V^H\left(ARA^H\right)^{-1}V\right]^{-1} \tag{15}$$

The optimal Lagrange multiplier $\hat{\gamma}^*$, namely equation (15) is brought into equation (14) to obtain the optimal spatial filter as follows.

$$\hat{H} = I_N - V\left[V^H\left(ARA^H\right)^{-1}V\right]^{-1}V^H\left(ARA^H\right)^{-1} \tag{16}$$

where I_N is a unit matrix of dimension $N\times N$. $\hat{H}$ is the stable point of $L(H,\lambda_1,...,\lambda_S,\delta_1,...,\delta_S)$, namely the globally optimal solution to the optimization problem.

This numerical solution $\hat{H}$ is the global optimal solution of the aforementioned optimization problem of the spatial matrix filter design. In the above formula, $R = \mathrm{diag}\left[w(\theta_1), w(\theta_2), \ldots, w(\theta_P)\right]$ is a diagonal matrix composed of the weighting values of the direction vector of the far-field plane wave signal by the filter.

2.2. Iterative Algorithm

By setting an appropriate weighting factor $w(\theta_p) = 1, p = 1, \ldots, P$ for the passband response error, the effect of equalizing the response of the spatial matrix filter to each direction of the far-field plane wave signal can be achieved under the condition of zero response constraint on the platform radiation noise. A concrete scheme for calculating the weighted spatial matrix filter by iterative method is given. The key of this scheme lies in the iteration of weighting coefficients and the calculation of the optimal spatial matrix filter.

Initial value:

$$w_1(\theta_p) = 1, p = 1, \ldots, P, \theta_p \in \Theta_P \tag{17}$$

Iteration:

$$R_k = \mathrm{diag}\left[w_k(\theta_1), w_k(\theta_2), \ldots, w_k(\theta_P)\right] \tag{18}$$

$$\hat{H}_k = I_{N\times N} - V\left[V^H\left(AR_kA^H\right)^{-1}V\right]^{-1}V^H\left(AR_kA^H\right)^{-1} \tag{19}$$

$$F_k(\theta_p) = \hat{H}_k a(\theta_p), p = 1, \ldots, P, \theta_p \in \Theta_P \tag{20}$$

$$\beta_k(\theta_p) = \frac{P\left\|F_k(\theta_p)\right\|_F^2}{\sum_{l=1}^{P} w_k(\theta_p)\left\|F_k(\theta_p)\right\|_F^2}, p = 1, \ldots, P, \theta_p \in \Theta_P \tag{21}$$

$$w_{k+1}(\theta_p) = \beta_k(\theta_p)\left[w_k(\theta_p) + o\right], p = 1, \ldots, P, \theta_p \in \Theta_P \tag{22}$$

where $w_k(\theta_p)$ is the error weighting factor of the filter response to the far-field plane wave signal direction vector during the kth iteration. o is a relatively small value that prevents $w_{k+1}(\theta_p)$ from going to zero when $w_k(\theta_p)$ goes to zero during a given iteration. $\hat{H}_k$ is the air-domain matrix filter obtained in the kth iteration, and $F_k(\theta_p)$ is the response of filter $\hat{H}_k$ to a far-field plane wave. $\beta_k(\theta_p)$ is the product vector of the weighted values of $w_k(\theta_p)$ for the kth iteration. R_k is the weighted coefficient matrix used for the kth iteration.

Termination conditions:

(1) $k = K$. At this point, the algorithm terminates after K iterations.

(2) $\max_p \frac{\left|\left\|F_k(\theta_p)\right\|_F^2 - \left\|F_{k-1}(\theta_p)\right\|_F^2\right|}{\left\|F_k(\theta_p)\right\|_F^2} < \varsigma_1, p = 1, \ldots, P$. After iteration, the change rate of response error of spatial matrix filter to all passband direction vectors is less than a constant ς_1, and the algorithm is terminated.

(3) Let $G_k(\theta_p)=10\lg(\|\hat{H}_k a(\theta_p)\|_F^2/N), p=1,...,P, \theta_p \in \Theta_P$, then the algorithm terminates when $\max_p \frac{|G_k(\theta_p)-G_{k-1}(\theta_p)|}{|G_k(\theta_p)|} \le \varsigma_2, p=1,...,P$, namely after iteration, the decibel change rate of the response of the null domain matrix filter for all passband direction vectors is less than or equal to a constant ς_2.

All the above conditions can be used as the termination conditions of iteration, which should be selected by considering the filtering effect and the timeliness of filter design.

3. Simulation Results and Data Analysis

3.1. Towed Linear Array Structure

In this section, the uniform hydrophone line array is consisted of 64 elements. The array element spacing is 4 meters, and the frequency of target signal is 192.5Hz. Refer to figure 1 for the relative position of ship-borne towed linear array sonar platform radiation noise propagation path and array, in which the seabed sound speed is 1682 m/s, the seabed medium density is 1.76 g/cm^3, and the platform radiation noise is located at a depth of 6 meters. The first hydrophone of horizontal towed linear array near the towed platform is located 400 meters behind the platform, with a towed depth of 25 meters. Based on this hydrophone, the interference-noise ratio (INR) of the platform radiation noise reaching this hydrophone is 10 dB.

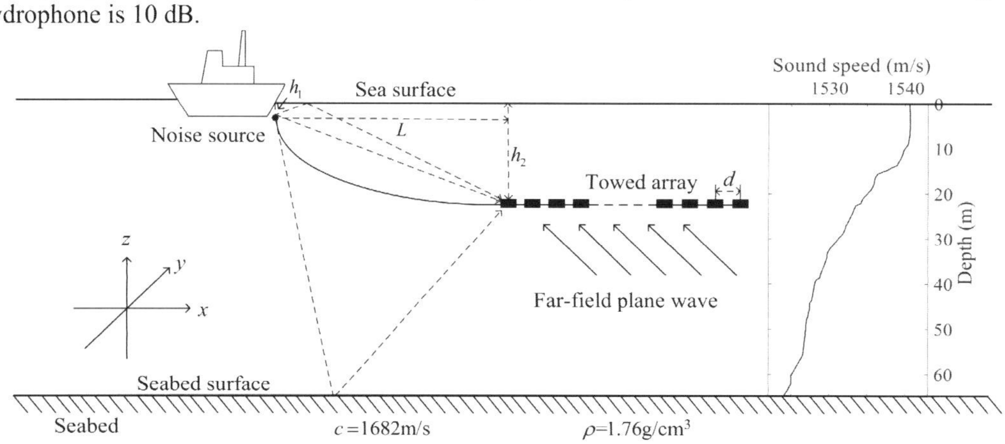

Figure 1. Schematic diagram of radiation noise propagation path of towed line array sonar platform.

3.2. Spatial Filtering Effect

The effect of the spatial matrix filter design is shown in figure 2. Figure 2a shows the filter response with two curves, where the black curve corresponds to the response of the initial spatial matrix filter $\hat{H}_0$ without iteration. The red curve corresponds to the response of the null matrix filter after 20 iterations, the weighted coefficient matrix is $R_{20}=\text{diag}\left[w_{20}(\theta_1), w_{20}(\theta_2),..., w_{20}(\theta_P)\right]$, corresponding to the response of the null matrix filter $\hat{H}_{20}$. The red curve corresponds to the response of the null matrix filter after 20 iterations. Figure 2b illustrates the filter response error, where the black curve is the response error corresponding to the null domain matrix filter $\hat{H}_0$ and the red curve is the response error corresponding to the null domain matrix filter $\hat{H}_{20}$. From figure 2a, it can be seen that after 20 iterations, the response of the filter corresponding to $\hat{H}_{20}$ is about -76° in the platform radiation noise orientation, and the response value is lower than the response value corresponding to $\hat{H}_0$. It means that after iteration, the filter is better for far-field plane waves than the result without iteration. As can be seen from figure

2b, the filter response error corresponding to $\hat{H}_{20}$, after 20 iterations, is lower than the filter response error of $\hat{H}_0$ in a large region of [-90°, 90°], indicating that the filter after iteration has lower distortion for far-field plane wave signals.

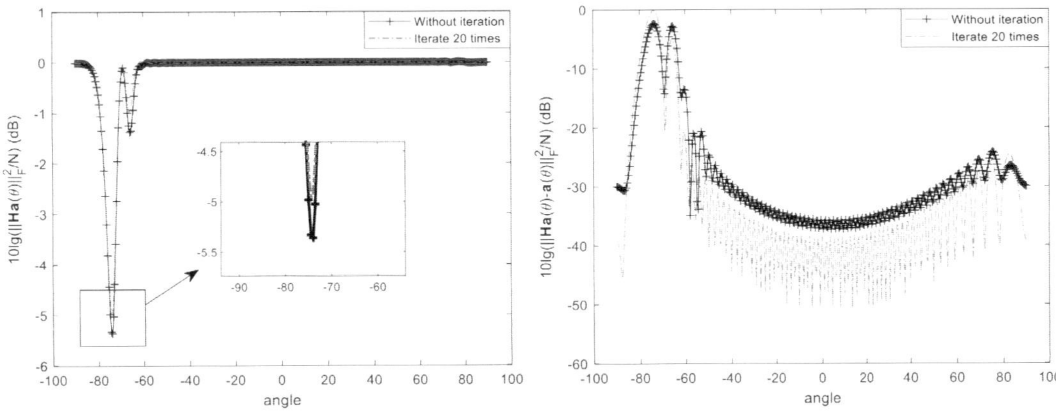

(a) Response of filter (b) Response error of filter

Figure 2. Response effect of spatial matrix filter.

Conventional beamforming (CBF) is a classic method in towed line array sonar signal processing. It has the advantages of simple structure, easy implementation, good adaptability to various environments and good stability in target azimuth estimation. Therefore, in this paper, the spatial matrix filtering technique is applied to CBF to test the ability of the filter to detect weak targets in the far field. It is assumed that the direction of continuous motion of the target is [-90°, -60°], the motion trajectory is oblique, and the position of the platform relative to the towed linear array remains unchanged. Figure 3 shows the CBF azimuth history charts before and after the weighted platform noise zero response suppression spatial filtering when the signal-to-noise ratio (SNR) of the weak target is -15dB (the abscissa is the beam search angle, the ordinate is the azimuth history angle of the weak target, and the scales on the right side of the two pictures are all CBF power values, in dB). Through the processing of this spatial filter, it is found that the platform noise interference with the azimuth of -75 is basically filtered out, and the strong interference with the azimuth of -66 is basically filtered out.

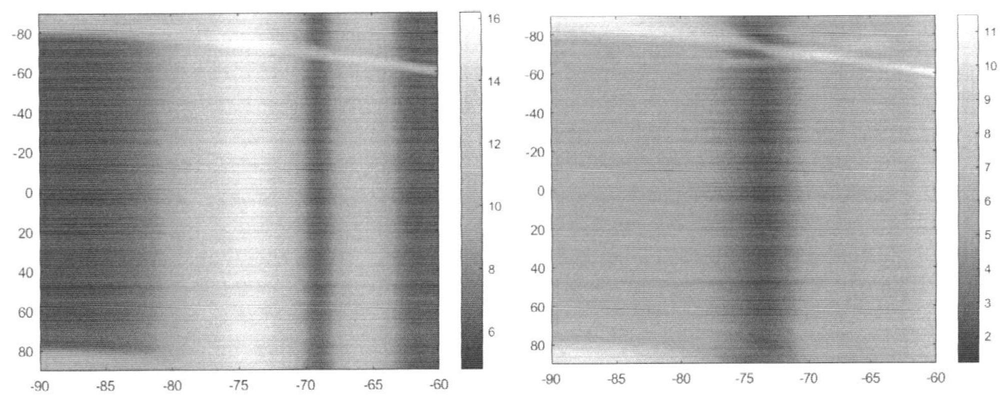

(a) Pre-filtering history (b) Post-filtering history

Figure 3. CBF azimuth history before and after weighted spatial filtering.

4. Conclusion

In this paper, the plane wave array manifold, the platform radiation noise copy vector and the weighting coefficient matrix are used to design the optimal spatial matrix filter. By iterating the weighting coefficient matrix, the spatial matrix filter can achieve the effect of balancing the response of the far-field plane wave signal in all directions under the constraint of zero response to the platform radiation noise. From the response effect of spatial matrix filter, the processing effect of the filter on far-field plane wave after iteration is better than that without iteration, and the distortion of the filter on far-field plane wave signal after iteration is lower.

References

[1] Ma Y L, Liu M A, Zhang Z B and Tong L 2002 Response of the towed line array to the noise of the tow ship in shallow water Acta Acust. 27 481-6
[2] Candy J V and Sullivan E J 2005 Cancelling tow ship noise using an adaptive model-based approach *Proc. IEEE/OES 8th Conf. on Current Meas. Technol. (Southampton)* vol 1 (Piscataway: IEEE) pp 14-8.
[3] Li Q H, Li S Q, Sun C Y and Yu H B 2007 The interference characteristics of platform and towed body noise in shallow water for active/passive towed array sonar *Acta Acust.* **32** 1-4.
[4] Maclnnes C S 2004 Source localization using subspace estimation and spatial filtering *IEEE J. Ocean. Eng.* **29** 488-97.
[5] Vaccaro R J, Chhetri A and Harrison B F 2004 Matrix filter design for passive sonar interference suppression *JASA* **115** 3010-20.
[6] Yan S F and Ma Y L 2004 Noise suppression for matched field: generalized spatial filtering method *Chin. Sci. Bull.* **49** 1909-12.
[7] Hassanien A, Elkader S A, Gershman A B and Wong K M 2006 Convex optimization based beam-space preprocessing with improved robustness against out-of-sector sources *IEEE Trans. Signal Process.* **54** 1587-95.
[8] Feng J, Yang Y X and Sun C 2007 Adaptive spatial matrix filter design with application to DOA estimation *J. Syst. Simul.* **19** 4798-802.
[9] Yuan Z T, Hu W D and Yu W X 2009 Spatial matrix prefilter DOA estimation based on four-order cumulate *Signal Process.* **25** 755-60.
[10] Han D, Zhang X H, Kang C Y and Li J 2010 Design of matrix filter with prescribed null *Acta Acust.* **35** 353-8.
[11] Han D, Zhang X H and Sun Y 2011 Design of optimal broad band spatial matrix filter *Acta Acust.* **36** 405-11.
[12] Han D, Li J, Kang C Y, Huang H N and Li Q H 2014 Towed line array sonar platform noise suppression based on spatial matrix filtering technique *Acta Acust.* **39** 27-34.
[13] Han D, Zhang H Y, Huang H N and Li Q H 2014 Towed line array sonar platform radiated noise spatial matrix filter based on far-field and near-field sound propagation characteristics *Acta Electro. Sinica* **42** 432-8.

Weighted Centroid Localization Algorithm Based on MEA-BP Neural Network and DBSCAN Clustering

Yi Li

Jiangsu Automation Research Institute, Lianyungang, Jiangsu, China
Email: zidanetorres@163.com

Abstract. In order to overcome RSSI ranging error and improve the accuracy of positioning results, a weighted localization algorithm based on MEA-BP Neural Network and DBSCAN clustering is proposed in this paper. This algorithm uses MEA-BP Neural Network (MEA-BP NN) model to optimize ranging information firstly, then it uses trilateral measurement method to get multiple initial localization results about unknown node and form a set. After clustering the results by DBSCAN and eliminating noise points, the estimated coordinate of unknown node in each cluster is obtained by using the weighted centroid localization algorithm based on collinearity. Next the number of core points in each cluster is regarded as weight value, the weighted centroid localization algorithm is used again, thus the final coordinates of unknown node can be got. Simulation results show that the localization accuracy of wireless sensor network can be improved significantly by using this algorithm in two-dimensional scene.
Keywords. Wireless sensor network; BP neural network (BP NN); mind evolutionary algorithm (MEA); DBSCAN clustering; weighted centroid localization algorithm.

1. Introduction

Wireless Sensor Network (WSN) is a kind of sensing network with easy deployment, low cost and flexible networking. Relying on the rapid development of System on Chip (SoC) and Internet of Things (IoT) technology, WSN is widely used in a wide spectrum of applications, for instance, environmental sensing [1], scene monitoring, target tracking and localization.

The algorithms for target localization using WSN can be broadly classified into two types: range-free algorithm and range-based algorithm. Range-free algorithm [2] estimates the coordinates of unknown node based on the connectivity of each node to the network, which is less accurate than the range-based algorithm. Range-based algorithm [3] firstly collects the distance or angle information of reaching the unknown node by anchor in WSN, and then performs geometric operations based on these measurement parameters to estimate location relationship between anchor and unknown node, and finally obtains the coordinates of unknown node. Common ranging methods include RSSI [4], TOA [5], TDOA [6], AOA [7], etc. As the research of machine learning technology and target localization technology keeps advancing, researchers have proposed different methods to improve the accuracy of WSN localization. Reference [8] introduces KNN and ANN into fingerprint localization algorithm to improve localization accuracy while effectively overcoming environmental interference. Reference [9] uses genetic algorithm and clustering to optimize localization process, and localization effect is significantly better than centroid algorithm and DV-hop algorithm. In Reference [10], a self-adaptive localization system using LSTM-based distance estimator by bluetooth is designed and localization experiment achieves better results.

In this paper, MEA-BP NN is applied to optimize ranging model of WSN, and the clustering analysis of the unknown node positioning coordinates in the positioning scenario, and a weighted localization algorithm combining MEA-BP Neural Network optimized ranging and DBSCAN clustering is proposed to further improve the positioning accuracy.

2. Optimizing RSSI Ranging Information

2.1. RSSI Ranging Model Analysis

Received Signal Strength Indication (RSSI) is the most commonly used and easily implemented measure of range, and the distance between anchor and unknown node can be calculated based on the fading of signal strength and transmission loss model when anchor communicates with unknown node. The commonly used transmission loss model is log-normal distribution model, as shown in equation (1).

$$PL(d) = PL(d_0) + 10\eta lg\left(d / d_0\right) + \chi \tag{1}$$

Suppose d_0 is reference distance, η is the path loss factor, χ is the Gaussian random variable, $PL(d)$ is path loss from unknown node to anchor . In addition, $PL(d_0)$ is path loss from unknown node to d_0.

Generally speaking, the wireless communication environment where WSNs are used for localization is variable and complex, and the presence of multipath effect, fading effect can lead to large fluctuations in RSSI ranging. Therefore, this paper uses improved BP NN to optimize ranging model and considers real-time path loss factor η in the RSSI transmission loss model as a variable, which is used as the input of the Neural Network together with the RSSI ranging information to reduce the effects of external environment on the ranging process.

2.2. BP NN Principle

BP NN is a multilayer feed-forward neural network trained by the error back propagation algorithm [11]. Let the number of neurons in input layer, hidden layer and output layer of BP NN be e, g and l respectively, weight value from input layer to hidden layer be ω_{ij} and activation function be f , and weight value from hidden layer to output layer be ω_{ki} and activation function be $f^{'}$. Then output of i-th neuron in hidden layer is as follows:

$$a_{1i} = f\left(\sum_{j=1}^{e} \omega_{ij} p_j + b_{1i}\right) \tag{2}$$

Output of h-th neuron in output layer is as follows:

$$a_{2h} = f^{'}\left(\sum_{i=1}^{g} \omega_{hi} a_{1i} + b_{2h}\right), \left(h = 1, 2, \cdots, l\right) \tag{3}$$

This model can be used to modify and optimize the RSSI ranging model because of its strong nonlinear mapping capability, self-adaptive capability and error-tolerance. Offline training phase and online ranging phase are most important steps of using this model. Measured RSSI value and real-time path loss factor between two training nodes are used as data input, and the real distance between two training nodes is used as output in offline training phase; and measured RSSI value and real-time path loss factor from unknown node to anchor are used as data input, and the data output is measured distance from unknown node to anchor in online ranging phase.

2.3. Application of MEA-BP NN

Easy to fall into local minima and slow convergence speed are disadvantages when using BP NN alone, so this paper uses Mind Evolutionary Algorithm (MEA) to improve BP NN and optimize the process of determining its initial weight value and threshold value to further improve RSSI ranging accuracy.

Mind Evolutionary Algorithm (MEA) [12, 13] is an algorithm for random searching optimization that simulates the evolutionary mechanism of human mind. Figure 1 shows the schematic diagram of BP Neural Network optimization using MEA. The specific steps of MEA are as follows.

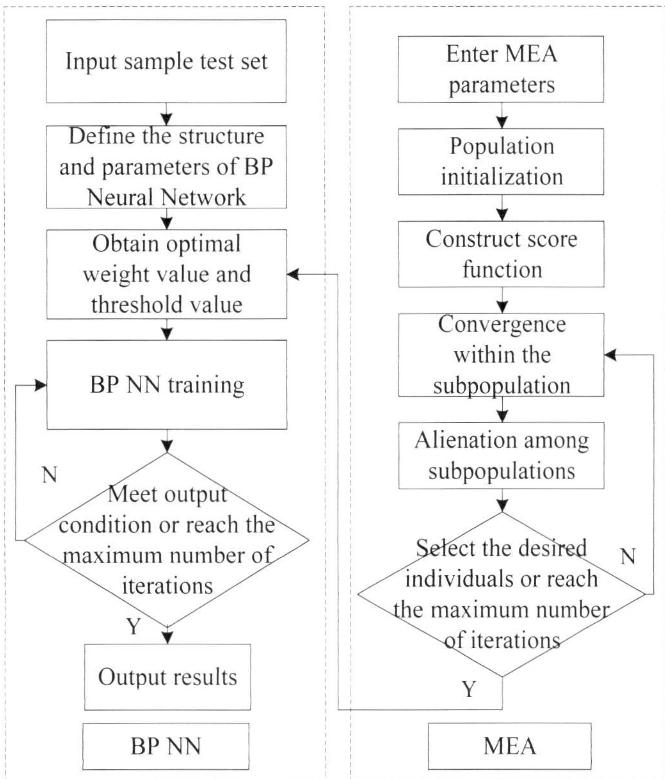

Figure 1. BP NN optimization using MEA.

(1) Population initialization. Let the number of initial weights and initial thresholds to be determined for the neural network be S. Then we have:

$$S=(e+1)\cdot g+(g+1)\cdot l \tag{4}$$

(2) Construction of score function. Score of individuals in the subpopulation is calculated and some of higher scorers are filtered out. Let score function be as follows:

$$score_j = \frac{1}{\frac{1}{Z}\sum_{j=1}^{Z}\left(T_j - O_j\right)^2} \tag{5}$$

where T_j and O_j are the true value and predicted value of BP Neural Network back propagation, respectively, and Z is the total number of samples. the larger the value of *score* function, the better the resulting individual.

(3) Convergence within the subpopulation. Each individual uses the score function as a criterion to compete with the surrounding winners, and the highest scoring individual is considered as the new winner. Multiple new individuals are derived from the winner and temporary individuals to form the winner subpopulation and temporary subpopulation. When the winner stops iterating, it means that this subpopulation has matured, at which point the convergence process ends, the score value of final winner is considered as the score value of this subpopulation.

(4) Alienation among subpopulations. All subpopulations are compared and the global range of high score groups are screened to release individuals in other subpopulations. The global individuals are searched again and new temporary subpopulations are generated when the higher score individuals are captured.

(5) The number of temporary subpopulations is kept constant to determine whether the accuracy meets the requirements, otherwise iterate Steps (3)-(4) until the desired individuals are selected or the maximum number of iterations is reached, and the optimal individuals are output as the optimal weight value and threshold value of BP NN.

3. Weighted Centroid Localization Algorithm Based on MEA-BP NN and DBSCAN Clustering

3.1. DBSCAN Clustering

Density-Based Spatial Clustering of Applications with Noise (DBSCAN) algorithm is different from K-means algorithm in that clusters of any shape can be partitioned and noise points can be filtered out using this algorithm [14, 15]. This algorithm has several main definitions as follows.

- Two input parameters: let neighborhood radius can be expressed as Eps and density threshold value can be expressed as $MinPts$.
- Three types of points: core points, boundary points and noise points. Core points contain more than or equal to $MinPts$ data points in Eps neighborhood; boundary points contain less than $MinPts$ data points in Eps neighborhood, but the points fall in Eps neighborhood of other core points; noise points refer to data points that are not classified into any of clusters.
- Direct density reachable: in data set D, sample point q is said to be directly density reachable from p if it is in Eps neighborhood of sample point p and p is the core point.
- Density reachable: in data set D, p_n is said to be density reachable from p_1 if a range of sample points p_1 , $p_2 \dots p_n$ are given and p_i is directly density reachable from p_{i-1}.
- Density connected: In data set D, if point o is the core point and both sample point q and sample point p are density reachable from o, then p and q are said to be density connected.

The pseudo code for the DBSCAN clustering algorithm is shown below:

Input: Eps, $MinPts$, data set D
Process:
1. Initialize all data points of data set D;
2. Determine status of the data point by two input parameters;
3. If the data point is a core point, create a new cluster and add all the points in Eps neighborhood of the data point to this cluster;
4. If the data point is a boundary point and no other point can be density reachable, then the next data point is examined;
5. Otherwise, the data point is judged as a noise point;
6. Repeat steps 2-5;
end

3.2. Algorithm Description

(1) For localization scenario with WSN, in offline training phase, MEA-BP Neural Network is trained using the RSSI ranging model described above;

(2) In the online localization stage, the measured RSSI values are first mean filtered. Let the number of measurements be r, then the mean filtering process is shown in the following equation. After that, the nodes use the trained MEA-BP-RSSI model for ranging and broadcast their own ID numbers.

$$RSSI(P)=\frac{1}{r}\sum_{i=1}^{r} RSSI(P_i) \tag{6}$$

(3) Unknown node establishes a database of ranging information for all anchors in communication range. Let there be n anchors that can communicate with unknown node, then these n anchors can form C_n^3 triangles (if $n<3$ then the localization process cannot be continued). If unknown node is located inside the triangle, such triangles will be called positioning triangles and establish the set. Let there be s positioning triangles to meet the requirements, then there is a set $\mathrm{A}=\{T_i \mid 1\le i\le s\}$. By the trilateral measurement method, we can find out localization coordinates of the positioning triangle to unknown node and build the set $\mathrm{B}=\{(x_i, y_i) \mid 1\le i\le s\}$.

(4) All data points in set B are clustered by DBSCAN algorithm. *Eps* and *MinPts* are input. In this paper, two parameters are associated and taken as:

$$Eps=\left(\frac{volume_D\cdot MinPts\cdot\Gamma\left(\frac{\dim}{2}+1\right)}{N\cdot\sqrt{\pi^{\dim}}}\right)^{\frac{1}{\dim}} \tag{7}$$

where $volume_D$ is the area of the data set D, dim is dimension of data point, N is the number of data points, and there are:

$$\Gamma(x)=\int_0^{\infty}\frac{t^{x-1}}{e^t}dt \tag{8}$$

Let data points in set B can be divided into k clusters and some of noise points can be filtered out.

(5) After removing the noise points, a weighted centroid localization algorithm is used for the localization results in each cluster obtained from Step (4). Let $t(1\le t\le k)$ -th cluster contain $m(t)$ available localization triangles. According to theory of collinearity, the minimum interior angle of each positioning triangle is taken as weight value, and let $w_{m(t)}=\theta_{min}$, and k coordinates of $P_t(X_t, Y_t)$, can be obtained as follows:

$$X_t=\frac{\sum_{i=1}^{m(t)} w_i\cdot x_i}{\sum_{i=1}^{m(t)} w_i},\ Y_t=\frac{\sum_{i=1}^{m(t)} w_i\cdot y_i}{\sum_{i=1}^{m(t)} w_i} \tag{9}$$

(6) The weighted centroid localization algorithm is used again for the resulting k localization coordinates. From the analysis of the DBSCAN clustering algorithm above, the core points reflect the real aggregation of data points in each cluster and are the most representative, so we consider the number of core points in each cluster as weight value. Let $t(1\le t\le k)$ -th cluster contain c_t core points

and all clusters have a total of c core points, the coordinates of the secondary localization $P(X,Y)$ for unknown node can be obtained:

$$X=\frac{c_1\cdot X_1}{c}+\frac{c_2\cdot X_2}{c}+...+\frac{c_t\cdot X_t}{c}...+\frac{c_k\cdot X_k}{c},\ Y=\frac{c_1\cdot Y_1}{c}+\frac{c_2\cdot Y_2}{c}+...+\frac{c_t\cdot Y_t}{c}...+\frac{c_k\cdot Y_k}{c} \tag{10}$$

4. Simulation Experiment Analysis

For verifying the effectiveness and reliability of the proposed algorithm, we use simulation experiments to analyse it. It is assumed that all nodes are randomly distributed in 50 m×50 m two-dimensional localization environment, and all simulation experiments are run 50 times to avoid chance errors, the final experimental results are averaged. Let actual coordinate of the unknown node be (x,y), and measured coordinate of the unknown node be $P(X,Y)$. We define the localization error as follows:

$$ER=\sqrt{(X-x)^2+(Y-y)^2} \tag{11}$$

The number of neurons in hidden layer of BP NN needs to be determined firstly before experiment, which can be derived from empirical equation (12):

$$g=\sqrt{e+l}+\lambda \tag{12}$$

where λ is a constant between 1 and 10. In this paper, g can be 5 in all experiments. We can also take the S-type tangent function as activation function of hidden layer and take the pure linear function as activation function of output layer. The parameters of the MEA are taken as follows: population size is 200, the number of winner subpopulations and temporary subpopulations is 5, the number of individuals in the subpopulation is 20, and the number of iterations is 10. *MinPts* of DBSCAN clustering algorithm is 2.

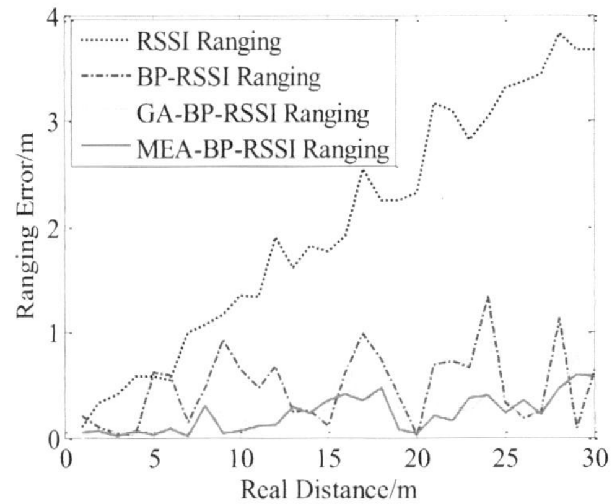

Figure 2. Relationship between ranging error and real distance.

As can be seen from figure 2, original RSSI ranging error increases with the increase of real distance, while ranging error of model trained by MEA-BP NN is smaller, and average ranging error from 1m to 30m is about 0.23m with higher stability.

Figure 3 shows sample distribution before and after clustering. Different shapes in figure 3b indicate different clusters, and • indicates noise. It can be seen that after DBSCAN clustering, which makes data points are divided into different clusters, different clusters are more obviously distinguished from each other, and noise points are filtered out.

Figure 4 shows effect of number of anchors on localization error. At this time, communication radius of all nodes is 20 m. From this figure, it can be seen that localization error of all four algorithms decreases with the increase of number of anchors, and localization accuracy of the proposed algorithm is higher. When the number of anchors is 30, localization error of proposed algorithm is about 0.18 m. This is because the increase of number of anchors can create more initial localization results, and localization results obtained after clustering weighting and noise elimination are more accurate.

Figure 5 shows effect of node communication radius on localization error. At this time, the number of anchors is 20. From this figure, it can be seen localization error of all four algorithms decreases with increase of communication radius of anchors, and localization effect of proposed algorithm is the best. Compared with other three algorithms under same conditions, localization accuracy can be improved by about 36.54%, 25.80% and 6.89%, respectively.

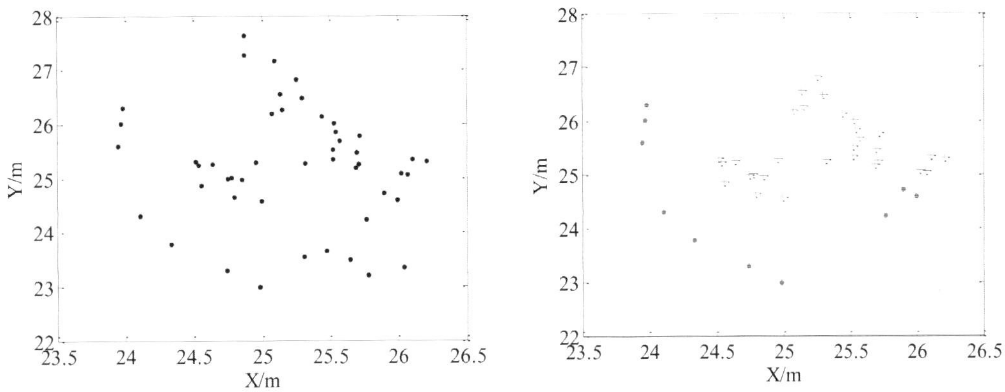

(a) Before DBSCAN clustering (b) After DBSCAN clustering

Figure 3. Sample distribution before and after DBSCAN clustering.

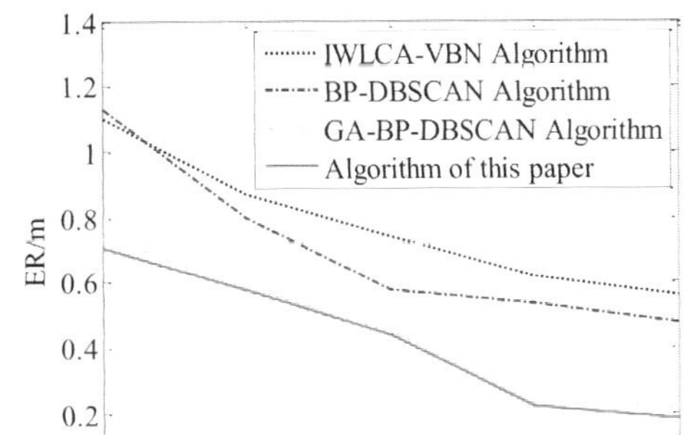

Figure 4. Effect of number of anchors on localization error.

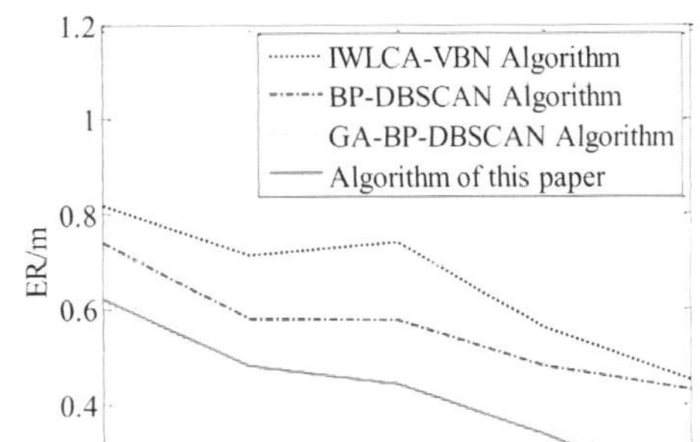

Figure 5. Effect of node communication radius on localization error.

Cumulative Distribution Function (CDF) directly reflects probability distribution of localization error in different algorithms. As shown in figure 6, the number of anchors is 30 and communication radius is 20 m. Localization error of proposed algorithm has 100% probability to fall within 0.5 m, and maximum localization error is only 0.45 m, which has better performance than other algorithms.

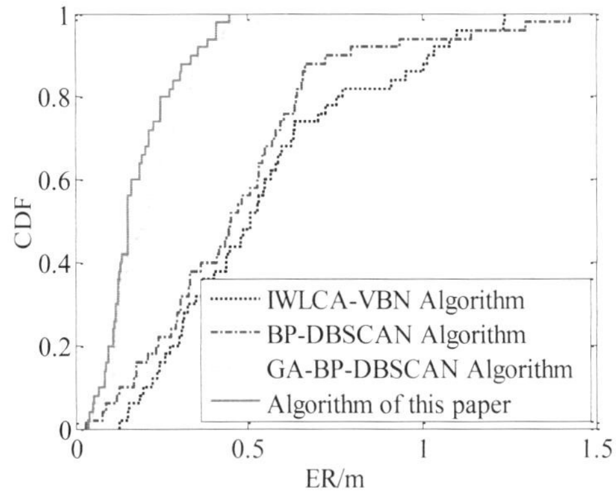

Figure 6. Cumulative distribution function.

5. Conclusion

This paper proposes a weighted localization algorithm based on optimized neural network ranging and clustering by using MEA-BP NN and DBSCAN clustering to improve the ranging model and adjust weighting factor. The simulation experiments show that, under the same conditions, proposed algorithm has higher accuracy and stability compared with other algorithms, and it has universal significance. Next step of research is to build target localization system, test algorithm performance further in different field scenarios, and debug it for improvement.

References

[1] Amutha J, Sharma S and Nagar J 2020 WSN strategies based on sensors, deployment, sensing models, coverage and energy efficiency: Review, approaches and open issues *Wireless Personal Communications* **111** (2) 1089-115.

[2] Kanwar V and Kumar A 2021 DV-Hop-based range-free localization algorithm for wireless sensor network using runner-root optimization *The Journal of Supercomputing* **77** (3) 3044-61.

[3] Liu X, Su S, Feng H, Liu Y and Pan Z 2018 A Range-Based Secure Localization Algorithm for Wireless Sensor Networks *IEEE Sensors Journal* 785-96.

[4] Li G, Geng E, Ye Z, Xu Y, Lin J and Pang Y 2018 Indoor positioning algorithm based on the improved RSSI distance model *Sensors* **18** (9) 2820.

[5] Zhang L, Yang Z, Zhang S and Yang H 2019 Three-Dimensional Localization Algorithm of WSN Nodes Based on RSSI-TOA and Single Mobile Anchor Node *Journal of Electrical and Computer Engineering* 1-8.

[6] Xiong H, Chen Z, Yang B and Ni R 2015 TDOA localization algorithm with compensation of clock offset for wireless sensor networks *China Communications* **12** (10) 193-201.

[7] Watanabe F 2021 Wireless sensor network localization using AoA measurements with two- step error variance-weighted least squares *IEEE Access* **9** 10820-8.

[8] D'Aloia M, Longo A, Guadagno G, Pulpito M and Rizzi M 2020 IoT Indoor Localization with AI Technique *2020 IEEE Inter. Workshop on Metrology for Industry 40 & IoT (MetroInd40&IoT)* pp 654-8.

[9] Sackey S H, Chen J, Henry A J and Zhang X 2019 A Clustering approach based on genetic algorithm for wireless sensor network localization *2019 15th Inter. Conf. on Computational Intelligence and Security (CIS)* pp 54-8.

[10] Li Z, Cao J, Liu X, Zhang J and Yao D 2022 A self-adaptive bluetooth indoor localization system using LSTM-based distance estimator *2020 29th Inter. Conf. on Computer Communications and Networks (ICCCN)* pp 1-9.

[11] Cui L, Tao Y, Deng J and Liu X 2021 BBO-BPNN and AMPSO-BPNN for multiple-criteria inventory classification *Expert Systems with Applications* **175** 114842.

[12] Qiao L, Liu Y, Zhu J and Wang Y 2021 Phase prediction of high carbon pearlitic steel: An improved model combining mind evolutionary algorithm and neural networks *Advanced Engineering Materials* **23** (8) 2100204.

[13] Cao H, Cui R, Liu W, Ma T and Shi Y 2021 Dual mass MEMS gyroscope temperature drift compensation based on TFPF-MEA-BP algorithm *Sensor Review* 41 (2) 162-75.

[14] Chen Y, Zhou L, Bouguila N, Wang C and Du J 2020 BLOCK-DBSCAN: Fast clustering for large scale data *Pattern Recognition* **109** 107624.

[15] Li S S 2020 An Improved DBSCAN algorithm based on the neighbor similarity and fast nearest neighbor query *IEEE Access* **8** 47468-76.

Global Mean Square Error Separation Loss

Yun Hou

Key Laboratory of Agile Intelligent Computing, Southwest China Institute of Electronic Technology, Chengdu, Sichuan, China
Email: yun.hou@foxmail.com

Abstract. Class imbalance problem greatly affects the accuracy of deep-learning-based object detectors. To weaken its influence, this paper proposes a global mean square error separation loss (GMSES Loss). GMSES Loss assigns a factor to each example, and then enhances the learning of hard-to-learn examples by reducing the factors of easy-to-learn examples. Meanwhile, the global mean square error separation method is introduced to enhance the learning of foreground classes. To verify the effectiveness of our algorithm, experiments are conducted with YOLOv4 and RetinaNet as the baselines, respectively. The results have shown that our loss function can improve the performance of one-stage object detectors without adding any extra hyperparameters.
Keywords. Object detection; deep learning; class imbalance.

1. Introduction

Object detection is an important visual recognition problem of computer vision. Traditional object detectors consist of three cascaded steps of proposal generation, feature extraction and region classification [1-4]. In the proposal generation stage, the detectors use a sliding window to find regions of interest (ROI). In the feature extraction stage, the feature vectors of ROIs are extracted by using the low-level vision descriptor, such as SIFT [1], Haar [2], HOG [3] or SURF [4]. In the region classification stage, classifiers such as SVM [5], decision tree [6] and neural network [7] are used to learn the feature vectors extracted in the previous stage, so as to obtain a training model to classify the ROIs. In recent years, the convolutional neural network algorithms have been introduced into the field of object detection, and the proposed object detectors have achieved the significantly better performance.

Different from the traditional ones, the object detector based on deep learning is end-to-end training model, which has achieved the significantly better performance. According to the mechanism of generating anchor boxes, the object detectors based on deep learning were categorized into two groups: two-stage detectors and one-stage detectors. Two-stage detectors (such as R-CNN [8], Fast-RCNN [9], Faster-RCNN [10], SPP-NET [11], etc) consist of two cascaded processes: proposal generation and region classification. The two-stage detectors first generate a sparse set of proposed regions, then use a backbone network to extract features from them, and feed the extracted features into a classifier to predict the categories of the regions. One-stage detectors (such as YOLO [12], SSD [13], RetinaNet [14], CornerNet [15], CenterNet [16], etc) directly classify and predict objects at each position of the feature map without a separate stage of generating proposal. Two-stage detectors usually have better performance, and one-stage detectors have better time efficiency and are more suitable for real-time tasks. The researchers believe that the class imbalance problem is the main factor for the different performance between these two types of detectors [14-22]. Class imbalance problems can be categorized into two groups: foreground-background examples imbalance and hard-easy examples imbalance. In recent years, the imbalance problem in object detectors based on deep learning has received more and more attention from researchers. Researchers proposed RPN [10], Focal Loss [14], Online Hard Example Mining (OHEM) [18], GHM [19], DR Loss [22], Selective Search [23], EdgeBoxes [24], DeepMask [25, 26], and ATSS [27] to deal with the class imbalance problem. Although the above algorithms for dealing with class imbalance problems can effectively improve the performance of one-stage detectors, these all algorithms have hyperparameters, which tuning them will be a time-consuming work.

To overcome the shortcomings of excessive hyperparameters in the related algorithms, a global mean square error separation loss is proposed in this paper. Inspired by Focal Loss, our loss function enhances the learning of difficult examples and foreground classes by introducing a modulating factor and the global mean square error separation method, respectively.

2. Related Works

2.1. Class Imbalance
Class imbalance problem exists in the training process of many classical one-stage object detectors. These detectors evaluate a large number of candidate locations in each image, but only a few of them contain objects. Most of candidate locations are easy background examples that contain no objects. Because the background examples dominate the training process, model degrades and training is inefficient. A common solution is hard negative mining [18] or example reweighting [14, 19]. In contrast, the reweighting scheme can make the model to efficiently train on all examples.

2.2. OHEM
Detection datasets contain a small number of examples and a large number of easy-to-learn examples. Automatically selecting these hard-to-learn examples for training can make the model more efficient and effective. Therefore, OHEM [18] selects a small subset contains hard examples with a large loss from all candidate boxes.

2.3. GHM
GHM [19] generalizes foreground-background examples imbalance and hard-easy examples imbalance as attribute imbalance. This attribute imbalance can be reflected in the gradient norm distribution. The easy-to-learn examples with very small gradient norm have a relatively large density, while hard-to-learn examples with large gradient norm have a relatively small density. Besides, the density of the medium examples is slightly smaller than the ones of outliers (very hard-to-learn examples). According to the above characteristics, GHM reduces the weight of outliers and easy-to-learn examples by embedding a harmonizing parameter in the classification loss.

2.4. DR Loss
DR Loss [22] addresses the imbalance problem in one-stage detectors by transforming the classification problem into a ranking problem. The model ranks the background candidates after the foreground ones. The worst-case loss distribution can be obtained by re-weighting the candidate box, so that the DR Loss can focus on the decision boundary between the foreground and background.

2.5. ATSS
The selection of positive and negative samples during training is the essential reason for the performance gap between anchor-free and anchor-based detectors [27]. Based on this theory, ATSS (Adaptive Training Sample Selection) [27] proposes the adaptive training sample selection according to the statistical characteristics of objects, which can automatically divide positive and negative training samples.

2.6. Focal Loss

Focal Loss [14] modifies the cross entropy loss by reweighting each example differently. It introduces a modulating factor to balance the learning of easy and hard examples, and a weighting factor to balance the learning of negative and positive classes.

The equation of cross entropy (CE) loss is expressed as.

$$CE = \begin{cases} -\log(p),\ y=1 \\ -\log(1-p),\ y=0 \end{cases} \tag{1}$$

where $y \in \{0, 1\}$ represents the ground-truth label and $p \in [0, 1]$ represents the predicted label. For notational convenience, we define p_t as the probability that an example is correctly classified.

$$p_t = \begin{cases} p,\ y=1 \\ 1-p,\ y=0 \end{cases} \tag{2}$$

The equation (1) can be rewritten as:

$$CE = -\log(p_t) \tag{3}$$

From the above equation, we can get that cross entropy loss does not consider the class imbalance problem, which usually results in the object detectors ignoring the foreground and hard-to-learn examples. In order to overcome this imbalance problem, Focal Loss (FL) is introduced into the network to increase the weight of foreground and hard-to-learn examples.

$$FL = -\alpha_t(1-p_t)^{\gamma}\log(p_t) \tag{4}$$

In equation (4), α_t is a balance factor, where the foreground examples will get larger values than background ones.

$$\alpha_t = \begin{cases} \alpha,\ y=1 \\ 1-\alpha,\ y=0 \end{cases} \tag{5}$$

$(1- p_t)^{\gamma}$ is a modulating factor, where hard-to-learn examples will get larger values than easy-to-learn ones.

3. Global Mean Square Error Separation Loss

Although many proposed algorithms can deal with class imbalance problem, these all algorithms have hyperparameters, which tuning them is a tedious task. To overcome the shortcoming of proposed algorithms, this paper proposes a global mean square error separation loss, called GMSES Loss. GMSES Loss (GL) is expressed as follows,

$$GL = -\delta((1-p_t)/p_t)\log(p_t) \tag{6}$$

where the sigmoid function δ is expressed as follows.

$$\delta(x) = \frac{1}{1+e^{-x}} \tag{7}$$

From figure 1, we can get that GMSES Loss reduces the modulating factors of easy-to-learn examples while keeping the ones of the hard-to-learn examples as constant as possible. By this way, GMSES Loss can enhance the network's learning ability for difficult examples.

In the traditional class loss function, global error is the average sum of the class errors of all examples. Since background classes account a large proportion of candidate locations, they dominate the training process. GMSES Loss introduces a global mean squared error separation method [28] to bias the network towards foreground examples. In GMSES, the global error is the mean value of all class losses.

$$E = \frac{1}{K}\sum_{k=1}^{K} E_k \tag{8}$$

where K represents the number of classes, E_k represents the error of k-th class.

$$E_k = \frac{1}{N_k}\sum_{i=1}^{N_k} GL(x_i) \tag{9}$$

where N_k represents the number of k-th class examples.

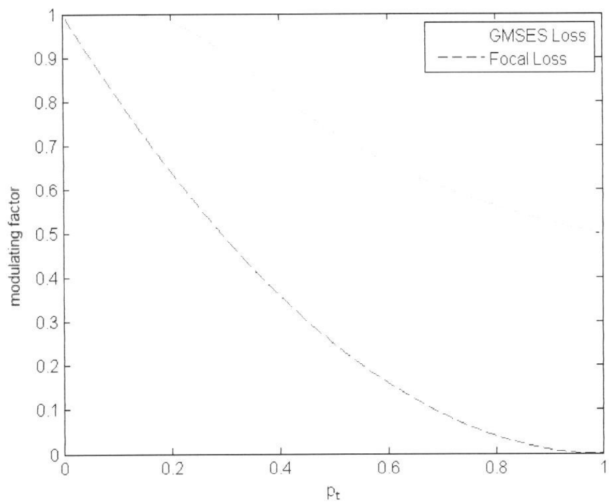

Figure 1. The modulating factors of Focal Loss and GMSES Loss.

Overall, in GMSES Loss, those easy-to-learn examples in each class are assigned smaller weights than other ones; hard background classes will receive a smaller class penalty factor than easy foreground classes. This makes the neural network pay more attention to hard-to-learn examples and foreground classes. Compared with Focal Loss, GMSES Loss modulation factor does not need to set any extra hyperparameters.

In our approach, since the global error is the weighted sum of each example, the gradient of each example on each neuron is expressed as follows.

The softmax function is expressed as:

$$p_i = \frac{e^{x_i}}{\sum_{k=1}^{K} e^{x_k}} \tag{10}$$

Among them, x is the output vector of the net, and x_i is the i-th element value of the output vector, K represents the number of classes. For GMSES Loss (GL), the gradient on each output neuron is expressed as:

$$\frac{\partial GL}{\partial x_i} = \left(-\frac{\mathrm{e}^{1-\frac{1}{p_t}} \cdot \log(p_t)}{p_t^2(1+\mathrm{e}^{1-\frac{1}{p_t}})^2} + \frac{1}{p_t(1+\mathrm{e}^{1-\frac{1}{p_t}})}\right) \cdot \frac{\partial p_t}{\partial x_i} \tag{11}$$

$$\frac{\partial p_t}{\partial x_i}=\frac{\frac{\partial(e^{x_t})}{\partial x_i}*\sum_{k=1}^{K}e^{x_k}-e^{x_t}*\frac{\partial(\sum_{k=1}^{K}e^{x_k})}{\partial x_i}}{(\sum_{k=1}^{K}e^{x_k})^2} \tag{12}$$

If $t=i$:

$$\frac{\partial p_t}{\partial x_i}=\frac{e^{x_t}*\sum_{k=1}^{K}e^{x_k}-e^{x_t}*e^{x_t}}{(\sum_{k=1}^{K}e^{x_k})^2}=p_t(1-p_t) \tag{13}$$

If $t\neq i$:

$$\frac{\partial p_t}{\partial x_i}=\frac{0-e^{x_i}*e^{x_t}}{(\sum_{k=1}^{K}e^{x_k})^2}=-p_i p_t \tag{14}$$

Then equation (11) can be expressed as:

$$\frac{\partial GL}{\partial x_i}=\begin{cases}-\frac{\mathrm{e}^{1-\frac{1}{p_t}}\cdot(1-p_t)\cdot\log(p_t)}{p_t(1+\mathrm{e}^{1-\frac{1}{p_t}})^2}+\frac{(1-p_t)}{(1+\mathrm{e}^{1-\frac{1}{p_t}})}, t=i\\ \frac{\mathrm{e}^{1-\frac{1}{p_t}}\cdot p_i\cdot\log(p_t)}{p_t(1+\mathrm{e}^{1-\frac{1}{p_t}})^2}-\frac{p_i}{(1+\mathrm{e}^{1-\frac{1}{p_t}})}, t\neq i\end{cases} \tag{15}$$

Noted that GMSES Loss is only used to compute the confidence and classification loss in the object detectors.

4. Experiments

4.1. Datastes

MS COCO is a famous database for segmentation and object detection, which contains about 118*k* images for training, 40*k* images for testing and 5*k* images for validation. Our experiments use Trainval35*k* as the training set and Minival set as the test set. The standard COCO evaluation criterion is used to compare the performance of different methods, which contains AP (average precision of all classes with the 10 IoU thresholds of 0.5:0.05:0.95), AP_{50} (AP with an IoU threshold of 0.50), AP_{75} (AP with an IoU threshold of 0.75), and AP_S (AP of small objects), AP_M (AP of medium objects), and AP_L (AP of large objects). Among them, the area of small-scale objects is less than 32^2, the area of medium-scale objects is greater than 32^2 and less than 96^2, and the area of large-scale objects is greater than 96^2.

4.2. Experimental Results

Firstly, the experiments use YOLOv4 [29] as the baseline to evaluate the proposed GMSES Loss. Unless explicitly stated, all experiments use CSPDarknet53 [30] as the backbone, and the input image size is 512×512. To be fairness, the rest of the experimental architecture is exactly the same as that in YOLOv4 except for the loss function. Specifically, the step-decay learning rate scheduling strategy is adopted in the experiments and the training step is 500,500. The initial learning rate is 0.01, which is reduced by a factor of 0.1 after the 400,000-*th* and 450,000-*th* iteration steps. For anchor, each location has 3 aspect ratios and 3 scales. In addition, the data augmentation method adopts the Mosaic data augmentation method, the weight decay parameter is 0.0005 and the momentum parameter is 0.9. Table 1 shows the experimental results, and the best results are shown in bold. In the table, S represents the elimination grid sensitivity method; M represents the Mosaic data augmentation algorithm; GA represents the genetic algorithm, so that the network selects the best hyperparameters in the first 10% of the training steps; IT represents the IoU threshold. When IoU (truth, anchor) > IoU threshold, several anchors are used for the single ground truth; GIoU, CIoU, DIoU [31], and MSE represent the regression loss function; OA represents the use of optimized anchors for the network training process with an input size of 512×512. Experimental results have shown that the combination of "S + M + IT + GA + OA + PANet + SPP + GIoU/CIoU + GMSES Loss" achieves the best detection results.

Table 1. The influence of different methods on YOLOv4 (%).

S	M	IT	GA	OA	PANet	SPP	Loss	AP	AP_{50}	AP_{75}
					✓	✓	MSE+Focal Loss	38.0	60.0	40.8
✓					✓	✓	MSE+Focal Loss	37.7	59.9	40.5
	✓				✓	✓	MSE+Focal Loss	39.1	61.8	42.0
		✓			✓	✓	MSE+Focal Loss	36.9	59.7	39.4
			✓		✓	✓	MSE+Focal Loss	38.9	61.7	41.9
	✓		✓	✓	✓	✓	CIoU+Focal Loss	36.1	56.5	38.4
✓					✓	✓	GIoU+Focal Loss	39.4	59.4	42.5
✓					✓	✓	CIoU+Focal Loss	39.6	59.2	42.6
✓					✓	✓	DIoU+Focal Loss	39.1	58.8	42.1
✓	✓	✓	✓		✓	✓	CIoU+Focal Loss	41.5	64.0	44.8
✓	✓	✓	✓	✓	✓	✓	MSE+Focal Loss	40.3	64.0	43.1
✓	✓	✓	✓	✓	✓	✓	GIoU+Focal Loss	42.4	64.4	45.9
✓	✓	✓	✓	✓	✓	✓	CIoU+Focal Loss	42.4	64.4	45.9
✓	✓	✓	✓	✓	✓	✓	GIoU+ GMSES Loss	**43.2**	**65.3**	**46.6**
✓	✓	✓	✓	✓	✓	✓	CIoU+ GMSES Loss	**43.2**	**65.3**	**46.6**

Secondly, the experiment inserts GMSES Loss into the YOLOv4 algorithm, and compares the improved algorithm with several other object detectors (including Faster RCNN [10], SSD [13], RetinaNet [14], CornetNet [15], Mask RCNN [32], CenterNet [33], Cascade RCNN [34], YOLOv2 [35], YOLOv3 [36], CenterMask [37], RefineDet [38], etc.). Table 2 shows the comparison results between these object detectors. Compared with the baseline YOLOv4, our algorithm achieves better detection performance.

Table 2. Comparison with other object detectors (%).

Algorithm	Backbone	AP	AP_{50}	AP_{75}	AP_S	AP_M	AP_L
Two-stage detectors							
Faster RCNN [10]	ResNeXt101-FPN	36.2	59.1	39.0	18.2	39.0	48.2
Libra RCNN [21]	ResNeXt101-FPN	43.0	64.0	47.0	25.3	45.6	54.6
Mask RCNN [32]	ResNeXt101-FPN	39.8	62.3	43.4	22.1	43.2	51.2
Cascade RCNN [34]	ResNet101	42.8	62.1	46.3	23.7	45.5	55.2
Anchor-free detectors							

Algorithm	Backbone	AP	AP_{50}	AP_{75}	AP_S	AP_M	AP_L
CornetNet511 [15]	Hourglass104	42.2	57.8	45.2	20.7	44.8	56.6
CenterNet [33]	Hourglass104	44.9	62.4	48.1	25.6	47.4	57.4
CenterMask [37]	ResNet101-FPN	44.0	—	—	25.8	46.8	54.9
One-stage detectors							
YOLOv2 [35]	DarkNet19	21.6	44.0	19.2	5.0	22.4	35.5
YOLOv3-SPP [36]	DarkNet53	36.2	60.6	38.2	20.6	37.4	46.1
SSD513 [13]	ResNet101-SSD	31.2	50.4	33.3	10.2	34.5	49.8
DSSD513 [39]	ResNet101-DSSD	33.2	53.3	35.2	13.0	35.4	51.1
RefineDet512 [38]	ResNet101	36.4	57.5	39.5	16.6	39.9	51.4
RetinaNet [14]	ResNeXt101-FPN	40.8	61.1	44.1	24.1	44.2	51.2
GHM [19]	ResNeXt101-FPN	41.6	62.8	44.2	22.3	45.1	55.3
DR.Retina [22]	ResNeXt101-FPN	42.5	62.8	45.9	25.2	45.8	53.5
YOLOv4-512 [29]	CSPDarkNet53	43.0	64.9	46.5	24.3	46.1	55.2
YOLOv4-512+GMSES Loss	CSPDarkNet53	43.2	65.3	46.6	24.7	46.3	55.4

Finally, the experiments use another one-stage detector, called RetinaNet [14], as the baseline to further verify the effectiveness of GMSES Loss. This experiment compares GMSES Loss with OHEM [18], Focal Loss [14] (FL), GHM-C [19], DR Loss [22], ATSS [27], and the results are shown in table 3. In OHEM, the hyperparameters are set as: N=2, B=128. In Focal Loss, the hyperparameters are set as: α=0.25 and γ=2. In GHM-C, the hyperparameters are set as: M=30 and α=0.75. In DR Loss, the hyperparameters are set as: h=4 and L=6. In ATSS, the hyperparameters are set as: k=9. Compared with several other methods for dealing with imbalance problem, GMSES Loss still achieves very good detection results without setting any hyperparameters. This is helpful for the model to improve training efficiency and detection accuracy.

Table 3. Comparison of algorithms dealing with imbalance problem (%).

Algorithm	Backbone	AP	AP_{50}	AP_{75}	AP_S	AP_M	AP_L
CE	ResNet50-FPN	28.6	43.3	30.7	11.4	30.7	40.7
OHEM [18]	ResNet101-FPN	31.1	47.2	33.2	—	—	—
FL [14]	ResNet50-FPN	35.6	55.6	38.2	19.1	39.2	46.3
GHM-C [19]	ResNet50-FPN	35.8	55.5	38.1	19.6	29.6	46.7
DR Loss [22]	ResNet50-FPN	37.2	56.7	39.8	20.1	40.0	48.3
ATSS [27]	ResNet101	39.3	57.5	42.8	24.3	43.3	51.3
GMSES Loss	ResNet50-FPN	36.7	56.3	39.1	19.7	40.2	47.8

5. Conclusion

Existing algorithms for dealing with class imbalance problems have a large number of hyperparameters, where tuning them is a tedious task. To overcome this drawback, this paper proposes a global mean square error separation loss. Reducing the weights of easy-to-learn examples, our loss enhances the model's learning of hard-to-learn examples. Meanwhile, a global mean square error separation method is introduced to enhance the learning of foreground classes. Experiments have shown that GMSES loss is helpful for improving the performance of one-stage object detectors without adding any additional hyperparameters. It can provide a technical support and theoretical basis for the model training process.

Acknowledgments
This work is supported by Key Research and Development Projects of Sichuan Science and Technology Program under Grant 2022YFG0231.

References

[1] Lowe D G 1999 Object recognition from local scale-invariant features *ICCV.*
[2] Lienhart R and Maydt J 2002 An extended set of HAAR-like features for rapid object detection *Inter. Conf. on Image Processing.*
[3] Dalal N and Triggs B 2005 Histograms of oriented gradients for human detection *CVPR.*
[4] Bay H, Tuytelaars T and Van Gool L 2006 Surf: Speeded up robust features *ECCV.*
[5] Haykin S 1994 *Neural Networks: A Comprehensive Foundation* (New York: Macmillan College Publishing Company).
[6] Vapnik V 1995 *The Nature of Statistical Learning Theory* (NewYork: Springer).
[7] Quinlan J R 1993 *C4.5: Programs for Machine Learning.*
[8] Girshick R, Donahue J, Darrell T, et al. 2014 Rich feature hierarchies for accurate object detection and semantic segmentation *CVPR.*
[9] Girshick R 2015 Fast R-CNN *ICCV.*
[10] Ren S, He K, Girshick R, et al. 2015 Faster R-CNN: Towards real-time object detection with region proposal networks *NeurIPS.*
[11] He K M, Zhang X Y, Ren S Q, et al. 2015 Spatial pyramid pooling in deep convolutional networks for visual recognition *TPAMI* **37** 1904-1916.
[12] Redmon J, Divvala S, Girshick R, et al. 2016 You only look once: Unified, real-time object detection *CVPR.*
[13] Liu W, Anguelov D, Erhan D, et al. 2016 SSD: Single shot multibox detector *ECCV.*
[14] Lin T Y, Goyal P, Girshick R, et al. 2017 Focal loss for dense object detection *IEEE Transactions on Pattern Analysis & Machine Intelligence* pp 2999-3007.
[15] Law H and Deng J 2018 CornerNet: Detecting Objects as paired keypoints *ECCV.*
[16] Zhou X, Wang D and Krahenbuhl P 2019 Objects as points *CVPR.*
[17] Lin T Y, Dollar P, Girshick R, et al. 2017 Feature pyramid networks for object detection *CVPR.*
[18] Shrivastava A, Gupta A, Girshick R 2016 Training Region-based Object Detectors with Online Hard Example Mining *IEEE Conf. on Computer Vision & Pattern Recognition* pp 761-769.
[19] Li B, Liu Y and Wang X 2018 Gradient harmonized single-stage detector *CVPR.*
[20] Chen K, Li J, Lin W, et al. 2019 Towards accurate one-stage object detection with AP-Loss *CVPR.*
[21] Pang J, Chen K, Shi J, et al. 2019 Libra R-CNN: Towards balanced learning for object detection *CVPR.*
[22] Qian Q, Chen L, Li H, et al. 2019 DR Loss: Improving object detection by distributional ranking *CVPR.*
[23] Uijlings J R, Van De Sande K E, Gevers T, et al. 2013 Selective search for object recognition *IJCV.*
[24] Zitnick C L and Dollar P 2014 Edge boxes: Locating object proposals from edges *ECCV.*
[25] Pinheiro P O, Collobert R and Dollar P 2015 Learning to segment object candidates *NIPS.*
[26] Pinheiro P O, Lin T Y, Collobert R, et al. 2016 Learning to refine object segments *ECCV.*
[27] Zhang S, Chi C, Yao Y, et al. 2020 Bridging the gap between anchor-based and anchor-free detection via adaptive training sample selection *CVPR.*
[28] Castro C L and Braga A P 2009 Artificial neural networks learning in ROC space *Inter. Joint Conf. on Computational Intelligence* pp 484-489.
[29] Bochkovskiy A, Wang C Y and Liao H Y M 2020 YOLOv4: Optimal speed and accuracy of object detection *Preprint* 2004.10934.
[30] Wang C Y, Liao H Y M, Wu Y H, et al. 2020 CSPNet: A new backbone that can enhance learning capability of CNN *CVPR.*

[31] Zheng Z, Wang P, Liu W, et al. 2020 Distance-iou loss: Faster and better learning for bounding box regression *AAAI Conf. on Artificial Intelligence.*
[32] He K, Gkioxari G, Dollar P, et al. 2017 Mask R-CNN *ICCV* pp 2980-2988.
[33] Duan K, Bai S, Xie L, et al. 2019 CenterNet: Keypoint triplets for object detection *ICCV* pp 6569-6578.
[34] Cai Z and Vasconcelos N 2018 Cascade R-CNN: Delving into high quality object detection *CVPR.*
[35] Redmon J and Farhadi A 2017 YOLO9000: Better, faster, stronger *CVPR.*
[36] Redmon J and Farhadi A 2018 YOLOv3: An incremental improvement *Preprint 1804.02767.*
[37] Lee Y and Park J 2020 CenterMask: Real-time anchor-free instance segmentation *CVPR.*
[38] Zhang S, Wen L, Bian X, et al. 2018 Single-shot refinement neural network for object detection *CVPR.*
[39] Fu C, Liu W, Ranga A, et al. 2017 DSSD: Deconvolutional single shot detector *CVPR.*

Accuracy Controllable SpMV Optimization on GPU

Jie Liu

School of Computer Science and Technology, Beijing Institute of Technology, Beijing, China.
Email: jieliu@bit.edu.cn

Abstract. Sparse matrix vector multiplication (SpMV) is a key kernel widely used in a variety of fields, and mixed-precision calculation brings opportunities to SpMV optimization. Researchers have proposed to store nonzero elements in the interval (-1, 1) in single precision and calculate SpMV in mixed precision. Though it leads to high performance, it also brings loss of accuracy. This paper proposes an accuracy controllable optimization method for SpMV. By limiting the error caused by converting double-precision floating-point numbers in the interval (-1, 1) into single-precision format, the calculation accuracy of mixed-precision SpMV is effectively improved. We tested sparse matrices from the SuiteSparse Matrix Collection on Tesla V100. Compared with the existing mixed-precision MpSpMV kernel, the mixed-precision SpMV proposed in this paper achieves an accuracy improvement.
Keywords. Mixed precision; SpMV; GPU; accuracy controllable.

1. Introduction

In many important fields of science and engineering, such as computational electromagnetic, the mathematical models established are often partial differential equations [1]. In general, finite volume, finite element and finite difference method are used for discretization, in which the discrete partial differential equations are transformed into sparse linear systems [2]. SpMV is the kernel frequently used in the iterative solution of sparse matrix linear systems (such as CG [3] and GMRES [4]), and it is also the main performance bottleneck [5]. Therefore, it is essential to optimize SpMV.

Ahmad et al. [6] proposed a data-driven mixed-precision SpMV algorithm, referred to as MpSpMV. The basic idea is that the distribution of floating-point numbers on the axis is uneven. Floating point numbers are denser near 0 and sparser near the ends of the axis. Therefore, this method partitions the nonzero elements of the matrix into two parts, the ones within the range (-1, 1) are represented in single precision, and those outside the range are represented in double precision. Although this method brings performance improvement, the accuracy loss is serious due to the conversion from double to single precision. This paper proposes an accuracy-controllable mixed-precision SpMV optimization method based on MpSpMV. Limiting the error caused by converting from double to single precision, a part of the nonzero elements in the (-1, 1) interval are selectively partitioned into single-precision sub-matrix, which has higher accuracy than MpSpMV and a little performance loss compared to MpSpMV.

2. Related Work

2.1. Mixed-Precision Numerical Method

For many problems in numerical calculation, it is possible to take advantage of the performance of low-precision arithmetic units, and at the same time perform double-precision operations in several stages of the algorithm to obtain double-precision results. Alfredo et al. [7] proposed a mixed-precision direct solution for sparse linear systems. However, the direct sparse method is affected by filling which results in high memory requirement, and the execution time is prolonged. Therefore, for large-scale problems, the direct sparse method is no longer feasible. Alfredo et al. [8] proposed a mixed-precision inner-outer iterative method, which replaced the preconditioner with an inner loop. The general attraction of this method is that the internal solver uses low precision to approximate the original matrix to achieve computational acceleration.

Mixed-precision linear algebra methods use reduced-precision data formats for calculations. The limitation of this method is that they are only suitable for iterative algorithms that can correct previous errors, such as iterative refinement [9], Jacobi solver and CG.

2.2. Decouple the Precision of Calculation and Storage

Considering that the communication overhead of applications occupies a large proportion in the CPU-GPU heterogeneous platform, the communication volume and the number of communication nodes should be reduced. Flegar et al. [10] proposed a new calculation method that separates calculation accuracy and storage accuracy for block-Jacobi preconditioner. A preconditioner is to transform the ill-conditioned linear system into a preconditioning system to speed up the convergence of the iterative solver of the linear system. The preconditioner is a rough approximation of the sparse matrix inverse. It is natural to question the necessity of using high precision for the preconditioner. Grützmacher et al. [11] proposed to split the mantissa part of the floating-point format to meet the accuracy requirements of different algorithms. Gallo et al. [12] designed an architecture based on in-memory computing for CG and GMRES. The computational task of these iterative algorithms is mainly composed of two parts: one is to obtain an approximate scheme, the other is to calculate the precise error according to the final goal. Zachariadis et al. [13] proposed a mixed precision SpGEMM based on GPU tensor cores, which supports low-precision input and performs high-precision arithmetic operations (half-precision input and single-precision multiplication and accumulation). Ahmad et al. [6] proposed a data-driven mixed precision SpMV algorithm, in which the nonzero elements within the interval of (-1, 1) are represented and calculated in single precision, while those outside the interval are represented and calculated in double precision. Although this method uses mixed-precision calculations to improve performance, the accuracy loss is serious.

3. Approach

3.1. Accuracy-Controllable Partition

Considering that storing and calculating nonzero elements in the interval (-1, 1) with single precision inevitably causes a large loss of accuracy of calculation results, we propose a mixed-precision SpMV with error limit. This method limits the accuracy loss caused by the conversion of nonzero elements in the interval (-1, 1) to single precision, thereby improving the accuracy of SpMV. Assuming that the tolerance of the absolute value of the error is represented by δ, a double-precision nonzero element $f \in (-1,1)$ and its single-precision representation is denoted as f'. If the nonzero element in the interval (-1, 1) satisfies equation (1), it can be converted to single precision. Otherwise, it is stored in double-precision format.

$$\left| f - f' \right| \le \delta \tag{1}$$

In general, the sparse matrix remains unchanged in an iterative solver, and the sparse matrix only needs to be processed once to be applied to an iterative solver all the time. Therefore, the overhead of this partition will be amortized by hundreds or thousands of iterations.

3.2. Accuracy Estimation

For a given precision format, the absolute error introduced by the calculation is proportional to the magnitude of the calculation result. In addition, floating-point error is highly uncombinable and does not obey the usual algebraic laws, such as distributive law [14]. Therefore, we use significant digits proposed in Reference [14] instead of absolute error for the accuracy analysis. The main idea of this metric is given by the equation (2):

$$\left|y' - y\right| \leq 5 \times 10^{-m-1} \tag{2}$$

where y is theoretical result and y' is experimental result. If $y' - y$ has magnitude less than or equal to 5 in the $(m+1)$ st digit of y', counting from the first nonzero digit in y' to the right, we say y' has m significant digits with respect to y, as shown in equation (2). For example, suppose $y' = 1.11056$, $y = 1.11055$, $\left|y' - y\right| = 0.00001 \leq 5 \times 10^{-4-1}$, hence y' has four significant digits with respect to y.

In order to predict the effective accuracy improvement of mixed-precision calculation brought by the accuracy-controllable partition after a given error limit δ, the sparse matrix is sampled by the equal-spaced sampling method. First, the sparse matrix is divided into k blocks by row. Then the first row of each block is taken as the sample data, and mixed-precision SpMV calculation is performed. Finally, the significant digits of the sample's result are counted. In this paper, 5% rows of the original sparse matrix are taken as the sample data, then k can be calculated by equation (3), where M represents the number of rows of the matrix.

$$k = M \cdot 5\% \tag{3}$$

3.3. Design and Implementation of SpMV Kernel

One of SpMV implementations based on the CSR format developed on the GPU is CSR-Scalar, in which each GPU thread processes a row of the matrix individually. The problem with this method is that the nonzero element distribution in each row may be uneven, which will lead to unbalanced load between threads in a warp. Moreover, the coalesced memory access cannot be used, which leads to more memory transactions. CSR-Vector solves these problems by processing the same row of matrix with a warp, which is the basic scheduling unit of GPU and has 32 threads, significantly improving the computational efficiency of SpMV. We designed and implemented our mixed-precision SpMV kernel on Tesla V100 based on the CSR-Vector provided by CUSP [15]. CUSP is a widely used sparse linear algebra and graph computing library [15]. Its SpMV has better performance in existing implementations, and the entire library is open source. Figure 1 shows the mixed-precision SpMV implementation.

In figure 1, the input includes the vector in single and double precision (*xS* and *xD*), single- and double-precision sub-matrices in CSR format. This format is widely used because of its low transformation cost. *THREADS_PER_BLOCK* is the block size of SpMV kernel, which is usually a multiple of 32. The outermost *for* loop (line 13) controls threads to traverse each row of the sparse matrix. A vector of cooperative threads processes the same row of the matrix, the number of cooperative threads is set by *THREADS_PER_VECTOR*. Threads which process the same row are renumbered by *thread_lane*. After renumbering, the thread whose *thread_lane is equal to 0* (denoted as thread0) reads the number of nonzero elements in *row*th row into the register *ptrs*, and the rest *(THREADS_PER_VECTOR-1)* threads call CUDA warp-level synchronization primitives *__shfl_sync* to receive data from thread0. The *for* loop (line 27) performs multiplication and accumulation for the single-precision sub-matrix and vector, and the *for* loop (line 30) calculates the double-precision sub-matrix. The *for* loop (line 33) calls the synchronization primitive *__shfl_down_sync* to merge the results of *THREADS_PER_VECOTR* threads to thread0, and this thread writes the result in register sum back to the result vector *y*. __shfl_sync and __shfl_down_sync exchange a variable between threads within a warp, data exchange is carried out between registers, which is more efficient than shared memory.

```
1  __global__ void spmv_GPU_SD(const unsigned int num_rows, const float *AxS, const float *xS,
2      const double *xD, const uint *AjS, const uint *ApS, double *y, const double *AxD, const uint *AjD,
       const uint *ApD){
3
4      const size_t VECTORS_PER_BLOCK = THREADS_PER_BLOCK / THREADS_PER_VECTOR;
5
6      const uint thread_id = THREADS_PER_BLOCK * blockIdx.x + threadIdx.x;
7      const uint thread_lane = threadIdx.x & (THREADS_PER_VECTOR - 1);
8      const uint vector_id = thread_id / THREADS_PER_VECTOR;
9      const uint vector_lane = threadIdx.x / THREADS_PER_VECTOR;
10     const uint num_vectors = VECTORS_PER_BLOCK * gridDim.x;
11     uint ptrsS0, ptrsS1, ptrsD0, ptrsD1;
12
13     for (uint row = vector_id; row < num_rows; row += num_vectors){
14         if (thread_lane == 0){
15             ptrsD0 = ApD[row + thread_lane];
16             ptrsD1 = ApD[row + thread_lane + 1];
17             ptrsS0 = ApS[row + thread_lane];
18             ptrsS1 = ApS[row + thread_lane + 1];
19         }
20
21         const uint row_startD = __shfl_sync(0xffffffff, ptrsD0, vector_lane * THREADS_PER_VECTOR);
22         const uint row_endD = __shfl_sync(0xffffffff, ptrsD1, vector_lane * THREADS_PER_VECTOR);
23         const uint row_startS = __shfl_sync(0xffffffff, ptrsS0, vector_lane * THREADS_PER_VECTOR);
24         const uint row_endS = __shfl_sync(0xffffffff, ptrsS1, vector_lane * THREADS_PER_VECTOR);
25
26         double sum = 0.0;
27         for (uint jj = row_startS + thread_lane; jj < row_endS; jj += THREADS_PER_VECTOR){
28             sum += AxS[jj] * xS[AjS[jj]];
29         }
30         for (uint jj = row_startD + thread_lane; jj < row_endD; jj += THREADS_PER_VECTOR){
31             sum += AxD[jj] * xD[AjD[jj]];
32         }
33         for (uint offset = THREADS_PER_VECTOR / 2; offset > 0; offset /= 2)
34             sum += __shfl_down_sync(FULL_MASK, sum, offset);
35         if (thread_lane == 0){
36             y[row] = sum;
37         }
38     }
39 }
```

Figure 1. Mixed-precision SpMV implementation with synchronization on Tesla V100.

Before the NVIDIA Volta architecture, 32 threads in a warp are in SIMT mode, which means that the GPU architecture only maintains scheduling resources for each warp. The threads in a warp execute the same instructions, and each thread uses its own data. However, starting from the Volta architecture, the feature of independent scheduling has been introduced, and the maintenance of scheduling resources has been added for each thread, including the program counter and call stack. Threads in a warp are no longer forced to synchronize, so manually control synchronization between threads if data is to be shared.

4. Evaluation

4.1. System Setup

We tested the mixed-precision SpMV using a machine equipped with a 2.4 GHz Intel Xeon E5-2680 v4 CPU, 128 GB main memory, and a Tesla V100 GPU with 16 GB of HBM2 memory. The operating systems are 64-bit Ubuntu 18.04, and the CUDA version is 11.2.

4.2. Parameter and Benchmark

- THREADS_PER_VECTOR: Considering the natural irregularity of sparse matrices, it should set different THREADS_PER_VECTOR for each sparse matrix. Reference [16] proposes the setting method shown in equation (4), which assigns a group of cooperative threads to a matrix row, called CSR-Coop for short. CSR-Coop has the best overall performance in CSR-based SpMV implementations, so we use double-precision CSR-Coop as the benchmark. We also evaluated the mixed-precision algorithm MpSpMV [6]. Since MpSpMV and our optimized mixed-precision SpMV generate two sub-matrices after partition, then equation (4) is no longer hold, so we set THREADS_PER_VECTOR to 32 for these two methods.

$$THREADS\_PER\_VECTOR = \min(2^{\lceil \log_2(sqrt(nnz/M)) \rceil}, 32) \quad (4)$$

- δ: We tested the accuracy under error limits in [10^{-15}, 10^{-6}]. The reason for choosing this interval is that single- and double-precision floating-point format can provide up to six and fifteen decimal digits of accuracy, respectively.

4.3. Data Set
The sparse matrices tested in our experiments are downloaded from the SuiteSparse Matrix Collection [16], which is a widely used set of sparse matrix benchmarks collected from a wide range of applications. There are a total of 2,870 independent matrices in the SuiteSparse Matrix Collection, including 48 complex matrices and 2,821 real matrices. Excluding complex matrices and some very large matrices, we tested 2,804 real matrices.

4.4. Accuracy
This subsection analyzes the accuracy of SpMV-SD and MpSpMV. We compare the result vector of SpMV-SD and MpSpMV to the result calculated by double-precision CSR-Coop under the same vector $\boldsymbol{x}$ to get the significant digits. There are five significant digit cases in this paper, including significant digits less than six, six significant digits, seven significant digits, eight significant digits and significant digits more than eight. The more significant digits, the higher accuracy of the results. We add up the number of elements of all SpMV result vectors with the same significant digits, and then divide by the sum of the dimensions of all result vectors to obtain the proportion of each significant number (referred to as p_m). p_m is calculated as equation (5).

$$p_m = \frac{\sum_{i=0}^{num\_matrix} num\_m_i}{\sum_{i=0}^{num\_matrix} M_i} \quad (5)$$

num_matrix represents the number of sparse matrices. The number of elements in result vector $\boldsymbol{y}$ that have *m* significant digits is recorded as num_m_i for the *i*th sparse matrix. M_i indicates the number of rows of the *i*th sparse matrix.

The proportions ($p_{m<6}$) of SpMV-SD and MpSpMV are approximately equal to 0 when the number of significant digits is less than six. Figure 2 shows the proportions of six significant digits (p_6). When the error limit is smaller than 10^{-8}, the proportions (p_6) is dramatically reduced compared to MpSpMV. Figure 3 shows the proportions ($p_{m>8}$) of SpMV-SD and MpSpMV when significant digits are more than 8. It can also be seen that the proportions ($p_{m>8}$) increases greatly when the error limit is smaller than 10^{-8}. After the accuracy-controllable partition, SpMV-SD proposed in this paper have higher computational accuracy than MpSpMV.

In order to analyze the accuracy improvement brought by SpMV-SD in more detail, the following is an analysis of the 17 matrices tested in [17] that can be optimized by mixed-precision calculations when δ=10^{-15}. Table 1 shows the nonzero elements in the single- and double-precision sub-matrices after performing partition corresponding to SpMV-SD and MpSpMV for these 17 matrices. nnzS and nnzD represent the number of nonzero elements of the single-precision sub-matrix and double-precision sub-matrix after partition, respectively, and the sum of the two is equal to the number of nonzero elements of the original matrix.

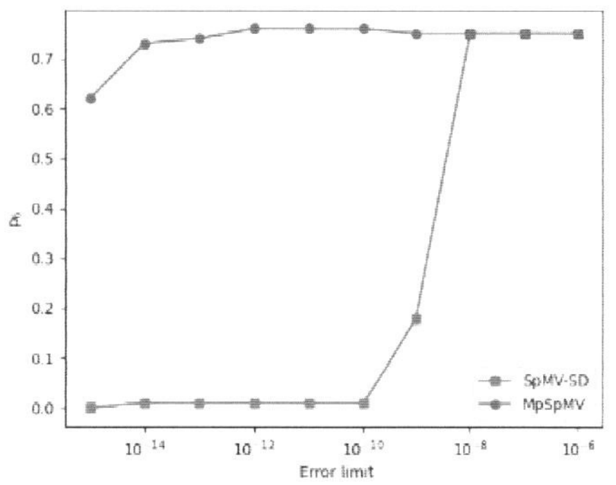

Figure 2. The proportion of six significant digits.

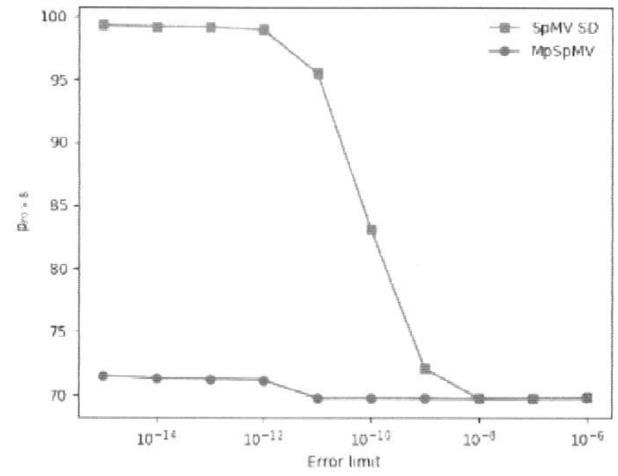

Figure 3. The proportion of significant digits which is less than eight.

Table 1. The nonzero elements in the single- and double-precision sub-matrices after partition.

Name	SpMV-SD		MpSpMV	
	nnzS	nnzD	nnzS	nnzD
mark3jac100sc	33,321	251,894	136,862	148,353
scagr7-2r	6,914	113,227	29,383	90,758
c-73	212	1,279,062	553,658	725,616
para-4	3,302,211	2,024,017	4,169,704	1,156,524
fs_760_3	997	4,979	5,382	594
oscil_dcop_42	406	1,138	662	882
PR02R	995,564	7,189,572	5,659,702	2,525,434
cavity01	1,832	5,495	6,760	567
c-54	34,369	357,324	255,955	135,738
freeFlyingRobot_3	476	12,446	6,005	6,917
bcsstk04	224	3,424	252	3,396
bcircuit	93,723	281,835	266,408	109,150
oscil_dcop_12	229	1,315	635	909
bcsstk03	4	636	4	636
ASIC_320ks	1,643,043	184,764	1,827,783	24
oscil_dcop_56	412	1,132	662	882
nasa2910	74,818	99,478	74,966	99,330

Table 2 shows the significant digits of SpMV-SD and MpSpMV versus double precision. Among them, the SpMV-SD calculation results of 15 matrices have an improvement in accuracy compared with MpSpMV. For the sparse matrix bcsstk03, there are fewer nonzero elements in the interval of (-1, 1). SpMV-SD and MpSpMV have the same partition for this matrix, so the accuracy seems to be the same. The absolute value of the result vector obtained by SpMV for nasa2910 matrix is very large, so the calculation results of SpMV-SD and MpSpMV have more than 8 significant figures. In general, SPMV-SD is promising in the case of high accuracy requirements.

Table 2. The accuracy of SpMV-SD and MpSpMV compared with CSR-Coop.

Name	<6(%)	6(%)	7(%)	8(%)	>8(%)
mark3jac100sc	0.00/0.00	≈0/1.73	5.34/25.67	9.82/36.42	84.84/36.18
scagr7-2r	0.00/0.00	0.00/0.05	7.67/10.82	5.52/15.61	86.81/73.52
c-73	0.00/0.00	0.00/0.01	0.00/0.34	0.00/2.40	100.00/97.24
para-4	0.00/0.00	0.00/0.17	0.00/3.84	0.00/7.80	100.00/88.20
fs_760_3	0.00/0.00	0.00/0.00	0.00/5.00	0.00/10.66	100.00/84.34
oscil_dcop_42	0.00/0.00	0.00/0.00	0.00/4.88	0.93/3.72	99.07/91.40
PR02R	0.00/0.00	0.00/1.21	0.00/12.25	0.00/23.04	100.00/63.49
cavity01	0.00/0.00	0.00/11.99	0.00/49.84	0.00/33.12	100.00/5.05
c-54	0.00/0.00	0.00/1.15	0.00/19.62	0.00/25.60	100.00/53.63
freeFlyingRobot_3	0.00/0.00	0.00/1.69	2.62/18.98	3.20/20.90	94.18/58.44
bcsstk04	0.00/0.00	0.00/0.00	0.00/0.00	0.00/0.76	100.00/99.24
bcircuit	0.00/0.00	0.00/0.29	0.00/19.49	0.00/25.31	100.00/54.91
oscil_dcop_12	0.00/0.00	0.00/0.47	0.00/4.65	0.93/5.11	99.07/89.77
bcsstk03	0.00/0.00	0.00/0.00	0.00/0.00	0.00/0.00	100.00/100.00
ASIC_320ks	0.00/0.00	0.00/≈0.00	0.00/0.09	0.00/0.27	100.00/99.64
oscil_dcop_56	0.00/0.00	0.00/0.00	0.23/5.58	0.23/4.42	99.54/90.00
nasa2910	0.00/0.00	0.00/0.00	0.00/0.00	0.00/0.00	100.00/100.00

Note: The left side of '/' is the accuracy of SpMV-SD, and the right side is the accuracy of MpSpMV.

4.5. Performance

Figure 4 shows the speedup of SpMV-SD and MpSpMV. It can be seen that when the accuracy is improved, the speedup of SpMV-SD gradually differs from that of MpSpMV, but the gap is relatively small.

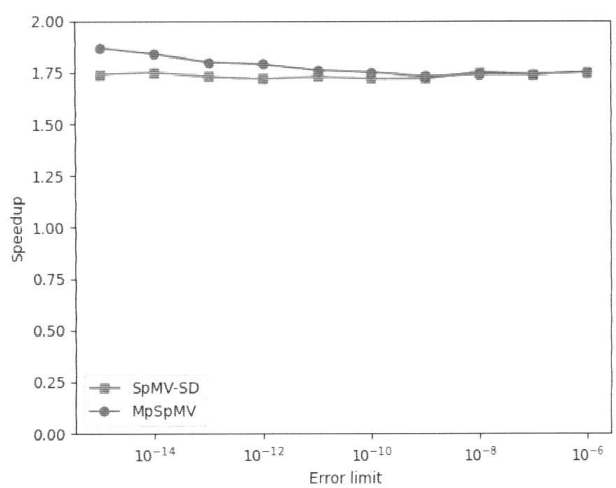

Figure 4. The average speedup of SpMV-SD and MpSpMV compared to CSR-Coop under different error limits.

4.6. Sample

Table 3 shows the significant digits of the mixed-precision computation for samples obtained by equidistant sampling of 17 matrices. It can be seen that the estimated results of the sample and the calculation results of the whole matrix have the same trend and similar proportions in significant digits. So, sampling can be used to estimate the significant digits of the mixed-precision SpMV under a given error limit.

Table 3. The significant digits of mixed-precision computation for 17 samples.

Name	<6(%)	6(%)	7(%)	8(%)	>8(%)
mark3jac100sc	0.00	0.00	5.33	9.85	84.82
scagr7-2r	0.00	0.00	7.56	5.65	86.79
c-73	0.00	0.00	0.00	0.00	100.00
para-4	0.00	0.00	0.00	0.00	100.00
fs_760_3	0.00	0.00	0.00	0.00	100.00
oscil_dcop_42	0.00	0.00	0.23	0.47	99.30
PR02R	0.00	0.00	0.00	0.00	100.00
cavity01	0.00	0.00	0.00	0.00	100.00
c-54	0.00	0.00	0.00	0.00	100.00
freeFlyingRobot_3	0.00	0.00	2.21	3.73	94.06
bcsstk04	0.00	0.00	0.00	0.00	100.00
bcircuit	0.00	0.00	0.00	0.00	100.00
oscil_dcop_12	0.00	0.00	0.47	0.23	99.30
bcsstk03	0.00	0.00	0.00	0.00	100.00
ASIC_320ks	0.00	0.00	0.00	0.00	100.00
oscil_dcop_56	0.00	0.00	0.47	0.47	99.06
nasa2910	0.00	0.00	0.00	0.00	100.00

5. Conclusion

To solve the problem that serious accuracy loss of the existing mixed-precision SpMV, this paper proposes an accuracy-controllable sparse matrix partition method to improve the calculation accuracy of mixed-precision SpMV. A mixed-precision SpMV kernel is designed and implemented in this paper, named SpMV-SD. It is proposed to perform equidistant sampling on matrices to estimate the accuracy of mixed-precision calculation after accuracy-controllable partition. We tested the performance of SPMV-SD on Tesla V100 and analyzed its accuracy. Taking CSR-Coop as a benchmark, SpMV-SD greatly improves the accuracy of the mixed-precision SpMV calculation results, and has similar performance to MpSpMV, demonstrating the superiority of the accuracy-controllable partition method. In our future work, we will apply our method to the optimization work of iterative solvers.

Reference

[1] Benzi M 2002 Preconditioning techniques for large linear systems: A survey *Journal of Computational Physics* **182** (2) 418-477.

[2] Dziekonski A, Lamecki A and Mrozowski M 2010 GPU acceleration of multilevel solvers for analysis of microwave components with finite element method *IEEE Microwave and Wireless Components Letters* **21** (1) 1-3.

[3] Hager W W and Zhang H 2006 A survey of nonlinear conjugate gradient methods *Pacific Journal of Optimization* **2** (1) 35-58.

[4] Couturier R and Domas S 2012 Sparse systems solving on GPUs with GMRES *Journal of Supercomputing* **59** (3) 1504-1516.

[5] Le Gallo M, Sebastian A, Mathis R, et al. 2018 Mixed-precision in-memory computing *Nature Electronics* **1** (4) 246-253.

[6] Ahmad K, Sundar H and Hall M 2019 Data-driven mixed precision sparse matrix vector multiplication for GPUs *ACM Trans on Architecture and Code Optimization (TACO)* **16** (4) 1-24.

[7] Buttari A, Dongarra J, Kurzak J, et al. 2008 Using mixed precision for sparse matrix computations to enhance the performance while achieving 64-bit accuracy *ACM Transactions on Mathematical Software (TOMS)* **34** (4) 1-22.

[8] Buttari A, Dongarra J J, Kurzak J, et al. 2006 Exploiting mixed precision floating point hardware in scientific computations *High Performance Computing Workshop* pp 19-36.

[9] Strzodka R and Goddeke D 2006 Pipelined mixed precision algorithms on FPGAs for fast and accurate PDE solvers from low precision components *14th Annual IEEE Symp. on Field-Programmable Custom Computing Machines* pp 259-270.

[10] Flegar G 2019 Sparse linear system solvers on GPUs: Parallel preconditioning, workload balancing, and communication reduction.

[11] Grützmacher T, Cojean T, Flegar G, et al. 2020 A customized precision format based on mantissa segmentation for accelerating sparse linear algebra *Concurrency and Computation: Practice and Experience* **32** (15).

[12] Le Gallo M, Sebastian A, Mathis R, et al. 2018 Mixed-precision in-memory computing *Nature Electronics* **1** (4) 246-253.

[13] Zachariadis O, Satpute N, Gómez-Luna J, et al. 2020 Accelerating sparse matrix–matrix multiplication with GPU Tensor Cores *Computers & Electrical Engineering* (88) 106848.

[14] Süli E and Mayers D F 2003 *An Introduction to Numerical Analysis* (Cambridge University Press, Cambridge) pp 1-403.

[15] Steven D, Nathan B, Luke O, et al. 2015 Cusp: Generic parallel algorithms for sparse matrix and graph computations. http://cusplibrary.github.io/

[16] Kolodziej S P, Aznaveh M, Bullock M, et al. 2019 The suitesparse matrix collection website interface *Journal of Open Source Software* **4** (35) 1244.

[17] Langr D and Tvrdik P 2015 Evaluation criteria for sparse matrix storage formats *IEEE Transactions on Parallel and Distributed Systems* **27** (2) 428-440.

Design of Unmanned Ship System with Two Engines and Two Propellers

Yongliang Zhang, Juanhui Yu, Cheng Dai and Tianshuang Fu

Tianjin Navigational Instrument Research Institute, Tianjin 300131, China
Email: 18041155481@163.com, 105436866@qq.com, 934605816@qq.com, 18920866592@qq.com.

Abstract. Unmanned ship is a kind of surface robot with integrated technology, which is composed of multiple disciplines, including intelligent remote control, wireless communication, autonomous navigation and obstacle avoidance algorithm.When the unmanned ship is sailing in the water, it is difficult to control its dynamic performance due to the influence of sea conditions. Therefore, the research on the unmanned ship technology has extremely important practical value. In order to improve its maneuverability and control accuracy, an unmanned ship based on two engines and two propellers is proposed. This paper designs an unmanned ship bottom system with two engines and two propellers. The system is mainly composed of two main engines, including two generators and two rudders and propellers. Two of them adopt Controller Area Network communication protocol and are controlled by two Controller Area Network communication interfaces respectively; The two generators adopt RS485 serial port communication protocol and are controlled by two RS485 serial ports respectively; The two-rudder propeller equipment adopt RS485 serial port communication protocol and are controlled by the other two RS485 serial ports respectively.Through the experiment, it is found that the unmanned ship equipped with two engines and two propellers can better control the navigation accuracy of the ship and have high maneuverability.
Keywords. Unmanned ship; bottom system; two engines and two propellers; Controller Area Network communication; RS485.

1. Introduction

In recent years, the rapid development and progress of computer technology, communication technology and navigation technology have promoted the further development of unmanned ships [1, 2] towards intelligence, networking and unmanned, making them widely used in marine environment monitoring, marine rescue, marine reconnaissance and other fields [3]. The western countries, led by the United States, have made early research in the field of unmanned ship technology, which is mainly used in the military field [4]. Although the research and development of unmanned ship technology in China is relatively late, with the enhancement of our national economic strength, we have invested more and more in unmanned ship technology [5]. At present, major enterprises, institutions, research institutions and universities have invested in the research and development of unmanned ship technology, forming a trend of flowers contending, and realizing the rapid breakthrough and development of unmanned ship technology [6]. In this paper, an unmanned ship bottom layer system based on STM32 [7] with two engines and two propellers is designed. Using the embedded programming and communication means of communication protocol [8], combined with the effective cooperation between each module system, the remote control and monitoring functions of the unmanned ship system [9] with two engines and two propellers are realized.

2. System Structure and Hardware Composition

A bottom layer module system based on ARM is designed. The bottom layer hardware design uses STM32F7 chip as microprocessor, and four RS485 serial ports are designed on the board to realize the control and feedback of multi-channel RS485 serial port equipment. The RS485 of the main chip_ 1 serial port connected to 1 serial port of generator, RS485_ 2 serial port connected to 2 serial port of generator, RS485_ 3 serial port connected to rudder propeller 1 serial port, RS485_ 4 serial port is connected with 2 serial port of rudder propeller, so as to realize RS485 control and feedback of dual generator and dual rudder propeller system. The master-slave chip communication is used to realize the dual channel Controller Area Network communication design. The UART serial port is used for two-way communication between the master and slave chips to realize the dual host Controller Area Network communication control and feedback.Figure 1 shows bottom layering system architecture.

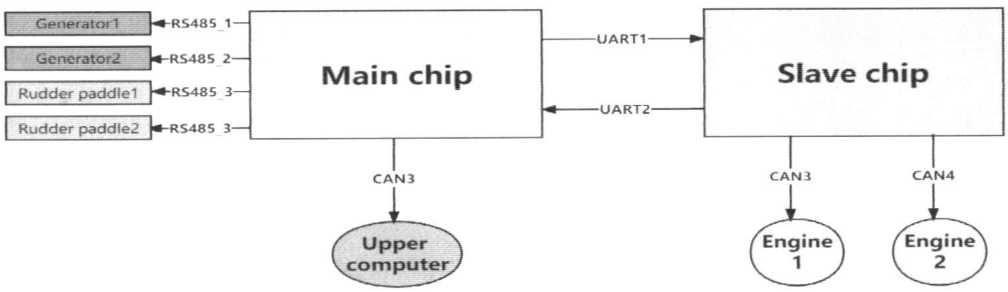

Figure 1. Bottom layer module system architecture.

3. Establishment of Communication

An unmanned ship bottom layer control system is designed. The control command transmission and information feedback are realized through Controller Area Network bus communication between shore based and ship bottom layer hardware. The system uses the shore based upper computer to send commands to the unmanned ship bottom board card through the radio through the Controller Area Network communication protocol. After receiving the commands, the unmanned ship bottom board card will be transferred to the designated equipment to perform relevant operations. The collected information of the relevant equipment will be converted into the Controller Area Network communication mode and sent back to the shore based to realize the establishment of communication.

In this paper, CTM1051(A)HQ can transceiver is selected as the interface between Controller Area Network controller and physical transmission line, which can realize message receiving and differential level signal conversion. Figure 1 shows the design of communication module.

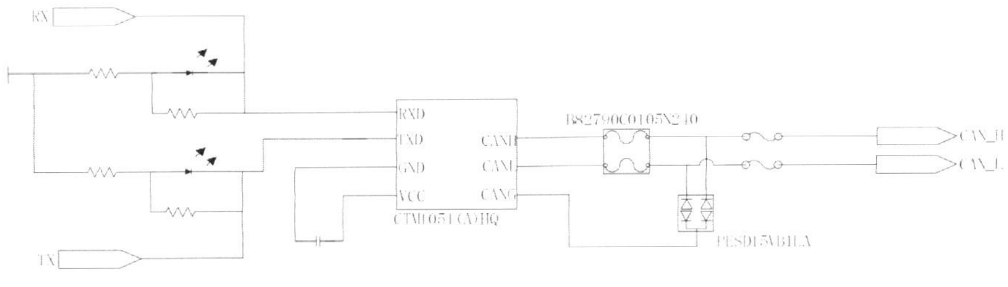

Figure 2. Design of Controller Area Network communication module.

The ground control station communicates with the unmanned ship through data transmission radio. The communication management unit and the bottom control unit are connected in series through Controller Area Network communication. The ground control station sends the control command to the communication management unit through the radio station, and then connects it to the bottom control board card through the Controller Area Network bus to realize the remote communication of the unmanned ship.Figure 3 shows overall communication architecture.

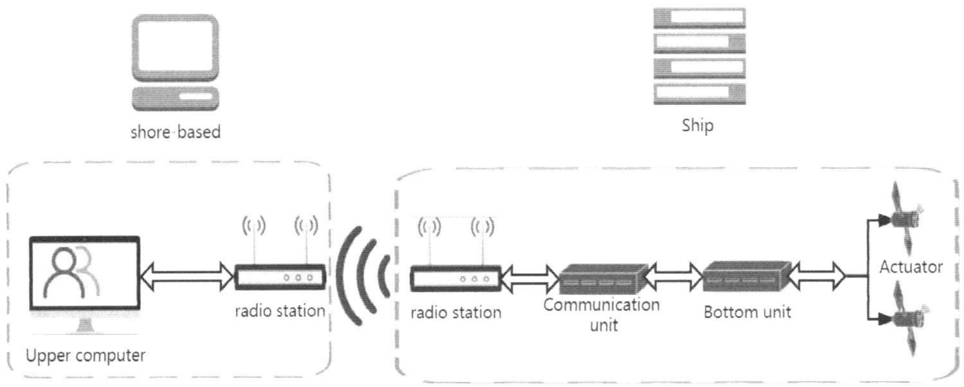

Figure 3. Overall communication architecture.

4. Design of Dual Main Engine Remote Control System

Field bus technology is mainly used to solve the communication problems between industrial field devices. Controller Area Network bus is a field bus developed by German automobile company to solve the internal communication problems of automobiles. Controller Area Network bus communication protocol is mainly divided into application layer, data link layer and physical layer. It has the characteristics of high data transmission efficiency, strong anti-interference ability, simple and reliable, two-way transmission, multi branch structure and so on.

The main engine is the main power device of the ship, and also the source of the ship power. With the development of the ship technology, the automation level of the main engine remote control technology is gradually improving. This system aims to design a host remote control system based on Controller Area Network communication network. The upper computer communicates with the ship radio through the radio station, and transmits it to the bottom control system through the communication unit. The bottom control system drives two hosts.As shown in Table 1, it is the host control command.

Table 1. Control command.

Start command	Stop command	Emergency stop command
0xFD	0xFD	0xFD
0x01	0x02	0x08
0x00	0x00	0x00
0x00	0x00	0x00
0x00	0x00	0x00
0x00	0x00	0x00
0x00	0x00	0x00
0x00	0x00	0x00

This design adopts an electric start host remote control system, which has a compact structure and can eliminate the hysteresis caused by pneumatic control. The host remote control system adopts two-way Controller Area Network communication to realize the separate control of two hosts. The control of the two main engines includes start, stop and emergency stop; Information feedback includes: engine speed, engine coolant temperature, engine fuel pressure, engine oil temperature, battery voltage, total operation time and fault alarm. Table 2 shows the feedback information of the host.

Table 2. Feedback information.

Data name	Data length	Remarks
Engine speed	Unit 16	0.125/bit
Coolant temperature	Unit 16	deviation-40
Engine fuel pressure	Unit 16	none
Engine oil temperature	Unit 16	none
Battery voltage	Unit 16	none
Total operation time	Unit 32	/100
Alarm fault	Unit 8	none

5. Design of Remote Control System for Twin Generators

With the development of unmanned ship technology, the ship power generation system needs to have a certain level of intelligent control to achieve the remote control of generators and real-time feedback of information.Figure 4 shows the design of generator remote control system.

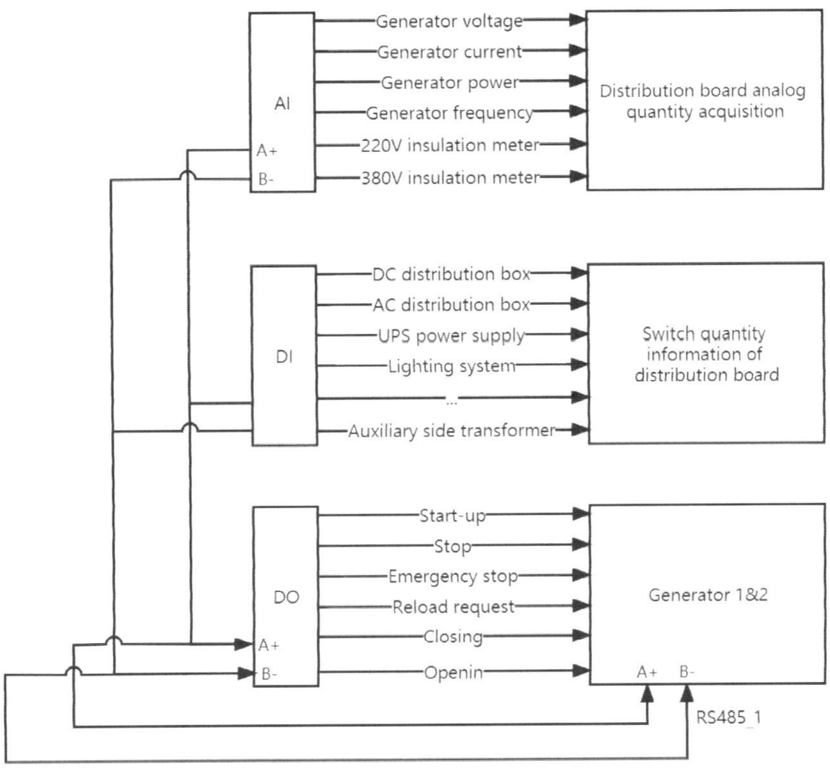

Figure 4. Design of generator remote control system.

The system uses two RS485 serial ports to control the two generators, including start, stop, emergency stop, opening, closing and heavy load start. The RS485 serial port is used to collect the power, frequency, current, insulation meter and other information of the generator on the distribution board through the analog quantity acquisition module AI; The switching value information on the distribution board is collected through the switching value acquisition module DI: DC distribution box switch, AC distribution box switch, UPS power switch, lighting system switch, switch information of the original and auxiliary side transformers, etc.; The start, stop, emergency stop, heavy load request, opening and closing of the generator are controlled by the switching value control module DO.

6. Design of double rudder propeller system

Polling is used to read the generator information. Including: generator speed, generator lubricating oil temperature, generator fresh water temperature, total operation time, battery voltage, fresh water pressure, and generator fault alarm.

A set of rudder propeller device is designed, which integrates propeller and rudder, and the underwater part including paddle and underwater gear box can rotate freely within 360°. Driven by the diesel engine, the blades form an all-round thrust, which greatly improves the flexibility of the ship. The full turn rudder propeller can realize 360° rudder angle control, and realize the advance, retreat and translation of the unmanned ship at a specific angle through the cooperation of the two propellers at different angles.

To ensure the safety and reliability of the rudder propeller system. AC and DC power supply is adopted, and the normal operation of the system can still be ensured after any power supply is lost.Figure 5 shows the design of rudder propeller remote control system.

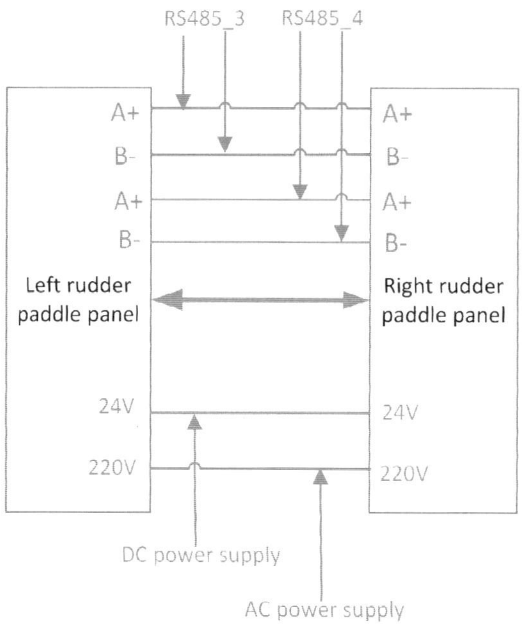

Figure 5. Design of rudder propeller remote control system.

Two RS485 serial ports are used to control the two rudder propeller systems, including 360° steering of rudder propeller, acceleration and deceleration of rudder propeller throttle, clutch release and connection. The feedback information includes: propeller speed, rudder angle information, clutch disconnection information, AC power loss alarm, DC power loss alarm, high temperature alarm of hydraulic pump station, high temperature alarm of rudder propeller oil temperature.

The left and right rudder propeller systems are controlled through two RS485 serial ports. The rudder propeller control functions mainly include:

(1) clutch control: clutch release and release.

(2) engine throttle control: propeller speed plus or minus throttle opening.

(3) steering control: 360 ° steering of steering gear.

7. Summary

With the rapid development of unmanned ship technology, it also plays an important role in the military and civil fields. This paper designs a bottom hardware board, which involves a variety of communication combinations. The host remote control uses dual Controller Area Network communication of master-slave communication to control and return information to the two hosts; The generator and power distribution system adopt RS485 serial port communication to monitor and control the ship power distribution system; The double RS485 serial port is used to realize the control and information collection of the double rudder propeller system. Finally, the design of a 360° full rotation unmanned ship system based on two engines and two propellers is realized.

References

[1] Xu T 2007 An intelligent navigation system for an unmanned surface vehicle *Fault Detection Supervision & Safety of Technical Processes* **6** (1) 1503-1508.

[2] Ataner E, Özdeş B, Öztürk G, Yasin T, et al. 2020 Design of communication and power systems in unmanned underwater vehicles *European Journal of Science and Technology* 262-267.

[3] Luo Y, Ai J, Zheng J and Wang J 2020 Control system design and thrust analysis of an unmanned surface cleaning vessel with a novel pump-valve propulsion system *IEEE Access* **8** (99) 46356 - 46372.

[4] Liu Y and Bucknall R 2015 Path planning algorithm for unmanned surface vehicle formations in a practical maritime environment *Ocean Engineering* **97** 126-144.

[5] Yong M, Zhao Y, Diao J, Gan L X, Bi H X and Zhao J M 2016 Design of sail-assisted unmanned surface vehicle intelligent control system *Mathematical Problems in Engineering* 1-13.

[6] Wang Z M, Yang D F, Yang K, Guo L Y and Tan J M 2014 Design and implementation of arm-based intelligent switchgear control system *Applied Mechanics & Materials* **644-650** 37-40.

[7] Xu X and Sheng L 2018 Design and implementation of USV tracking controller based on STM32 *Journal of Wuhan University of Technology (Transportation Science and Engineering)* **42** (4) 626-630.

[8] Liu Q, Fu G H, Lian X G and Wang J 2013 Based on ARM embedded remote data acquisition system *Advanced Materials Research* **614-615** 1566-1569.

[9] Luo Z, Qian T, Ye X, Huang J H and Yu L W 2021 Research on a course control strategy for unmanned surface vessel *Journal of Physics: Conference Series* **1948** 012106.

Improvement of Vulnerable Code Dataset Based on Program Equivalence Transformation

Dejiang Jing

School of Computer Science and Technology, Beijing Institute of Technology, Beijing, China
Email: jdj@bit.edu.cn

Abstract. Code vulnerability dataset plays an important role in the development and evaluation of vulnerability detection tools. Aiming at the problem that programs in existing vulnerability datasets usually have simple structures and small scales, which is not satisfying for testing, we proposed a generator of complex code vulnerability dataset based on program equivalence transformation in this paper. It improves the complexity of program structure and code size while preserving the labels of the original case. Based on the PHP vulnerability test suite in Software Assurance Reference Dataset (SARD), a large number of complex cases were constructed and tested using two open-source PHP vulnerability detection tools RIPS and WAP. Experimental results show that the cyclomatic complexity and code size of the generated cases increase by about five times after transformation. The detection accuracy of the tools on the generated dataset decreases significantly compared with the results on the original dataset. The false positive rate increases by about 30%, and the false negative rate increases by about 10%.
Keywords. Vulnerability detection; program equivalence transformation; dataset generator.

1. Introduction

In the process of software development and maintenance, it is often necessary to use code analysis tools to detect defects and vulnerabilities in software programs, so as to ensure the stability and security of programs. However, the availability and effectiveness of a vulnerability detection tool often require a large number of correctly labelled vulnerability dataset for testing, so the construction of vulnerability datasets plays an important role in the field of code analysis. In order to evaluate the real analysis ability, an effective dataset should cover possible vulnerability categories in software programs and simulate the complex structure of real software as much as possible. However, there are many kinds of program vulnerabilities with complex syntax structure, and it is time consuming and labor consuming to write test cases manually or collect them from existing programs. Therefore, many large-scale vulnerability datasets are constructed automatically or semi-automatically.

PHP is one of the most popular website programming languages. The demand for vulnerability detection of PHP programs is increasing in recent years, but it is difficult to find a perfect vulnerability dataset of PHP in the market currently. The PHP vulnerability test suite contained in SARD is a widely used vulnerability dataset, which covers a variety of common PHP vulnerability categories and contains some simple syntax of PHP language. However, the cases in the dataset have small scale and simple structures, which is hardly to meet the evaluation needs of vulnerability detection tools.

In order to construct a better PHP vulnerability dataset, this paper designs a generator of complex code vulnerability dataset based on program equivalence transformation. This paper applies a variety of equivalent transformation strategies to existing programs and after incorporating a variety of nested syntax structures into the code, the newly generated cases can greatly increase the program complexity while maintaining vulnerability information labelled in the original case.

In the experiment, taking the PHP vulnerability dataset in SARD as the input, three equivalent program transformation strategies are designed and implemented to increase the code complexity. The three strategies are "inserting multiple loops", "using variables instead of constants", and "inserting useless assignments" respectively. The generator finally constructs a large number of complex cases through a mixture of the three strategies, and the cyclomatic complexity and code size of the generated cases increase by about five times compared with the original cases in SARD dataset.

This paper uses two existing open-source PHP vulnerability detection tools: RIPS [1] and WAP [2] to evaluate the generated dataset. The results show that the detection effect of various tools on the dataset generated in this paper is far lower than that on the SARD dataset. Specifically, the false positive rate increases by about 30%, while the false negative rate increases by about 10%. Besides, the time consuming also increases significantly.

2. Related Work

Common methods of building vulnerability datasets include automatic generation according to vulnerability patterns, generation from simple code and extraction from open-source projects.

Software Assurance Reference Dataset (SARD) is a vulnerability sample code base developed for Software Assurance Metrics and Tool Evaluation (SAMATE), which is a project at the National Institute of Standards and Technology (NIST) with the goal of improving software assurance. SARD dataset can be accessed and downloaded. PHP vulnerability test suite in SARD is a PHP vulnerability dataset which was automatically constructed by the case generator designed in [3]. It contains 42,212 test cases, including 29,258 safe cases and 12,954 vulnerable unsafe cases, covering 12 common CWE (Common Weakness Enumeration) vulnerability categories. The case generator focuses on the three parts of vulnerability injection, including taint source introduction, tainted information filtering and vulnerability triggering. The generator constructs code components for each part respectively. It also inserts simple syntax structures such as object-oriented, loop and conditional branch to each code component. The automatically generated vulnerability dataset has become the evaluation standard for some research work [4, 5] related to code analysis, because it covers a wide range of vulnerability categories and contains a large number of cases.

In addition, the Juliet test suite generator [6] includes 57,099 C/C++ test cases and 23,957 Java test cases.

Although the vulnerability datasets automatically constructed by the generator have a large scale and high coverage of vulnerability categories, the cases in the datasets usually have small code scales and simple syntax structures, which is difficult to characterize the complexity of real software.

Some researchers focus on online programming websites and intend to build large-scale vulnerability datasets from simple code segments. Puri et al. [7] constructed a large-scale dataset, CodeNet, using code submitted from two online sites, AIZU and AtCode. Although this method can easily obtain a large amount of data, the code complexity collected is low and does not fit the real project requirements.

Some researchers [8-10] extracted defect codes by searching submitted information with CVE id or keyword. In order to improve accuracy, further manual verification is usually required. However, due to the large scale of the open-source projects and the complexity of the vulnerability, it undoubtedly increases the difficulty and cost of manual inspection. At the same time, some researchers [11-13] built vulnerability datasets by analyzing CVE (Common Vulnerabilities & Exposures), NVD (National Vulnerability Database) and other databases and their linked libraries. Although this method ensures the existence of vulnerabilities in the collected code to a certain extent, it is limited by the number of open-source projects provided by CVE and NVD.

Zhou et al. [14] selected four large C programs: Linux Kernel, QEMU, Wireshark and FFMPEG, and found the vulnerability-related functions by analyzing the modification log information of these large-scale programs. They spent 600 hours carrying out two rounds of manual labelling and cross verification. In addition, Walden et al. [15] sorted out 223 vulnerabilities from 71 versions of the program Moodle, 95 versions of the program PHPMyAdmin and 30 versions of the program Drupal. Through manual analysis and annotation, a collection of PHP security vulnerability datasets was constructed.

In SARD, there is also a vulnerability dataset obtained by manually analyzing and labelling the program WordPress of version 2.0, which contains 23 vulnerabilities totally.

The way of relying only on manual labelling and verification has high requirements for personnel professionalism. Due to the limited number and types of vulnerabilities in the software, the number of program test cases is usually small, and it cannot cover all vulnerability categories and patterns well.

3. Approach and Implementation

3.1. Overall Framework

Figure 1 shows the overall framework of the generator of complex code vulnerability dataset based on program equivalence transformation proposed in this paper. The input of the generator is the original PHP vulnerability dataset, and the output is the newly generated complex dataset. Firstly, the framework analyzes the original code to get the corresponding abstract syntax tree. In this process, the information in the source code is analyzed and extracted to facilitate the subsequent construction of cases through program equivalence transformation. The syntax structure information to be collected depends on the subsequent transformation strategy, mainly including the type and position of each statement, the constant information in the statement and the assignment information. Next, according to this syntax structure information, the original code is transformed by applying the three program equivalence transformation strategies designed in this paper, and a new program case is constructed. Compared with the original code, the new code constructed after transformation has more complex syntax structure and larger scale.

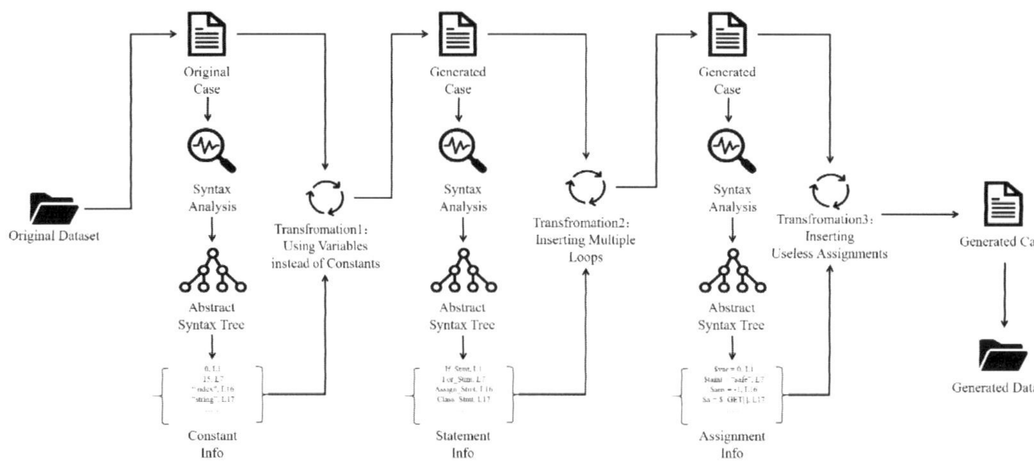

Figure 1. Overall framework of test case generator.

3.2. Program Equivalence Transformation Strategies

Three program equivalence transformation strategies are designed to insert and replace the code on the basis of maintaining the vulnerability information labelled originally. The transformation strategies are described as follows.

3.2.1. Strategy 1: Using Variables instead of Constants. Most of the existing cases use constants as the parameters of function call and the indices of array indexing. However, in real software program, due to the complex functional logic, variables are often used as the indices to access the array elements. For code analysis tools, the analysis of variables is much more difficult than constants. Therefore, in order to improve the complexity of existing programs and make them better characterize real software programs, we design an equivalent transformation strategy of "using variables instead of constants". We find each constant in the original code according to the syntax tree, create a new variable to replace the constant, and assign the constant value to the newly created variable before the location where the constant is referenced, so as to ensure the invariance of program semantics.

Figure 2 shows a simple example of the "using variables instead of constants" transformation strategy. As shown in the bold part, the transformation process uses the variable *$variable0* to replace the original constant 1 as the index of array *$array*.

```
<?php
  $array = array() ;
  $array[] = 'safe' ;
  $array[] = $_GET['userData'] ;
  $array[] = 'safe' ;
  $variable0 = 1 ;
  $tainted = $array[ $variable0 ] ;
  $query = "find / size '". $tainted . "'" ;
  $ret = system($query) ; // flaw
?>
```

Figure 2 Code example of "using variables instead of constants" strategy.

3.2.2. Strategy 2: Inserting Multiple Loops. Most of the existing test cases can cover the conditional branch jump structure (such as if-else statement block) and loop structure (such as for statement block), but it is usually a single-layer if-else or single-layer loop structure, which is relatively simple. However, in the real code, in order to realize some complex functional logic, loops and branches are often nested.

```
<?php
  $tainted = $_GET['userData'] ;
  $re = "/^.*$/" ;
  for($var0=0; $var 0<1; $var 0++)
  {
    for($var 1=0; $var1<1; $var 1++)
    {
      if(preg_match($re, $tainted) == 1)
      {
        $tainted = $tainted ;
      }
      else
      {
        $tainted = "" ;
      }
    }
  }
  $query = "cat ' $tainted '" ;

  //flaw
  $ret = system ($query) ;
?>
```

Figure 3. Code example of "inserting multiple loops" strategy.

In order to more effectively evaluate the ability of a code analysis tool for complex syntax structures, this paper designs a program equivalence transformation strategy of "inserting multiple loops". We select a node in the syntax tree, that is, a statement or statement block in the original case program, or nest multiple loops in its outer layer. On the one hand, if the original code is a single statement, the syntax structure of multiple loops is added to the current case. On the other hand, if the included code is an *if-else* statement block, the current case introduces the nested structure of loops and conditional branches, which can effectively improve the structural complexity of the code. Notably, we need to ensure that the program is semantically equivalent before and after the transformation, and it does not affect the correctness of the program and the vulnerability information. Therefore, we set each layer of multiple loops to iterate only once, so as to ensure that the original code in the loop body after the transformation runs only once which is the same as before the transformation.

Figure 3 presents a simple example of the "inserting multiple loops" transformation. The bold codes are the new double loops after the transformation.

3.2.3. Strategy 3: Inserting Useless Assignments. Injection vulnerability means that a potentially unsafe pollution value (such as user input) from external input will propagate to an operation (such as system function call) that may affect system security without any security judgment and processing, thus triggering a security vulnerability.

In a simple program, the data flow of taint information is very intuitive. because after one variable is assigned tainted value, usually no other assignment will be made to the variable. In the real PHP code, developers are likely to assign different values to a variable in different code branches, which increases the complexity of data flow analysis. Therefore, we propose a transformation strategy of "inserting useless assignments". After one variable is assigned taint in the original test case, another assignment to this variable is inserted to cause confusion and increase the complexity of data flow. In order to ensure that the program has the same semantics before and after transformation, the new assignment statement is embedded in a loop that executes 0 times.

Figure 4 presents a simple example of "inserting useless assignments" transformation strategy. The bold part of the code contains a new while statement block. The loop condition is always evaluated to false, so the new assignment statement will never be executed, and the variable *$tainted* is still tainted.

```
<?php
  $tainted = $_GET['userData'] ;
  $a = 1 ;
  while( $a != 1 ) {
    $tainted = 'safe' ;
  }
  $query = "cat ' $tainted '" ;
  $ret = system ($query) ; // flaw
?>
```

Figure 4. Code example of "inserting useless assignments" strategy.

3.3. Implementation

We choose the PHP vulnerability test suite in SARD as the input of our dataset generator. During the experiment, through the manual verification of original cases, we found that a small part of the cases in SARD dataset were incorrectly labelled.

After removing hundreds of cases involving the errors in the original dataset, the remaining 40,244 correct cases are selected as the input of the generator.

For the first program equivalence transformation strategy "using variables instead of constants", the specific steps are as follows:

- Step 1: Depth first search is performed to traverse the syntax tree and find a constant.
- Step 2: The location of the statement where the constant is used according to syntax tree is found.
- Step 3: A new variable is defined before the statement and it is assigned the constant value, and then we jump to Step 1.

For the second strategy "inserting multiple loops", in the implementation process, we insert double loops for each statement block in the source code except the definition statement of class or function, which is all the first level child nodes under the root node in the syntax tree. The specific implementation steps are as follows:

- Step 1: The first child node of the next layer under the root node of the syntax tree is scanned.
- Step 2: We check whether the node type is a class definition or function definition. If so, we directly jump to Step 5.
- Step 3: The location of the statement block is found according to the node information from the syntax tree.
- Step 4: Two layers of loops outside the statement block are added. It is noted that the variable name used by the loop condition cannot be reused.
- Step 5: we scan the next node in the syntax tree and jump to Step 2.

For the transformation of the third strategy "inserting useless assignments", it is necessary to identify each assignment statement *S* in the original program according to the syntax tree. Then its starting and ending locations as well as the content of the assigned object *O* are found. Next, a new assignment is added after *S*, and a new value *V* is assigned to object *O*. Finally, the new assignment statement is embedded into a loop that will not be executed. As the new value for assignment, *V* has four categories: external user input, constant value, non-existing variable and existing variable. For the first category, external input is considered unsafe, that is, tainted information may be introduced through that assignment. The second and third kinds of value are safe. For the last one, since it is unknown whether the variable in the program is tainted before, it is uncertain whether the assignment will introduce tainted information.

In the implementation of our tool, in order to achieve better effect, the value selection for *V* depends on the value in the original assignment statement *S*. If the original assigning value is a safe constant, the new confusion assignment will use an unsafe external input *$_GET['userData']*. If the original value is an external input, the new confusion assignment uses constant. Random assignment in other cases.

We mix the three transformation strategies, perform the transformation process of the three strategies on the original dataset one by one, and finally generate the mixed transformed dataset.

4. Evaluation

4.1. Research Questions

In this section, we evaluate the proposed approach by investigating the following research questions:

- RQ1: How much increase in code scale and complexity of the new cases generated by program equivalence transformation compared with the original cases?
- RQ2: How does the result of the code analysis tool detect the newly generated dataset compared with the original dataset?
- RQ3: How does the detection efficiency of the code analysis tool detect the newly generated dataset compared with the original dataset?

4.2. Experiment Design

In order to answer RQ1, the experiment counts and compares the number of code lines before and after transformation. In addition, the cyclomatic complexity of the programs is calculated and compared. In order to answer RQ2, open-source PHP vulnerability detection tools RIPS and WAP were used to analyze the datasets before and after transformation. After analysis, we count the detection results, and calculate the FN (false negative) and FP (false positive) rates for comparison.

In order to answer RQ3, the running time of each detection tool applied to each dataset is recorded and compared.

4.3. Results and Analysis

The experiment analyzes the original dataset SARD and the dataset MIX constructed by using the hybrid program equivalence transformation strategy.

Table 1. The statistical results of code lines and cyclomatic complexity of two datasets.

Dataset	Average code lines	Average cyclomatic complexity
SARD	15.00	1.00
MIX	93.65	5.06

Table 1 shows the average code lines and cyclomatic complexity of the two vulnerability datasets, in which the number of code lines is the actual number of effective lines in the program except comments and empty lines. It can be seen that the average number of lines of the program in the original SARD dataset is only 15, which means the code scale is relatively small. In addition, the average cyclomatic complexity of SARD is only 1, indicating that the logical structure of the program is relatively simple. The average number of code lines of the MIX dataset generated after transformation is 93.65, and the average cyclomatic complexity is 5.06. Both values have increased by more than five times. The scale and the complexity of the program have been significantly improved, which can better represent the characteristics of real software programs.

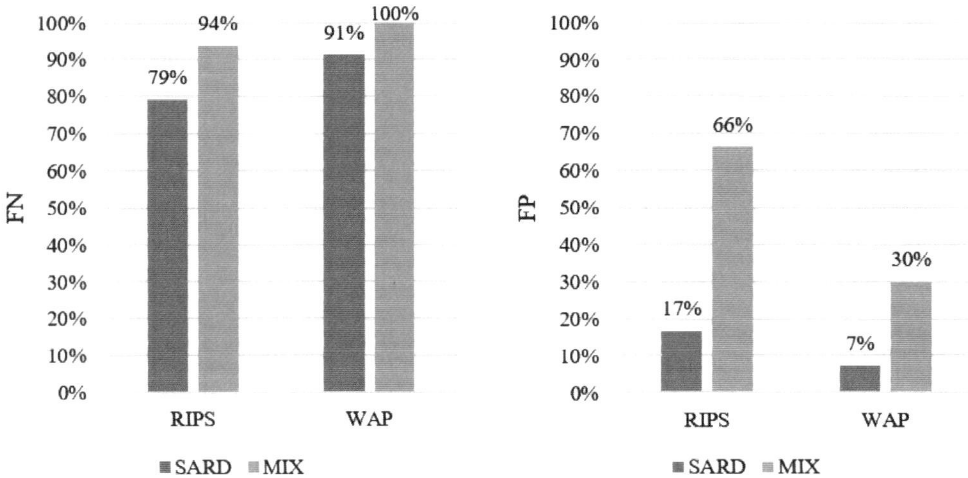

Figure 5. The detection results of two tools on two datasets.

Figure 5 shows the detection results of detection tools RIPS and WAP applied to two datasets. FN is the proportion of cases detected as safe by the tool in the set of cases labelled as vulnerable, and FP is the proportion of cases with vulnerabilities detected by the tool in the set of cases labelled as safe.

In general, the detection effect of the transformed dataset on the two tools is obviously behind the original dataset before transformation. In summary, the FN of RIPS increases by 10%-20%, and the FN of WAP tool is up to 100% when detecting MIX dataset. It indicates that the new complex structure and useless code after transformation have brought greater challenges to the detection of the tool, and it is difficult for the tools to detect vulnerabilities in the program. In addition, the FP detected by RIPS increases from 17% to 66%, and that detected by WAP increases from 7% to 30%. Their FP increases by about four times, indicating that the transformed dataset makes it easier for the detection tools to misidentify the vulnerability characteristics and mistake the security code as a vulnerability.

Table 2. The detecting time consumption results of two tools on two datasets.

Dataset	Time consumption (s)	
	RIPS	WAP
SARD	40.69	69.23
MIX	191.50	116.80

Table 2 shows the detection time on two datasets by RIPS and WAP. It can be seen that RIPS takes 40.69 seconds to detect the original dataset. It increases by about four times compared with 191.50 seconds to detect the transformed dataset. The time taken by WAP to detect the two datasets is 69.23 seconds and 116.80 seconds respectively. It shows that for the cases with improved code scale and complexity after transformation, the analysis process of detection tools is more complex and the detection efficiency is significantly reduced.

5. Conclusion

This paper proposes an approach to generate complex code vulnerability dataset based on program equivalence transformation. We design three equivalent transformation strategies to improve the complexity of program structure and code scale, so as to better characterize the real software programs. Based on the PHP vulnerability test suite in SARD, a large number of complex cases are constructed. This paper counts and compares the code size and complexity of the vulnerability dataset before and after the transformation, and uses the existing two open-source PHP code vulnerability detection tools RIPS and WAP for vulnerability detection. The results shows that the complexity of the test cases increases by about five times, and the detection accuracy of the tools decreases significantly. Additionally, the detection time of the two tools increases significantly. It proves the effectiveness of the complex code vulnerability generation approach based on program equivalence transformation proposed in this paper.

References

[1] Dahse J and Holz T 2014. Simulation of built-in PHP features for precise static code analysis *NDSS* **14** 23-6.

[2] Medeiros I, Neves N and Correia M 2015 Detecting and removing web application vulnerabilities with static analysis and data mining *IEEE Transactions on Reliability* **65** (1) 54-69.

[3] Stivalet B and FongE. 2016 Large scale generation of complex and faulty PHP test cases *2016 IEEE Inter. Conf. on Software Testing, Verification and Validation (ICST)* pp 409-15.

[4] Kronjee J, Hommersom A and Vranken H 2018. Discovering software vulnerabilities using data-flow analysis and machine learning *Proc. of the 13th Inter. Conf. on Availability, Reliability and Security* pp 1-10.

[5] Fang Y, Han S, Huang C and Wu R 2019 TAP: A static analysis model for PHP vulnerabilities based on token and deep learning technology *PloS one* **14** (11) e0225196.

[6] Boland T and Black P E 2012 Juliet 1. 1 C/C++ and java test suite *Computer* **45** (10) 88-90.

[7] Puri R, Kung D S, Janssen G, Zhang W, Domeniconi G, Zolotov V, et al. 2021 CodeNet: A large-scale AI for code dataset for learning a diversity of coding tasks *arXiv preprint arXiv:2105.12655*.

[8] Perl H, Dechand S, Smith M, Arp D, Yamaguchi F, Rieck K, et al. 2015 Vccfinder: Finding potential vulnerabilities in open-source projects to assist code audits *Proc. of the 22nd ACM SIGSAC Conf. on Computer and Communications Security* pp 426-37.

[9] Rahman F, Khatri S, Barr E T and Devanbu P 2014 Comparing static bug finders and statistical prediction *Proc. of the 36th Inter. Conf. on Software Engineering* pp 424-34.

[10] Tóth Z, Gyimesi P and Ferenc R 2016 A public bug database of github projects and its application in bug prediction *Inter. Conf. on Computational Science and Its Applications* pp 625-38.
[11] Fan J, Li Y, Wang S and Nguyen T N 2020 AC/C++ code vulnerability dataset with code changes and CVE summaries *Proc. of the 17th Inter. Conf. on Mining Software Repositories* pp 508-12.
[12] Gkortzis A, Mitropoulos D and Spinellis D 2018 VulinOSS: A dataset of security vulnerabilities in open-source systems *Proc. of the 15th Inter. Conf. on Mining Software Repositories* pp 18-21.
[13] Ponta S E, Plate H, Sabetta A, Bezzi M and Dangremont C 2019 A manually-curated dataset of fixes to vulnerabilities of open-source software *2019 IEEE/ACM 16th Inter. Conf. on Mining Software Repositories (MSR)* pp 383-7.
[14] Zhou Y, Liu S, Siow J, Du X and Liu Y 2019 Devign: Effective vulnerability identification by learning comprehensive program semantics via graph neural networks *Advances in Neural Information Processing Systems* **32**.
[15] Walden J, Stuckman J and Scandariato R 2014 Predicting vulnerable components: Software metrics vs text mining *2014 IEEE 25th Inter. Symp. on Software Reliability Engineering* pp 23-33.

A Deep Facial BRDF Estimation Method Based on Image Translation

Lulu Feng, Dongdong Weng* and Bin Liang

Beijing Engineering Research Center of Mixed Reality and Advanced Display, School of Optics and Photonics, Beijing Institute of Technology, Beijing, China
Email: fll_tty@163.com; crgj@bit.edu.cn; 18203416228@163.com

Abstract. The reconstruction of photorealistic 3D face geometry, textures and reflectance (BRDF) is one of the most popular fields in computer vision, graphics and machine learning. However, the acquisition of facial reflectance remains a challenge. In this article, we propose a method for estimating the facial reflection properties of a single portrait image based on image translation. From a RGB face image, we obtain the BRDF with a large amount of detail. To achieve it, we perform a reverse engineer, which renders face images with the obtained texture map to form training data pairs based on the Blinn-Phong illumination model. We also apply random rotate-and-crop and sliding-window-crop to augment the data and optimize the network weights by minimizing the generated adversarial loss and reconstruction loss. As demonstrated in a chain of quantitative and qualitative experiments, our method achieves superior performance compared to the state-of-the-art methods.
Keywords. 3D face reconstruction; BRDF estimation; cGAN; image translation.

1. Introduction

In the past few years, with the rapid development of VR and AR, photorealistic 3D face reconstruction has gradually become a popular research area in computer vision, graphics and deep learning. However, estimating the facial reflectance properties of a human face is a tricky problem due to the complexity of its optical property.

The pioneering work in this area is the 3D Morphable Model fitting algorithm proposed by Blanz in 1999 [1]. The algorithm represents the geometric and texture with the linear combination of vectors which is obtained by Principal Component Analysis. Although this algorithm performs well in reconstructing the basic features of the face shape, it does not perform well in reconstructing the high frequency details of the texture, which results in a lack of realism in rendering human faces.

In addition, classical methods like the photometric stereo vision [2] can recover the orientation of the object's surface and reconstruct the normal information. This method collects multiple images under varying lighting conditions to fit photometric or geometric models and calculates the normal vector of surface. This method is relatively simple and suitable for any approximate diffuse reflection surface. But an ideal Lambertian reflection model is assumed, which leads to anomalous parts of non-Lambertian reflections in image.

With the rise of deep learning, the inverse rendering methods based on deep convolutional networks have been proposed to regress the face geometry, illumination and reflectance from single image [3-7]. These methods demonstrate that deep learning has already shown excellent performance on reflectance inference, lighting estimation and intrinsic image decomposition. Nevertheless, many of them use the basic Lambertian reflectance models or unrealistic synthetic data, so they fail to reconstruct highly-detailed texture and geometry. Some of them even lose the identity characteristics. Given the excellent performance of generative adversarial networks (GAN) [8] in image inpainting, style transfer, and image super-resolution reconstruction, many GAN-based methods have begun to emerge, especially the application of GAN in image translation.

Considering the excellent performance of conditional generative adversarial networks in style transfer and image translation, we propose a method to infer the diffuse albedo, specular albedo and normal map from a RGB face image. Firstly, we implement a reverse engineer to synthesize face images under specific illumination condition based on Blinn-Phong illumination model with collected face reflectance map. Then we construct training data pairs with rendered face image and the corresponding BRDF. We regard the estimation of BRDF as the domain-to-domain image translation, so we input the training pairs into the networks and optimize it by the generated adversarial loss and specific reconstruction loss for each task. To maximize the diversity of data, we also apply random rotate-and-crop and sliding-window-crop to augment the data, which tremendously upgrades the performance of our network.

In conclusion, our contributions are:

(1) A deep facial BRDF estimation method is proposed. Compared with some previous work, our method is non-Lambertian and the training data is high fidelity scan data which is captured in real scene.

(2) For normal inference, we propose a cosine angle loss function, which improves the quality of normal maps estimated from a single image.

(3) Our method directly infers high-resolution BRDF without texture reconstruction from low-resolution textures by super-resolution network.

2. Related Work

2.1. Classical Methods

The traditional method to calculate the surface reflection properties of materials mainly by fitting geometric or photometric models. Blanz et al. first proposed the 3D Morphable Model [1] (3DMM) fitting algorithm in 1999. This algorithm represents the geometry and texture of human face with the linear combination of high-dimensional space vectors, including the shape and texture vector. Each group of shape and the texture vectors form a standardized face. Blanz et al. calculated the average value of shape and texture vectors based on 200 sets of face models in the database and used principal component analysis (PCA) to further extract features. Although this algorithm performs well in reconstructing the basic features of the face shape, it can not reconstruct the high frequency details of the texture very well, which results in a lack of realism in rendering human faces.

Photometric stereo vision [2] captures multiple images with fixed viewpoint under different illumination conditions and fits photometric models to calculates the normal vector of surface. Many variants of this method also take spatially-varying BRDFs [9] and statistical BRDF models [10] into account. From multiple images the structure-from-motion [11, 12] is able to recover the dense mesh models although the lumination effects of the texture are included. Thus, some methods [13-15] seek to combine the strengths of geometric and photometric methods. Haber et al. [16] apply the multiview stereo to compute the geometry and infer the reflectance and illumination from collected images.

Some work directly captures the reflectance properties based on special designed optical system. Debevec et al. [17] first proposed an optical system called lightstage to capture the reflectance light field of human faces. They also applied the image-based rendering for relighting application with the captured data. In the next few years, the team led by Debevec had completed the updates on light stage optical system several times. In 2007, Ma et al. [18] proposed the polarized spherical gradient illumination for acquiring high-precision face highlight and diffuse reflection components and photometric normal of human face with just eight pictures. However, this method is view-dependent and it is available only in the frontal viewpoint. Then Ghosh et al. [19] proposed multiview face capture using polarized spherical gradient illumination that broke the single viewpoint limitation with two proposed orthogonal spherical polarization illumination modes. Recently, Kampouris et al. [20] introduced the binary spherical gradient illumination pattern without polarized light field to separate highlight and diffuse reflection components and photometric normal of human face. Although these methods can obtain high-fidelity facial reflectance, they are all based on spherical setup similar to the light stage, which are costly and sophisticated.

2.2. Deep Learning Based Approaches

Recently, with the rise of deep learning and convolutional neural networks (CNNs), numerous deep learning based methods have emerged to acquire the reflectance of materials from a single image. The network structure of most methods adopts encoder-decoder structure [21, 22], which usually with an environmental constraint. Guo et al. [23] leveraged the power of convolutional neural networks and 3DMM to synthesize images to solve the image translation problem. However, the synthesize images from 3DMM were generally unreal and some facial details were failed to recover due to the convolution operation.

Sengupta et al. [7] inferred facial reflectance from an "in-the-wild" image by image translation networks while the resolution of generated image is low. Current approaches [3] attempt to utilize extra facial normals or displacement maps to produce result with high frequency detail. This method performs well in geometry synthesis but fails in some extreme condition with harsh illumination. The closest approach to our work is the "AvatarMe" proposed by Lattas et al. [24]. This method uses the 3DMM algorithm to fit the face geometry and texture from a image, and obtains a high-resolution image by feeding a low-resolution image into a super-resolution network. Then it obtains the diffuse albedo through the de-lighting network. Finally the diffuse albedo and geometric information are concatenated together to generate specular albedo and normal maps through an image translation network.

Unlike this method, we directly synthesize the training data by rendering under specific illumination conditions without applying the 3DMM fitting algorithm for the texture it produces is generally unrealistic. Thus, our data is naturally high-resolution and without using super-resolution networks. Moreover, we train three independent image translation networks to directly estimate the diffuse albedo, specular albedo and normal from a single face image with data augmented by random rotate-and-crop and sliding-window-crop to improve the diversity of data. Figure 1 shows the overview flow of our work.

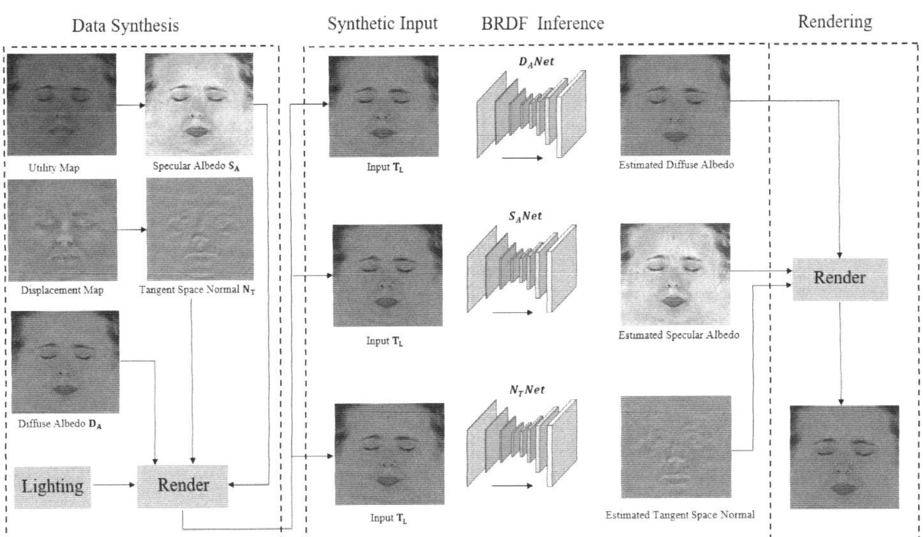

Figure 1. Overview of authors' method.

3. Method

To estimate the diffuse albedo (D_A), specular albedo (S_A) and tangent space normal (N_T) from a given single face image, as illustrated in figure 1, we perform a reverse engineering to synthesize face images under specific illumination conditions based on Blinn-phong [25] illumination model. And the face image and corresponding ground truth are used as training data pairs. Then we train three independent image translation networks to estimate diffuse albedo, specular albedo and tangent space normal directly from a single face image with training data pairs augmented by random rotate-and-crop and sliding-window-crop. The following chapters will describe these steps in detail.

3.1. Dataset

We employ the high-fidelity scan data from Reference [26] as our ground truth. The pore-level reflectance data from Reference [26] is captured by device like the light stage [27] with high resolution over 4K. The multi-channel facial texture assets from Reference [26] is currently recognized as the best texture asset and has been widely used by many special effects companies and artists in their standard workflow. We can directly obtain the diffuse albedo without any processing, while the specular albedo is acquired by obtaining the blue channel of the utility map. And the tangent space normal can be calculated from the displacement map through differentiation. Part of the ground truth are displayed in figure 2.

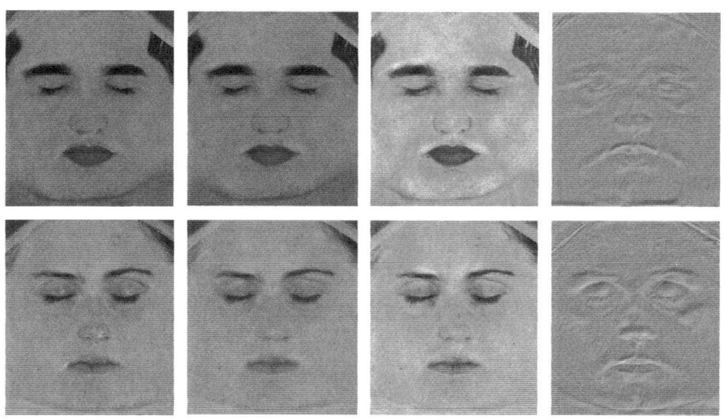

(a) Synthetic Face Image (b) Diffuse Albedo (c) Specular Albedo (d) Tangent Space Normal

Figure 2. Two subjects' rendered face images and reflectance.

The utility map, displacement map and diffuse albedo map are collected as our raw data. We separate the B channel of the utility map to obtain the specular albedo and calculate the tangent space normal from displacement map. The Blinn-phong illumination model is utilized for rendering to synthesize the face image on specific lighting condition, which is to form the training data pairs with ground truth. And three sub-networks are trained independently for inferring diffuse albedo, specular albedo and tangent space normal. Finally, the estimated BRDFs are input into the render to obtain the face image and compare with the input image T_L.

To construct training data pairs, we perform a reverse engineering as illustrated in figure 1. We build a rendering system in python based on the Blinn-phong illumination model [27] to synthesize face images under specific illumination conditions. The reason why we choose this model is that it is not only computationally efficient but also close enough to the physical facts to get reasonable results.

We select 25 individuals (including 13 women and 12 men) of different ages and characteristics. Then we synthesize face images under specific parallel lighting direction based on our rendering system. These face images and their corresponding reflectance maps shown in figure 2 constitute the training data pairs.

Although the resolution of our data is about 5K, we cannot directly feed the entire image into the network for training due to the hardware limitations. So we divided the original high-resolution data into small chunks of size 256×256.

To increase the amount of data and promote the generalization of our model, we design two data augmentation patterns as illustrated in figure 3. We call the two patterns: "Slide Crop" and "Random Rotate Crop". For pattern "Slide Crop", we generate the image data with partial area repetition by sliding a window of size 256×256 horizontally or vertically along the UV direction using a stride of size 128. For pattern "Random Rotate Crop", we rotate the original image at a random angle, then we randomly select a 256×256 square area on the rotated original image for cropping.

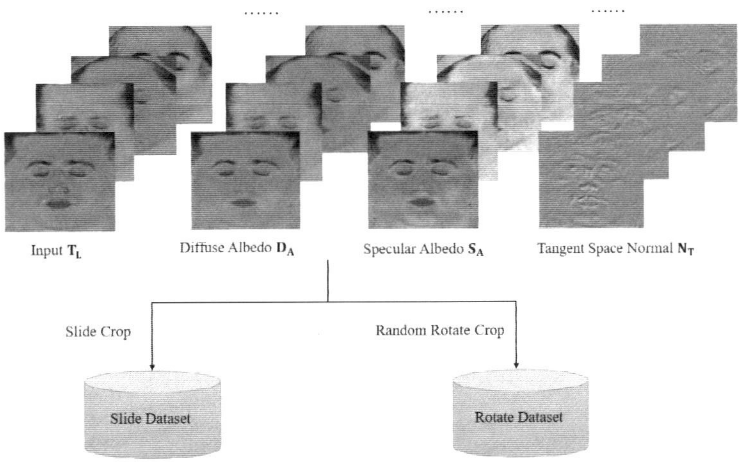

Figure 3. Two data augmentation patterns.

3.2. BRDF Inference Networks

Given the light direction described in Section 3.1, we now have the dataset consisting of the single face image (T_L) under specific illumination condition and the corresponding diffuse albedo D_A), specular albedo (S_A) and tangent space normal (N_T), where L represents the illumination conditions.

We regard the estimation of facial BRDF as a problem of domain adaptation, so we choose pix2pix [28] as our model, which shows impressive performance in image translation. And we independently train three image translation networks for estimating diffuse albedo, specular albedo and tangent space normal. Our view is based on the fact that although the human face skin shows different lighting effects under different lighting conditions, the information of color, highlight and shape of the skin can still be extracted.

So for the extraction of diffuse albedo, we train the image translation network D_ANet with patches: $T_L \rightarrow D_A \in R^{256\times256\times3}$, which can be represented as:

$$D_A = D_ANet(T_L) \tag{1}$$

For the extraction of specular albedo, we train the S_ANet with patches: $T_L \rightarrow S_A \in R^{256\times256\times3}$, which can be represented as:

$$S_A = S_ANet(T_L) \tag{2}$$

Similarly, we train the N_TNet with patches: $T_L \rightarrow N_T \in R^{256\times256\times3}$, which can be represented as:

$$N_T = N_TNet(T_L) \tag{3}$$

The "BRDF Inference" part in figure 1 illustrates the entire deep BRDF inference framework.

During training, we first normalize T_L channels to [-1, 1] and then divide the original high-resolution image into many small image patches of size 256×256 pixels by data augmentation methods described in Section 3.1. to augment the quantity of training data to prevent overfitting.

3.3. Loss Function

For D_ANet, S_ANet and N_TNet, since they all adopt the pix2pix structure [28], so the loss functions of these three networks all contain GAN loss. Particularly, considering the N_TNet generates normal maps where each pixel represents a normal vector, we add a cosine angle loss function for the N_TNet to better supervise the generation of normal and ensure the generated normal maintain its physical meaning during the generation process. The cosine angle loss function guarantees the generated normal and the ground truth are as close as possible in direction consistent.

Let us denote our training image pair (x, y), where $x \in R^{256\times256\times3}$ is the input image and $y \in R^{256\times256\times3}$ is the ground truth. In our work, x represents the face image T_L, and y represents diffuse albedo (D_A), specular albedo (S_A) and tangent space normal (N_T) in D_ANet, S_ANet and N_TNet respectively, and z is random noise.

So for D_ANet and S_ANet, the loss function can be formulated as:

$$G^* = arg \min_G \max_D L_{GAN}(G, D) + \lambda_1 L_{L_1}(G) \tag{4}$$

where:

$$L_{GAN}(G, D) = E_{x,y}[log\, D\,(x, y)] + E_{x,z}[log(1 - D(G(x, z), x)] \tag{5}$$

$$L_{L_1}(G) = E_{x,y,z}[\|y - G(x, z)_1\|] \tag{6}$$

Particularly, we add the cosine angle loss function for the N_TNet to maintain the physical meaning of the normal map, which can be utilized to measure the difference in direction between the normal vector represented by each pixel of generated normal map and the normal vector of ground truth.

So for N_TNet, the loss function is formulated as:

$$G^* = arg \min_G \max_D L_{GAN}(G, D) + \lambda_1 L_{L_1}(G) + \lambda_2 L_N\big(1 - cos(N_{gt}, N_{est})\big) \tag{7}$$

where N_{gt} represents the ground truth and N_{est} represents the generated normal vector.

4. Experiments

4.1. Training Setup

The D_ANet, S_ANet and N_TNet networks are implemented with the Pytorch library on a PC with an Intel Core i7 CPU, 32GB of RAM and NVIDIA GeForce RTX 3080 GPU. We utilize the traditional image translation network pix2pix [28]. More specifically, the networks adopt Adam optimizer with 0.5 for β1 and 0.999 for β2. The learning rate we adopted is 0.0002 and the batch size is 1. Finally, we utilize 3 channel single facial image as input and output the inferred BRDF. And the L_1 loss weight λ_1 is 100 and the cosine angle loss weight λ_2 is 50.

4.2. Evaluation

4.2.1. Qualitative Comparisons. For the qualitative comparison, we crop a high-resolution face image in test set into tiles of 256×256 size and feed them into the trained network and output the inferenced diffuse albedo, specular albedo and tangent space normal. And all the predicted BRDF patches are concatenated to form the high-resolution BRDF, which is applied to a topological face model for rendering. Arguably, our method produces high-fidelity details without artifacts compared with References [3, 24, 29] as shown in figure 4.

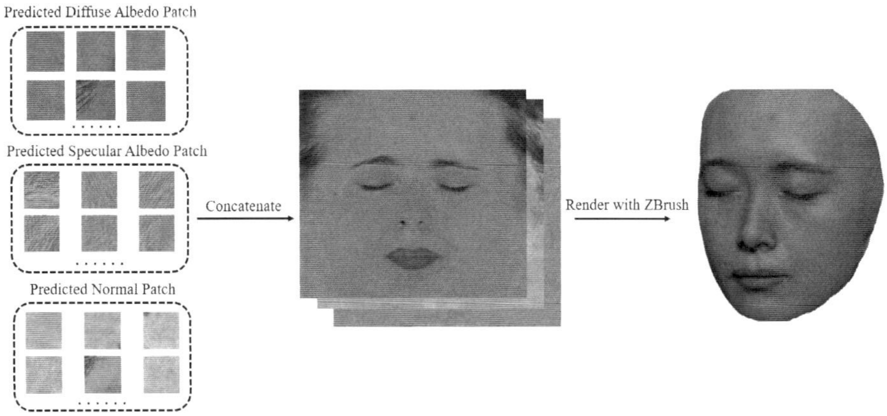

Figure 4. The rendered result with predicted BRDFs.

4.2.2. Quantitative Comparisons. For the quantitative comparisons, we randomly selected 800-1000 images in test set and adopt the widely used peak signal to noise ration (PSNR) metric 0 and we output the indicator calculation results to table 1. To confirm the validity of our method from various aspects, we also utilize mean square error (MSE), structural similarity (SSIM) and image perceptual similarity metric LPIPS0 to evaluate the performance of our method and we record the results under the four metrics in table 2. Obviously, our method outperforms References [3, 24, 29].

Table 1. Comparison with References [3, 24, 29].

Algorithm	Reference [29]	Reference [3]	Reference [24]	Ours
PSNR(Albedo)	11.23	14.37	24.05	**33.76**
PSNR(Normal)	21.89	17.32	26.97	**27.22**

Table 2. Metric results of our BRDF inference networks.

Metric	MSE	SSIM	PSNR	LPIPS
$D_A Net$ (diffuse albedo)	5.33	0.99	44.07	0.07
$S_A Net$ (specular albedo)	644.41	0.96	23.46	0.11
$N_T Net$ (normal)	87.96	0.83	27.99	0.37

4.3. Ablation Study

For the normal inference network, we compare the performance of with the cosine angle loss function and without the cosine angle loss function which is shown in figure 5. For quantitative comparison, we randomly select 800 images in test set and compute the peak signal to noise ration (PSNR), mean square error (MSE), structural similarity (SSIM) and image perceptual similarity metric LPIPS with test data from the same person who never appeared in the training set. And we record the metric results in table 3.

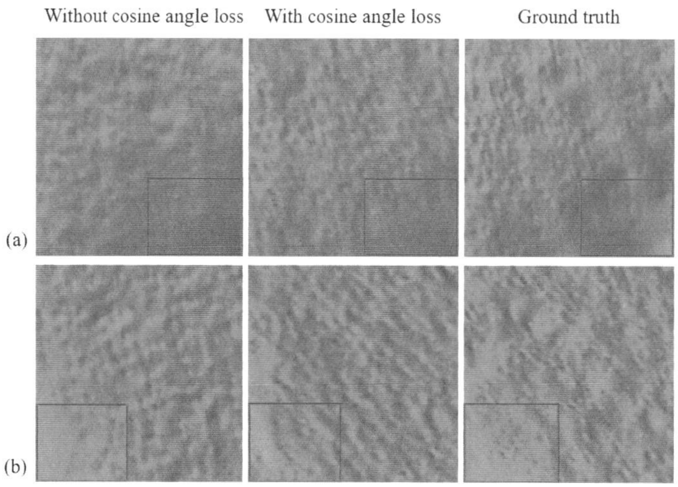

Figure 5. Patch based normal inference comparison between N_TNet with (the second column) and without (the first column) the supervision of cosine angle loss.

Note: In the regions marked by the red box, the normal inference network without the supervision of cosine angle loss produces more blur and less detail.

Table 3. Comparison between N_TNet with and without the supervision of cosine angle loss.

Metric	MSE	SSIM	PSNR	LPIPS
w/o cosine angle loss	99.85	0.82	27.22	0.39
w/ cosine angle loss	87.96	0.83	27.99	0.37

For qualitative comparisons, we demonstrate the inferenced normal results by the N_TNet with cosine angle loss supervision and without cosine angle loss for face images in test set. Obviously, the N_TNet with the supervision of cosine angle loss is better than the network without cosine angle loss in various metrics.

4.4. Limitations
Although our data has a very high resolution and high quality, it does not include enough samples from different ethnicities. Our data are mostly people with light skin tone. Therefore, our approach does not perform well when we infer the BRDFs of ethnicity with dark skin. Though we use some state-of-the-art data augmentation methods to fill this gap, but for objects not included, our method performs not that well.

5. Conclusion
In this paper, we propose a method to estimate the BRDF from a single face image, which is based on an image translation network. We train three independent networks for inferring diffuse albedo, specular albedo and tangent space normal. In order to promote the generalization of our network, we use the random rotate-and-crop and sliding-window-crop for data augmentation. Also we propose the cosine angle loss function to supervise the generation of normal. We demonstrate that it is possible to infer the high-resolution BRDFs from a single face image on a dataset with a relatively small sample size. And our results preserve high level of detail and realism in photorealistic rendering and achieve superior performance compared to state-of-the-art approachs.

Acknowledgments
This work was supported by the Key-Area Research and Development Program of Guangdong Province (No. 2019B010149001) and the National Natural Science Foundation of China (No. 62072036, No. 61902026) and the 111 Project (B18005).

References

[1] Blanz V and Vetter T 1999 A morphable model for the synthesis of 3D faces *Proc. of the 26th Annual Conf. on Computer Graphics and Interactive Techniques* pp 187-94.
[2] Furukawa Y and Ponce J Accurate, dense, and robust multiview stereopsis 2009 *IEEE Transactions on Pattern Analysis and Machine Intelligence* **32** (8) 1362-76.
[3] Chen A, Chen Z, Zhang G, Mitchell K and Yu J 2019 Photo-realistic facial details synthesis from single image *Proceedings of the IEEE/CVF Inter. Conf. on Computer Vision* pp 9429-39.
[4] Li Z, Shafiei M, Ramamoorthi R, Sunkavalli K and Chandraker M 2020 Inverse rendering for complex indoor scenes: Shape, spatially-varying lighting and SVBRDF from a single image *Proceedings of the IEEE/CVF Conf. on Computer Vision and Pattern Recognition* pp 2475-84.
[5] Gao D, Li X, Dong Y, Peers P, Xu K and Tong X 2019 Deep inverse rendering for high-resolution SVBRDF estimation from an arbitrary number of images *ACM Trans Graph* **38** (4) 134:1-134:15.
[6] Yu Y, Smith W A P. 2019 Inverserendernet: Learning single image inverse rendering *Proc. of the IEEE/CVF Conference on Computer Vision and Pattern Recognition* pp 3155-64.
[7] Sengupta S, Kanazawa A, Castillo C D and Jacobs D W 2018 SFSNET: Learning shape, reflectance and illuminance of FACESIN the wild *Proc. of the IEEE Conf. on Computer Vision and Pattern Recognition* pp 6296-305.
[8] Goodfellow I, Pouget-Abadie J, Mirza M, Xu B, Warde-Farley D, Ozair S, Courville A and Bengio Y 2014 Generative adversarial nets *Advances in Neural Information Processing Systems* **27**.
[9] Goldman D B, Curless B, Hertzmann A and Seitz S M 2009 Shape and spatially-varying BRDFS from photometric stereo *IEEE Transactions on Pattern Analysis and Machine Intelligence* **32** (6) 1060-71.
[10] Alldrin N, Zickler T and Kriegman D 2008 Photometric stereo with non-parametric and spatially-varying reflectance *IEEE Conf. on Computer Vision and Pattern Recognition* pp 1-8.
[11] Furukawa Y and Ponce J 2009 Accurate, dense, and robust multiview stereopsis *IEEE Transactions on Pattern Analysis and Machine Intelligence* **32** (8) 1362-76.
[12] Kazhdan M and Hoppe H 2013 Screened poisson surface reconstruction *ACM Transactions on Graphics* **32** (3) 1-13.
[13] Barron J T and Malik J 2014 Shape, illumination, and reflectance from shading *IEEE Transactions on Pattern Analysis and Machine Intelligence* **37** (8) 1670-87.
[14] Jeon J, Cho S, Tong X and Lee S 2014 Intrinsic image decomposition using structure-texture separation and surface normals *European Conf. on Computer Vision* Springer Cham pp 218-33.
[15] Kim K, Torii A and Okutomi M 2016 Multi-view inverse rendering under arbitrary illumination and albedo *European Conf. on Computer Vision* pp 750-67.
[16] Haber T, Fuchs C, Bekaer P, Seidel H.P, Goesele M and Lensch H P 2009 Relighting objects from image collections *IEEE Conf. on Computer Vision and Pattern Recognition* pp 627-34.
[17] Debevec P, Hawkins T, Tchou C, Duiker H.P, Sarokin W and Sagar M 2000 Acquiring the reflectance field of a human face *Proceedings of the 27th Annual Conference on Computer Graphics and Interactive Techniques* pp 145-56.

[18] Ma W C, Hawkins T, Peers P, Chabert C F, Weiss M and Debevec P E 2007 Rapid acquisition of specular and diffuse normal maps from polarized spherical gradient illumination *Rendering Techniques* (9) 10.

[19] Ghosh A, Fyffe G, Tunwattanapong B, Busch J, Yu X and Debevec P 2011 Multiview face capture using polarized spherical gradient illumination *Proceedings of the 2011 SIGGRAPH Asia Conf.* pp 1-10.

[20] Kampouris C, Zafeiriou S and Ghosh A 2018 Diffuse-specular separation using binary spherical gradient illumination *EGSR (EI&I)* pp 1-10.

[21] Deschaintre V, Aittala M, Durand F, Drettakis G and Bousseau A 2018 Single-image svbrdf capture with a rendering-aware deep network *ACM Transactions on Graphics* **37** (4) 1-15.

[22] Li X, Dong Y, Peers P and Tong X 2017 Modeling surface appearance from a single photograph using self-augmented convolutional neural networks *ACM Transactions on Graphics* **36** (4) 1-11.

[23] Guo Y, Cai J, Jiang B and Zheng J 2018 CNN-based real-time dense face reconstruction with inverse-rendered photo-realistic face images *IEEE Transactions on Pattern Analysis and Machine Intelligence* **41** (6) 1294-307.

[24] Lattas A, Moschoglou S, Gecer B, Ploumpis S, Triantafyllou V, Ghosh A and Zafeiriou S 2020 AvatarMe: Realistically renderable 3D facial reconstruction "in-the-wild" *Proceedings of the IEEE/CVF Conference on Computer Vision and Pattern Recognition* pp 760-9.

[25] Blinn J F 1977 Models of light reflection for computer synthesized pictures *Proc. of the 4th Annual Conf. on Computer Graphics and Interactive Techniques* pp 192-8.

[26] High-end content for Artists. https://texturing.xyz/

[27] Wenger A, Gardner A, Tchou C, Unger J, Hawkins T and Debevec P 2005 Performance relighting and reflectance transformation with time-multiplexed illumination *ACM Transactions on Graphics (TOG)* **24** (3) 756-64.

[28] Isola P, Zhu J Y, Zhou T and Efros A A 2017 Image-to-image translation with conditional adversarial networks *Proc. of the IEEE Conf. on Computer Vision and Pattern Recognition* pp 1125-34.

[29] Yamaguchi S, Saito S, Nagano K, Zhao Y, Chen W, Olszewski K, Morishima S and Li H 2018 High-fidelity facial reflectance and geometry inference from an unconstrained image *ACM Transactions on Graphics (TOG)* **37** (4) 1-14.

Target Detection Network for Underwater Image Based on Adaptive Anchor Frame and Re-parameterization

Huipu Xu and Yiteng Wang

College of Marine Electrical Engineering, Dalian Maritime University, Dalian 116026, Liaoning, China
Email: hpx1212@dlmu.edu.cn; wyiteng163@163.com

Abstract. For low target detection rate and inaccurate localization of most target detection networks in special underwater environments, we propose an underwater target detection algorithm with adaptive anchor frames. The adaptive anchor frame selection strategy differs from the traditional methods of manually designing anchor frames and obtaining anchor frames using the *K*-means clustering algorithm. Our method is based on the *K*-means clustering algorithm and consists of a grouping strategy and a dynamic *K*-value strategy. It can also automatically calculate the appropriate numbers of anchor boxes and layers of network output based on the characteristics of label data. And performance of the network is guaranteed while avoiding the computation caused by redundant structures. The simple but powerful RepVGG classification network is used as the backbone network. Therefore, the feature extraction capability of the network is enhanced and the detection performance of the network on fuzzy targets is improved. The experimental results show that the algorithm in this paper achieves better detection results in the underwater target detection task compared to other algorithms.
Keywords. Target detection; underwater image; adaptive anchor frame; re-parameterization.

1. Introduction

Targeted detection technologies play an important role in tasks such as aquatic growth detection, aquatic fishing, and marine biodiversity statistics. Deep convolutional neural networks can automatically acquire more representative features by learning from a large number of samples, and the structure of multiple convolutional layers stacked makes the network acquire stronger semantic information [1]. Among the deep learning-based target detection algorithms, single-shot multiscale structures [2-4] have made a breakthrough in recent years because of their relatively simple structures, low computational cost, and good detection accuracy.

Compared to the land-based optical environment, the underwater optical imaging environment is more unusual. Photos taken underwater are often affected by suspended matter in the water and the attenuation effects of water on light, resulting in more noise, blurred texture, low brightness, chromatic aberration, and other problems [5]. These make it difficult to collect the characteristics of the target. In addition, a large number of small targets and complex backgrounds in the image are also challenging. Therefore, target detection in the underwater environment is a challenging task.

To address the problems of poor imaging quality and the complex background of underwater images, we use the RepVGG [6] network to extract target features from the images. The RepVGG network combines the advantages of single-branch and multi-branch structures to improve the feature extraction capability of the network while maintaining a fast operation speed. The anchor frame is an important factor influencing the performance of the sliding window-based target detection algorithm.

IOP Publishing
2363 (2022) 012012 doi:10.1088/1742-6596/2363/1/012012

Therefore, an adaptive anchor box selection strategy is designed in order to adapt to the size characteristics of different targets in images. The size and number of anchor frames as well as the network structure are adaptively adjusted for the size characteristics of the detected targets to improve the fitting performance of network on the data during training.

2. Related Work

A lot of previous research has been done in the field of underwater target detection. Zhao et al. [5] introduced an auxiliary learning network and improved multi-scale feature fusion structure based on the cascaded R-CNN model, and obtained high detection accuracy in the fish detection task. But the real-time performance is poor and also has high hardware requirements. Compared to two-stage networks, single-stage networks are better in real time, and their accuracy has been greatly improved in recent years. Zhang et al. [7], introduced an attention module and used a multiscale feature fusion structure based on the SSD algorithm, however, the algorithm did not pay attention to the importance of the backbone network for feature extraction. Fan et al. [8] used two backbone networks with composite connections as feature extractors and improved the multiscale feature fusion structure to get the underwater detection of the network better. But there is no consideration of anchor boxes manually set matching to underwater targets.

In deep learning-based target detection algorithms, the backbone network provides the feature map of the image which is an important part of the algorithm. VGG16 [9] was a classic backbone network designed with a single-way model, which has the advantages of simplicity and fast computing speed. However, the number of convolutional layers stacked in the single-way model is too large, and the network degrades during the training process, so the expression capability of model is limited. ResNet [10] was proposed as a viable solution to this problem by incorporating jump connections into the network to avoid the degenerative phenomenon. Compared with the single-way model, the multiple-way models have higher accuracy, but they are also relatively more time consumin.

For target detection networks using anchor boxes, the selection strategy of anchor box is one of the important influences on the network performance. In SSD [2], a set of anchor boxes with dimensions and aspect ratios manually set based on experience were used. This design can satisfy the need to detect targets with common size, but it is difficult to cope with targets of special size. YOLOv2 [3] and YOLOv3 [4] used a clustering algorithm to analyze the data to obtain a set of anchor boxes as priori, and then they are assigned to different detection output layers in order of size sorting. However, for data with only large or small targets, this method may lead to the problem that the anchor box size does not match the feature map receptive field.

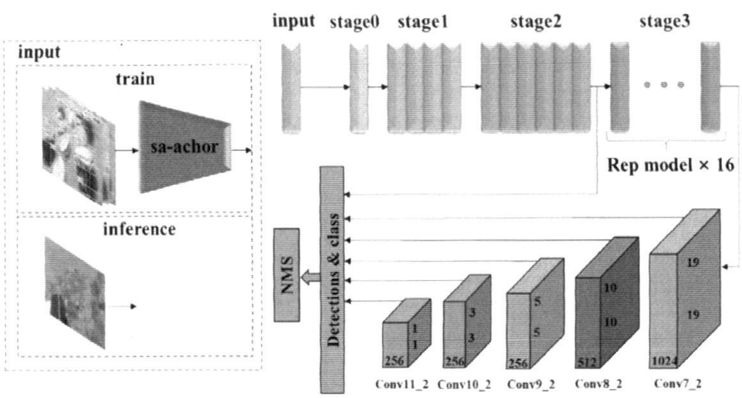

Figure 1. Diagram of network structure.

3. Our Method

The overall structure of the network is shown in figure 1. The network is an end-to-end model consisting of three parts: an anchor box extraction module, a feature extraction module, and a multi-scale detection module. The network runs in training mode and detection mode with different structures. In the training mode, the preselected frame extraction module analyzes the size of the detected target and adjusts the distribution of the preselected frame and the structure of the network. The feature learning module consists of a multi-branch structure to improve the performance of network training. In the detection mode, the anchor frame extraction module does not work, and the feature extraction module switches to a single-branch structure to improve the detection speed of the network and reduce the memory occupation.

3.1. The Feature Extraction Module

For target detection, feature extraction is an important part of the algorithm. We create RepNet as a feature extraction module using the structural re-parameterization module [6], combining the benefits of single-way and multi-way models.

The structural re-parameterization module is shown in figure 2 and contains both training state and inference state. In the training state, the input is divided into three branches: a 3×3 branch, a 1×1 branch, and a residual branch. BN in the figure indicates the batch normalization layer. Multi-branch structures can reduce the difficulty of training the network. In the inference state, the training parameters of each branch are equivalently transformed into the form of a 3×3 convolution. Thus, the detection performance of the network remains unchanged, but has a cleaner structure and faster operation speed compared to the pre-transformation one.

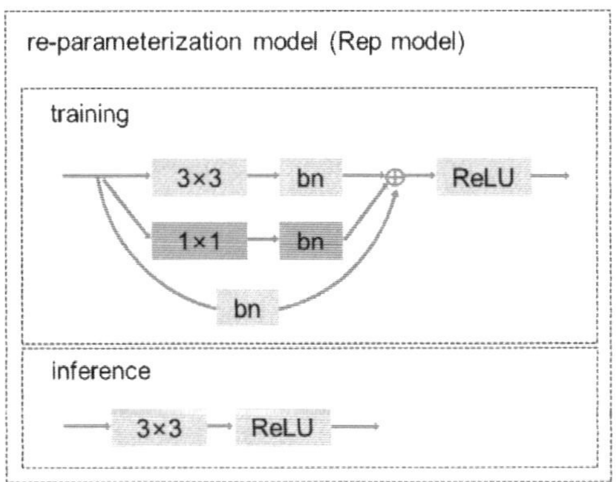

Figure 2. The structural re-parameterization module.

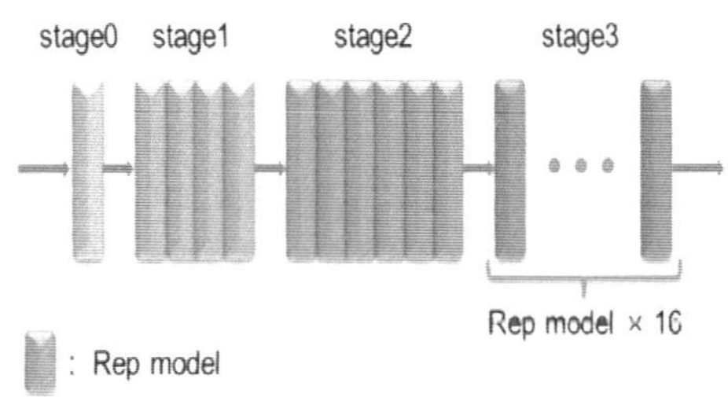

Figure 3. The feature extraction module.

The feature extraction module is built as shown in figure 3 using the structural re-parameterization module. Referring to the classical VGG16 and ResNet101, we divide the feature extraction module into four stages. Considering the large resolution of the feature map in the first stage and the time-consuming processing, we use only one layer to downsample the feature map. The subsequent three layers consist of 4, 6, and 16 structural re-parameterized modules, respectively. In the first layer of each stage, we perform a down-sampling operation with step size 2 and double the number of feature maps to maintain the complexity of the network layer.

3.2. Strategy for Grouping Anchor Frames

In order to improve the ability of network to detect different sizes of target, we optimize the method of finding anchor boxes using the clustering algorithm. Unlike the assignment method of clustering first and sorting later, we divide the truth boxes into different regions by size before clustering. According to the perceptual fields of different detection output layers, we use a set of parallel lines with slope of -1 to divide the 300×300 pixel range into multiple regions, as shown in figure 4, and assign a suitable region to each detection output layer. Then, the data in each region is analyzed separately using a clustering algorithm. Our grouping strategy not only avoids the problem of mismatch between the size of anchor frame and perceptual field of feature map, but also largely improves the computational efficiency in the grouping and clustering process.

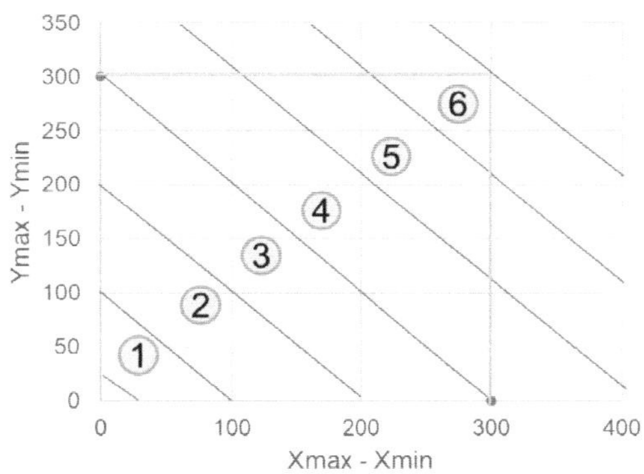

Figure 4. Anchor box grouping strategy and area division method. where 1-6 correspond to six outputs and a set of parallel diagonal lines with a slope of -1 are used as splitting lines.

3.3. Dynamic K-Value Strategy
The number of clustering centers K is an important parameter in the process of finding anchor boxes using the k-means algorithm. In order to determine the optimal K value, we design the dynamic K-value strategy.

During the training process of the network, Anchor boxes that overlapping with the truth box by more than a threshold value will be matched with that true value box, while the rest of the anchor boxes will be discarded. Therefore, different combinations of sizes and aspect ratios are usually chosen in the design process of anchor boxes to avoid similar or duplicate anchor boxes. Therefore, we avoid similar anchor boxes by removing the data from the true value box in the label with a large overlap with the known anchor box, and Intersection over Union (IoU) is used as the basis for judging the overlap. The optimal value of K is determined gradually by continuously reducing the number of truth boxes. Considering the problem that the value of K may be too large during the calculation for data with more dispersed sizes, and lead to a significant increase in the network computation, we set an upper limit for K for each detection layer. When the upper limit is reached, the generated anchor boxes can still represent the primary distribution properties of the sizes of target due to the characteristics of the clustering algorithm. When K is 0, It means that the target to be detected has few corresponding sizes, so we will turn off the corresponding detection output layer to reduce the computational load of network.

4. Experiment
In this paper, two underwater biological datasets, URPC2018 [11] and URPC2019 [12], are combined as our dataset. The dataset mainly includes four categories: holothurian, echinus, scallop, and starfish. The images in the dataset are characterized by underwater blur, color distortion, and complex backgrounds, and contain a large number of aggregated targets and small targets. In the combined dataset (8444 images), 800 images are randomly selected as the test set, and 7644 images in the training and validation sets, which are randomly divided according to the ratio of 9:1 during the training.

The algorithm in this paper is implemented based on the PyTorch framework. The structure of proposed network is pre-trained on the VOC2007 and VOC2012, and the obtained parameters is used as the initialization parameters of the network. To ensure fairness, we also used the corresponding pre-training weights when training other comparison algorithms. The input image size of our model is 512×512 pixels, using the Adam optimizer with momentum parameters $\beta 1=0.9$ and $\beta 2=0.999$. In training, we first freeze the backbone network for 15 epochs, and learning rate is 5×10^{-4}. Learning rate decay strategy is used with a fixed step, a decay rate of each epoch is 0.95, and the size of a batch is

32. The experimental hardware includes an Intel (R) Core (TM) i9-10900X CPU running at 3.70GHz, 32GB of RAM, and a NVIDIA GeForce RTX3060 GPU.

Table 1. Ablation experiments.

Method	AP (%)				mAP (%)
	holothurian	echinus	scallop	starfish	
SSD+VGG16	45.20	57.46	26.53	61.58	47.69
SSD+RepNet	69.24	80.25	40.24	81.34	67.77
SSD+RepNet +sa-anchor (ours)	76.46	86.83	46.14	86.06	73.87

In this section, we investigate the effectiveness of each component, including the structural re-parameterization feature extraction module and the proposed anchor box selection strategy. The experimental results are shown in table 1. The SSD+VGG16 target detection network as the benchmark achieves an average accuracy of only 47.69% in the underwater target detection task. while the SSD algorithm using RepNet has a significant advantage over the VGG16 backbone network. This is mainly due to the multi-branch residual structure used in the training process of RepNet, which is able to provide the network with more adequate feature information during target detection in complex underwater environments. The detection performances of the network are further improved by using the proposed anchor box selection strategy based on SSD+RepNet. As shown in figure 5, the algorithm using the automatic anchor box selection strategy has better performance in the detection task of small target compared to the method with anchor boxes manually set. This is because the proposed method adjusts the anchor boxes according to the characteristics that more small targets are included in the dataset, thus improves the relevance of the network to the detected targets.

Figure 5. Comparison of the detection performances of the manually set anchor box and the automatically set anchor box.

To further validate the performance of the proposed model, it was compared with several advanced target detection algorithms, including Faster-RCNN [13], RFBNet [14], EfficientDet [15], Retinanet [16], Centernet [17], and M2det [18]. From table 2, we can see that the detection task for underwater target is challenging for most target detection networks, and Faster-RCNN, as a classical two-stage network, only achieves an average accuracy of 44.72%. This is because Faster-RCNN only uses the feature map from the last layer of the backbone network with a large sensory field, so it cannot be well adapted to the application environment with more underwater small targets. Subsequently, several methods use multi-scale feature maps, but the anchor boxes artificially set limit their performance. Among them, M2det obtains better detection results due to its excellent feature fusion structure and the introduction of an attention mechanism. Centernet discards the use of anchor boxes, which allows the network to better adapt to different sizes of detection targets. However, the poor quality of underwater images and blurred targets have greater impacts on the ability of Centernet to locate targets, so the final average accuracy of 64.36% is obtained. Compared with other models, the proposed method in this paper achieved significant results, and the highest average detection accuracy are obtained on the underwater target detection dataset.

Table 2. Comparison results on the underwater target detection dataset.

Method	AP (%)				mAP (%)
	holothurian	echinus	scallop	starfish	
Faster-RCNN	60.73	41.24	31.51	45.41	44.72
EfficientDet	48.44	60.16	29.20	65.57	50.84
RFBNet	73.99	62.82	44.84	69.30	62.74
M2det	68.53	73.72	46.00	76.65	66.22
RetinaNet	59.30	70.05	36.27	70.01	58.91
Centernet	55.55	81.87	44.51	75.50	64.36
Ours	76.46	86.83	46.14	86.06	73.87

5. Conclusions

This paper presents an end-to-end neural network model for underwater target detection. According to the characteristics of underwater environmental imaging, a RepNet feature extraction network is built using a structural re-parameterization module to improve the feature extraction capability of the model while maintaining the faster detection speed of the network. In addition, an automatic selection method of anchor box is designed based on the *K*-means algorithm, which enhances the adaptability of the model to different detected targets. The effectiveness of the modules is verified by ablation experiments on the underwater dataset, and the superiority of the proposed method in the task of detection of underwater targets is verified by comparison with other advanced methods. No further design for the multiscale feature fusion structure has been done in this paper, which will be the focus of the next work. Meanwhile, it is our focus to further improve the detection speed of the model while ensuring detection accuracy.

References

[1] Chen S L, Yang C, Ma J W, Chen F and Yin X C 2019. Simultaneous end-to-end vehicle and license plate detection with multi-branch attention neural network *IEEE Transactions on Intelligent Transportation Systems* **21** (9) 3686-95.

[2] Liu W, Anguelov D, Erhan D, Szegedy C, Reed S, Fu C Y and Berg A C 2016 SSD: Single shot multibox detector *European Conf. on Computer Vision* pp 21-37.

[3] Redmon J and Farhadi A 2017 YOLO9000: better, faster, stronger *IEEE Conf. on Computer Vision and Pattern Recognition* pp 7263-71.

[4] Redmon J and Farhadi A 2018 Yolov3: An incremental improvement *arXiv preprint arXiv:1804.02767*.

[5] Zhao Z, Liu Y, Sun X, Liu J, Yang X and Zhou C 2021 Composited FishNet: Fish detection and species recognition from low-quality underwater videos *IEEE Transactions on Image Processing* **30** 4719-34.

[6] Ding X, Zhang X, Ma N, Han J, Ding G and Sun J 2021 RepVGG: Making VGG-style convnets great again *IEEE/CVF Conf. on Computer Vision and Pattern Recognition* pp 13733-42.

[7] Zhang J, Zhu L, Xu L and Xie Q 2020 MFFSSD: An Enhanced SSD for Underwater Object

Detection *2020 Chinese Automation Congress (CAC)* pp 5938-5943.
[8] Fan B, Chen W, Cong Y and Tian J 2020 Dual refinement underwater object detection network *European Conf. on Computer Vision* pp 275-291.
[9] Simonyan K and Zisserman A 2014 Very deep convolutional networks for large-scale image recognition *arXiv preprint arXiv:1409.1556.*
[10] He K, Zhang X, Ren S and Sun J 2016 Deep residual learning for image recognition *IEEE conference on computer vision and pattern recognition* pp 770-778.
[11] 2018 underwater robot picking contest 2018.cnurpc.org/index.html
[12] 2019 underwater robot picking contest 2019.cnurpc.org/index.html
[13] Ren S, He K, Girshick R and Sun J 2015 Faster r-CNN: Towards real-time object detection with region proposal networks *Advances in Neural Information Processing Systems* **28**.
[14] Liu S and Huang D 2018 Receptive field block net for accurate and fast object detection *European Conference on Computer Vision (ECCV)* pp 385-400.
[15] Tan M, Pang R and Le Q V 2020 Efficientdet: Scalable and efficient object detection *IEEE/CVF Conf. on Computer Vision and Pattern Recognition* pp 10781-90.
[16] Lin T Y, Goyal P, Girshick R, He K and Dollár P 2017 Focal loss for dense object detection *IEEE International Conf. on Computer Vision* pp 2980-88.
[17] Zhou X, Wang D and Krähenbühl P 2019 Objects as points *arXiv preprint arXiv:1904.07850.*
[18] Zhao Q, Sheng T, Wang Y, Tang Z, Chen Y, Cai L and Ling H 2019 M2det: A single-shot object detector based on multi-level feature pyramid network *AAAI Conf. on Artificial Intelligence* **33** (01) 9259-66.

Detection of Defects in Power Grid Inspection Images Based on Multi-scale Fusion

Jia Wu, Zhenbo Liu and Qi Ren

State Grid Jibei Eiectric Power Company Limited, Beijing 100054, China
Emails: jbdl_wu@hotmail.com; jbdl_liu@hotmail.com; jbdl_ren@hotmail.com

Abstract. With the increasing application of visual application scenarios in various industries, power grid inspection system based on vision is widely used. However, the power grid intelligent inspection system has the problem of unsatisfactory accuracy of small target detection. To address this problem, this paper proposes a multi-scale fusion-based defect detection model for grid inspection images. The object features are first acquired by improving the CSPDarknet53 network using the attention mechanism. Second, the multi-scale feature fusion module is constructed to obtain more feature information. Finally, the defect detection model of the grid inspection image is constructed based on the multi-scale feature fusion module. The experimental results of our proposed algorithm are compared and analyzed with the current mainstream algorithms in all aspects. The optimal performance accuracy of our model reaches 0.927, which is improved by 6.06%.
Keywords. Multi-scale fusion; power grid detection; image processing; defect detection.

1. Introduction

The traditional power inspection work mainly relies on manual completion, which requires a lot of human and material resources and is poor in real-time. For the inspection personnel, such work has the characteristics of a difficult inspection environment, inspection labor intensity, and high danger index. With the increasing application of visual application scenarios in various industries, the demand for intelligent inspection instead of manual inspection to achieve real-time monitoring of hidden areas is becoming increasingly strong [1].

The main task of target recognition is to realize the location and perception of interested targets, which is a hot topic at present. The detection model is divided into one stage or two stages. These two different processing methods are the two major research strategies for constructing short hair models based on deep learning. Comparatively speaking, the strategy of constructing the model in two stages is generally superior to the strategy of constructing the model in one stage in terms of accuracy. [2]. The more authoritative early research on the one-stage model includes YOLO [3], SSD [4], YOLO-v2 [5], YOLY-v3 [6], etc. Other typical works based on R-CNN model and its improved algorithm are shown in documents [7-10].

The intelligent inspection technology based on target recognition mainly includes image acquisition, feature extraction and matching, signal enhancement and recognition, etc. The power intelligent inspection equipment detects foreign objects or abnormalities of power equipment through computer vision technology. However, the defects in power equipment are mostly small targets such as pin defect storage and insulator breakage, and the existing general target detection models are not ideal for small target detection.

For the practical problems of power industry, this paper presents an image defect perception algorithm based on different scale information fusion strategies. The overall framework of the algorithm is shown in Figure 1. Firstly, the improved CSPDarknet53 network using the attention mechanism is used to improve the backbone feature extraction network in the small target attention level. Second, the multiscale feature information acquisition model is constructed to obtain more target features through multi-scale information fusion. Finally, the defect detection model for grid inspection images is constructed based on the multi-scale feature fusion module. Accurate identification of small targets will help improve the performance of the grid intelligent inspection system. This has a positive effect on the development of intelligent power grids.

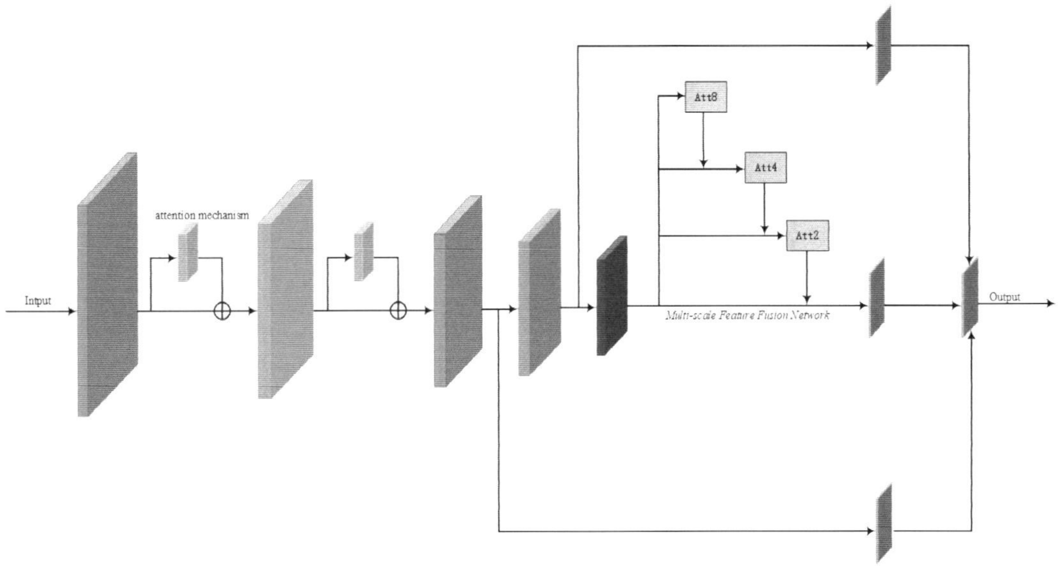

Figure 1. Multi-scale fusion model framework for power grid inspection defect detection.

The rest of this paper is presented below. Part II describes the related work. The construction idea and mathematical principle of this algorithm will be highlighted in part III. In order to verify the feasibility of the practical application of the proposed method in the power industry, the part IV focuses on the relevant experimental results and representative data analysis basis, and uses quantitative data to verify the advancement of the proposed algorithm. The part V concludes and general description of possible future work.

2. Related Work

At present, relevant research institutions and related research work of many scholars in the early stage have achieved many remarkable results for the intelligent inspection of power grids, and we will briefly review the relevant contents.

2.1. Intelligent Inspection of Power Grid

The electric inspection unmanned aerial vehicle (UAV) technology takes pictures on the line through the camera and realizes the real-time automatic monitoring of line faults and hidden dangers through image processing and analysis, which greatly improves the inspection efficiency. A Japanese research and development team has developed a highly integrated, compact and low-power drone inspection system [11]. The UAV inspection system has a high recognition rate for various common faults, such as insulator pollution flashover and wire icing. North China Power Group mentions visual inspection, thermal image inspection technology, for the first time in the country to achieve a manually driven helicopter patrol line [12]. Southern Network Guangdong also invested a lot of manpower and material resources in the UAV power line inspection research, from the actual site, research many practical technical solutions, do a lot of useful exploration [13].

2.2. Small Target Detection

Deep learning-based target recognition technology is fast maturing and widely used, however, it is still challenging for small target detection and recognition. To address the problem that small targets are difficult to recognize and the recognition accuracy is low, Zhao et al. [14] used the generator of the GAN network to learn the corresponding feature mapping and mapped small target features to large target features before detection. Zhang et al. [15] use a combination of pooling and deconvolution to improve recognition accuracy by using a unique "rainbow-connected" cascade approach. Li et al. [16] expanded an ROI network based on Faster R-CNN, which was used to backtrack and retrain difficult samples screened through loss, and improve the detection accuracy of difficult samples without resetting the positive and negative sample ratio.

2.3. Intelligent Inspection of Power Grid Based on Target Detection

The object detection model based on intelligent processing algorithm has also been gradually applied to the inspection and defect detection of transmission lines to realize intelligent inspection of transmission lines [17]. To solve the problem of short-distance identification on transmission lines, Qi et al. [18] proposed a hardware based transmission line detection model for power industry. The problem of too few samples can be solved by the method of rotation and clipping. The appearance of dense occlusion will seriously affect the recognition effect. The repulsion loss function is constructed to eliminate this effect. Henriques et al. [19] Organically combine with dense spatiotemporal context combined, the tracking process is completed by means of probability theory. Furthermore, according to the feature that foreign bodies always appear in the image sequence outside the region, significance calculation is used to evenly extract foreign bodies on the transmission line. Zhao et al. [20] proposed a YOLOv3 method for typical components of transmission lines, including composite insulators, glass insulators, pressure equalizing rings, shielding rings and shock absorbers.

Through the above analysis, intelligent inspection of power grid with target recognition has achieved rapid development, but there are still some shortcomings in small target recognition and real-time detection. This paper further studies the accuracy of small target recognition.

3. Algorithm

A defect detection model of power grid inspection image based on multi-scale fusion is proposed to enhance the ability to acquire subtle target features. The following explains the construction strategy of the model from three aspects.

3.1. Target Feature Extraction

The CSPDarknet53 network is introduced as the backbone feature extraction network. CSPDarknet53 adds Cross-Stage-Partial-connections to each residual cell of the original Darknet53. Its convolution block is composed of single convolution, normalization and Mish activation function. There is a large residual edge in this network, which is directly connected to the output without operation, making the backbone part easier to optimize. Meanwhile, the Mish activation function can guarantee the gradient descent effect, whose mathematical principle is described as follows.

$$Mish = x * \tanh(\text{In}(1+\text{e}^{x})) \tag{1}$$

where x denotes the value of the input node Mish activation function also retains the advantage of unbounded activation function and cleverly avoids the saturation problem caused by the limit gradient tending to 1 on both sides.

The end-to-end target detection model is difficult to detect because the low-level details and location information are ignored. This is mainly due to two main reasons: the lack of contextual information and the lack of superficial features of semantic information. This paper uses attention mechanism to improve the feature extraction backbone network.

The visual attention mechanism allows attention to be focused on one part of the image. The structure of attention module designed in this paper is shown in figure 2. The designed attention module is placed after the CSPRes1 and CSPRes2 modules and consists of a trunk branch and a branch. The trunk branch has two attention mechanisms with three convolutional layers per block. The branch outputs the attention map by down-sampling and up-sampling the residual connections and then uses sigmoid activation for final processing. The principle is as follows:

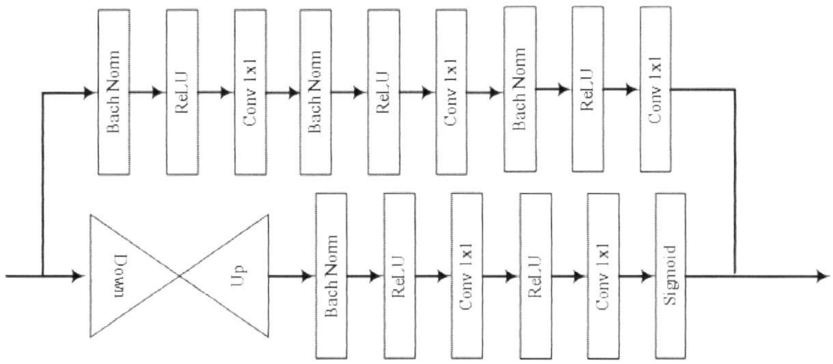

Figure 2. Network model of attention mechanism.

$$T = Soft\max(\mathrm{A}_{ba} \otimes \mathrm{A}_{br}) \tag{2}$$

where, A_{ba} represents the attention weight of the main branch, and A_{br} represents the attention weight of the branch. T Represents the weight of attention after recalibration

In this paper, the attention recalibration weight T was multiplied by the corresponding scale feature Figure F_i to obtain the attention feature map, which was integrated with the backbone network. The text of your paper should be formatted as follows:

$$F = [F_i : F_i \cdot T] \tag{3}$$

where F represents the fused features and [:] represents the fusion operation.

3.2. *Multi-scale Feature Fusion Network*

As the network deepens, the lost information will grow in multiples, and the later the hierarchical structure is, the more effective information will be lost.This paper designs a multi-scale feature fusion solution. The principle is as follows:

$$X_B = C^0\left(X_A\right) \tag{4}$$

$$X_C = C^2\left(X_A\right) \tag{5}$$

$$X_D = C^4\left(X_A\right) \tag{6}$$

$$X_E = C^8\left(X_A\right) \tag{7}$$

where $C^n(\cdot)$ represents the feature scaling operation. n indicates scaling. Then, the compressed features are input into the multilevel attention module.

The multi-scale fusion module used at the input end is described in detail. The process is shown in equations (8)-(11).

$$X_B' = ATT^0\left(X_C' \oplus X_B\right) \tag{8}$$

$$X'_C = Att^2\left(X'_D \oplus X_C\right) \tag{9}$$

$$X'_D = Att^4\left(X'_E \oplus X_D\right) \tag{10}$$

$$X'_E = \mathrm{Att}^8\left(X_E\right) \tag{11}$$

where $\oplus$ represents channel dimension superposition. $Att^n(\bullet)$ represents the attention mechanism feature fusion operation for the feature with compression rate of n.

3.3. Target Detection

The coordinates of candidate frames are generated at the regression layer, and the confidence degree of candidate frames is calculated at the classification layer. To facilitate expression, G is used in this paper to represent the coordinate of the candidate box. Formally, the network predicts a probability distribution instead of only bounding box location. Assuming that the coordinate of candidate box is independent, the principle of obtaining probability distribution is as follows:

$$P(\mathrm{G}) = \frac{1}{\sqrt{2\pi\sigma^2}} e^{-\frac{(\mathrm{G}_g - \mathrm{G})^2}{2\sigma^2}} \tag{12}$$

where G_g represents the coordinates of the real boundary box. σ represents the standard deviation, which measures the uncertainty of the estimate. When $\sigma \to 0$ indicates that the candidate box for identifying framework proposal in this paper is worthy of consideration. Not all candidate boxes obtained through the above procedure are valid. Therefore, further screening is needed. We set the IOU threshold at 0.5 to obtain accurate target recognition results.

4. Experiment and Analysis

The experimental data and results based on the proposed algorithm will be highlighted in this section, which will more authoritatively illustrate the practical application value of the algorithm in the power industry. The specific experimental results are as follows.

4.1. Experimental Setup

This experiment is run on the server based on Windows 10 system. Run Python3.7. The deep learning platform software used to compile the tests was configured as PyTorch and CUDA V1.0. Hardware configuration: GPU is RTX 1080 Ti. In this paper, 1000 images including insulators, shock hammers and spacers were collected to construct a dataset. These images are scaled to 227x227. This paper will choose four methods to compare with our model. These four methods are SSD, Faster R-CNN and YOLO-v4. Use the SGD optimizer to train all models separately. Models are trained in a batch size of 10. The initial learning rate of the model is set to 0.01. When the training epoch is 20th epoch and 40th epoch, the learning rate of the model is attenuated by 0.1 times. The weight decay and momentum respectively set to 0.0001 and 0.9.

4.2. Qualitative Experimental Analysis

To verify the proposed grid inspection image defect detection model based on multi-scale fusion is to solve the problem of low accuracy of small target recognition in complex scenes. Set the test image set to 1000, and use this test set to evaluate the implementation of the algorithm, the improved model can well identify the position and classification of insulators, shock hammers and spacers.

Figure 3 shows the detailed comparison results between the proposed algorithm and the comparison algorithm in the effects of under different backgrounds, different lights and different shelters. The algorithm proposed can accurately identify targets. Compared with other methods, the corresponding confidence degree of the method proposed in all experiments. The contrast experiment shows that the designed model can accurately identify the insulators, shock hammers and spacers in transmission lines, and can also identify the target object well under different backgrounds.

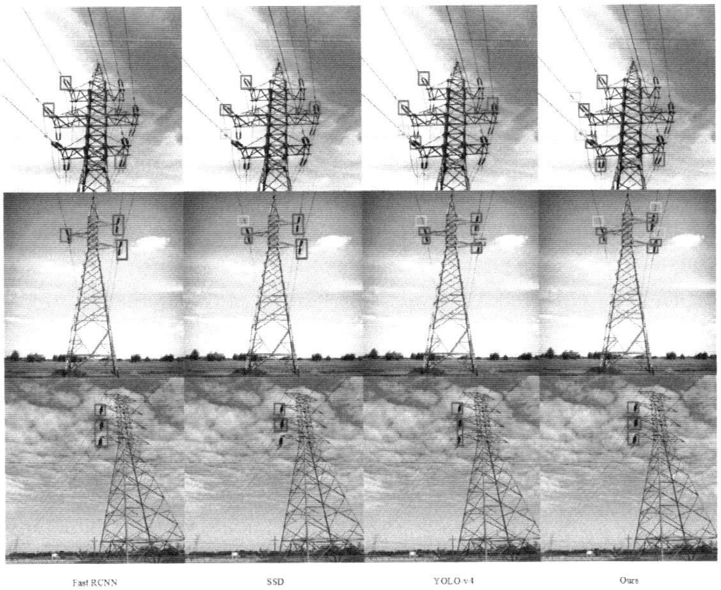

Figure 3. Test renderings.

4.3. Quantitative Experimental Analysis

Different evaluation individuals will have different views on the visual evaluation results from different perspectives. In order to avoid the impact of such artificial sensory differences, quantitative measurement data are further provided. This article will qualitatively analyze the object recognition models of SSD, Faster R-CNN, YOLO-v4 and Ours from the objective data level. This article introduces AP and mAP object recognition evaluation standards to analyze the object recognition results of each object recognition model from objective data. Table 1 shows the objective data corresponding to the different scene simulation experiments.

Table 1. Experimental results.

	AP			mAP
	Nsulators	Shock hammers	Spacers	
Faster R-CNN	0.899	0.902	0.893	0.898
SSD	0.853	0.865	0.845	0.854
YOLO-v3	0.906	0.886	0.852	0.874
Ours	0.931	0.904	0.896	0.927

The AP, and mAP values for the different scene simulation experiments are shown in the table above. From the data point of view, the algorithm in this paper is the best, and the AP is the highest in the Insulators, Shock Hammers and spacers experiments. The optimal performance accuracy of our model reaches 0.927, which is improved by 6.06%.

5. Conclusion

The multi-scale fusion-based defect detection model for power grid inspection images is proposed. Firstly, the attention mechanism is used to improve the CSPDarknet53 network to obtain target features. Secondly, The multi-scale feature fusion module is constructed to obtain more feature information. Finally, the grid inspection image defect detection model is constructed based on the multi-scale feature fusion module. The feasibility of the proposed recognition model is verified by extensive simulation experiments. Regardless of qualitative or quantitative comparisons, the algorithm proposed in this paper has certain advantages. Although the proposed algorithm has some advantages, it cannot be applied in engineering. In future research, underwater object recognition in more complex environments will be further explored. At the same time, the object recognition model is further optimized to realize engineering applications.

References

[1] Li Z, Liu Y, Walker R, et al. 2010 Towards automatic power line detection for a UAV surveillance system using pulse coupled neural filter and an improved Hough transform *Machine Vision and Applications* **21** (5) 677-686.

[2] Xu C D, Zhao X R, Jin X, et al. 2020 Exploring categorical regularization for domain adaptive object detection *2020 IEEE Conf. on Computer Vision and Pattern Recognition* pp 11721-11730.

[3] Redmon J, Divvala S, Girshick R, et al. 2016 You only look once: Unified, real-time object detection *2016 IEEE Conf. on Computer Vision and Pattern Recognition* pp 779-788.

[4] Liu W, Anguelov D, Erhan D, et al. 2016 SSD: Single shot multibox detector *European Conf. on Computer Vision* pp 21-37.

[5] Redmon J and Farhadi A 2017 Yolo9000: better, faster, stronger *2017 IEEE Conf. on Computer Vision and Pattern Recognition* pp 6517-6525.

[6] Redmon J and Farhadi A 2018 YOLOv3: An incremental improvement *2018 IEEE Conf. on Computer Vision and Pattern Recognition* pp 13674-13682.

[7] Girshick R, Donahue J, Darrell T, et al. 2014 Rich feature hierarchies for accurate object detection and semantic segmentation *2014 IEEE Conf. on Computer Vision and Pattern Recognition* pp 580-587.

[8] He K M, Gkioxari G, Piotr D, et al. 2017 Mask R-CNN *IEEE Transactions on Pattern Analysis and Machine Intelligence* **42** (2) 386-397.

[9] Girshick R 2015 Fast R-CNN *2015 IEEE Inter. Conf. on Computer Vision* pp 2380-7504.

[10] Ren S Q, He K M , Girshick R, et al. 2017 Faster R-CNN: Towards real-time object detection with region proposal networks *IEEE Transactions on Pattern Analysis and Machine Intelligence* **39** (6) 1147-1149.

[11] Shi J T, Gu C Y and Sun H 2019 Data expansion for foreign object detection in power grid *11th IEEE PES Asia-Pacific Power and Energy Engineering Conf.* pp 69-73.

[12] Girshick R, Donahue J, Darrell T, et al. 2014 Rich feature hierarchies for accurate object detection and semantic segmentation *2014 IEEE Conf. on Computer Vision and Pattern Recognition* pp 580-587.

[13] Julien B, Kristopher T and Nicolas P 2010 On the application of VTOL UAVs to the inspection of power utility assets *Inter. Conf. on Applied Robotics for the Power Industry* pp 620-631.

[14] Zhao P, Deng C and Yuan Y C 2002 Patrol of transmission line with helicopter *North China Electric Power* **25** (10) 23-26.

[15] Zhang W F. Peng X Y and Zhong Q 2014 Status quo and expectation of safety inspection technology for electric power lines based on remote sensing *Guangdong Electric Power* **27** (02) 163-169.

[16] Li J, Liang X, Wei Y, et al. 2017 Perceptual generative adversarial networks for small object detection *2017 IEEE Conf. on Computer Vision and Pattern Recognition* pp 1320-1333.

[17] Cai Y, Luan T, Gao H, et al. 2020 YOLOv4-5D: An effective and efficient object detector for autonomous driving *IEEE Transactions on Instrumentation and Measurement* **70** (99) 93-105.

[18] Qi Y, Jiang A, Zhao Z, et al. 2019 Fittings detection method in patrol images of transmission line based on improved SSD *Electrical Measurement & Instrumentation* **56** (22) 7-12+43.
[19] Henriques J F, Rui C, Martins P, et al. 2014 Iligh-speed tracking with kernelized correlation filters *IEEE Transactions on Pattern Analysis&Machine Intelligence* **37** (3) 583-596.
[20] Zhao Y, Jia J, Liu D, et al. 2021 He-yolo: aerial target detection based on improved YOLOv3 *International Journal of Pattern Recognition and Artificial Intelligence* **35** (13) 583-596.

Pose Measurement Technology of Roadheader Body based on Fusion of Visual and SINS*

Lixin Wang[1], Chengjun Hu[1], Gege Pan[1], Chao Zhang[2], Bin Liu[1], Jicheng Wan[2], Hongqiang Yang[2], Shuo Shi[2], Zhen Dong[2], Yuyang Du[2], Wenjuan Yang[2] and Xuhui Zhang[2]

[1] China Coal Underground Engineering Intelligent Research Institute Co. Ltd. and China Coal Design Engineering Co. Ltd., Tianjin 300000, China
[2] Mechanical Engineering College,Xi'an University of Science and Technology, Xi'an 710050, Shaanxi, China
Email: wanglixin@chinacoal.com; 289926387@qq.com; pangege@chinacoal.com; 1354277569@qq.com; bin.liu@chinacoal.com; 1981724247@qq.com; 1757307011@qq.com; 1278817233@qq.com; 460757110@qq.com; 1090973350@qq.com; yangwj@xust.edu.cn; zhangxh@xust.edu.cn

Abstract. In order to solve the problem of pose measurement of roadheader body in low illumination and high dust environment in coal mine, a pose measurement technology of roadheader body based on visual and strapdown inertial navigation system (SINS) is proposed. By extracting the point and line features and using the monocular vision positioning principle based on the gate three lines, the spatial pose of the roadheader body measured by vision is obtained according to the spatial matrix transformation. At the same time, the spatial pose of the roadheader body measured by SINS is calculated by gyroscope and accelerometer. Then the least square method is used to align the two-sensor data, and the registration data is fused based on the extended Kalman filter fusion algorithm to obtain the final fusion roadheader pose data. The experimental results show that the measurement error of the roadheader body position in the x axis direction of the roadway coordinate system is less than ± 20 mm, the error of the y axis direction is less than ± 35 mm, the error of the z axis direction is less than ± 10 mm, the yaw angle error is between-0.4° to 0.7°, the roll angle error is between -0.2° and 0.6°, and the pitch angle error is between -0.02° and 0.5°. This method can be used to measure the pose of the roadheader stably and accurately in the complex environment of high dust, miscellaneous light and low illumination.
Keywords. SINS; roadheader; extended Kalman filter; monocular vision; pose measurement.

1. Introduction

The working conditions of underground roadways in coal mines are complex, and they have the characteristics of high dust and low illumination, which will make it difficult to guarantee the forming quality of roadway excavation [1, 2]. And there are risks such as gas outburst and coal rock collapse in roadway [3]. It is necessary to realize automation, intelligence and unmanned of coal mine. The measurement technology of roadheader fuselage posture is its core problem [4].

In recent years, experts at home and abroad have carried out corresponding research on the pose measurement of roadheader fuselage. Fu et al. [5] designed UWB pose detection system. The system uses four mobile base station robots equipped with UWB modules to measure the range of the fuselage nodes, and the pose parameters of the roadheader are obtained by calculating the ranging information.

The system uses four mobile base station robots equipped with UWB modules to measure the range of the fuselage nodes, and the pose parameters of the roadheader are obtained by calculating the ranging information. Based on the principle of monocular vision measurement [6, 7], using the visual measurement model based on "a gate with three lines", according to the laser characteristics, the roadway image collected by the explosion-proof industrial camera is processed. Through the combination of Hough transform and least-square method. The Point and line characteristics of flare and beam of the two laser pointing instruments are obtained. The spatial pose of the roadheader fuselage is calculated by spatial matrix transformation. Zhang et al. [8] used binocular stereo vision to compose the infrared LED light source into an image measurement target, the Gaussian fitting method is used to extract the feature of the facula center quickly and accurately, and the 3D-3D motion estimation method is used to solve the pose parameters. Zhang et al. [9] used passive broadband harmonics to generate available harmonic signals. According to the phase Difference Ranging Algorithm of transmitting broadband multi-frequency continuous wave Coherently, the pose of the roadheader is identified. Tao et al. [10] installed the laser receiver at the fixed position of the roadheader fuselage, and measured the three-dimensional coordinates of the laser receiver in the launch station coordinate system. Hargrave [11] designed for mine equipment positioning based on array ranging sensors. However, the application stability of single sensor positioning method in mine is relatively low, so the combination positioning method has become a research hotspot. Zhang et al. [12] measured the position parameters of the roadheader by total station, and measured the pose parameters of the roadheader with the strapdown inertial navigation system (SINS), and calculated and fused them to obtain the pose data of the roadheader. Hua et al. [13] used SINS to measure the attitude information of intelligent tunneling robot in real time, and combined with the displacement data of the moving cylinder to calculate the position. Then the pose fusion is carried out to improve the pose measurement accuracy, but this method is difficult to obtain the relative position of the fuselage. Xu et al. [14] obtained the absolute position information with UWB, and estimated the absolute scale of monocular vision odometer by using the least square method. Then, the displacement increment of monocular vision odometer and the ranging information of UWB are taken as the measurement values, and the extended Kalman filter algorithm is used for data fusion to realize equipment positioning. Lu et al. [15] designed the positioning system of roadheader robot with total station and compass. However, total station and UWB have great influence on water mist and dust, and inertial navigation can also produce drift [16], which is difficult to be used remotely.

2. Frame of Position and Attitude Measurement System for Roadheader Fuselage

Vision and SINS fusion of roadheader fuselage pose measurement technology scheme as shown in figure 1. The scheme mainly includes three modules, the roadheader visual pose measurement module, the SINS positioning module and the combination of vision and SINS positioning module. Vision Pose Detection of Roadheader is to collect the light flare and light image of two lasers of laser pointer through explosion-proof camera on the fuselage, the connection between the two laser spots, the light on the spot and the spot is simplified as the feature of "two points and three lines", and the pose relationship between the camera and the laser pointer is calculated by visual measurement method. Then the posture of the roadheader in the roadway coordinate system is calculated by the established posture calculation model. SINS is rigidly connected with the fuselage of the roadheader. Instantaneous acceleration and angular velocity information of roadheader can be obtained by SINS. Before the pose measurement of the SINS, the initial alignment of the SINS is carried out to obtain the zero drift data of the sensor. The quaternion attitude calculation model is established by the SINS, and the position, attitude and speed information of the roadheader are obtained by the three-sample algorithm.

The coordinate transformation relationship between the camera and the SINS can be obtained parameter calibration. Coordinate transformation between camera and SINS can be obtained by calibrating external parameters. Realized the unification of space coordinates, unified the coordinates to the navigation coordinate system. Because the output frequency of vision is much lower than that of SINS, the original data of vision sensor and SINS should be aligned before data fusion. Secondly, both of the two positioning methods can output the pose information of the roadheader, but the SINS has the characteristics of good short-term performance and independent measurement. Therefore, the process model is mainly constructed based on the output pose data of SINS, and the extended Kalman filter is designed with the visual data as the observation value. Finally, the data fusion of binocular vision and SINS is realized to achieve the purpose of accurate positioning of roadheader fuselage.

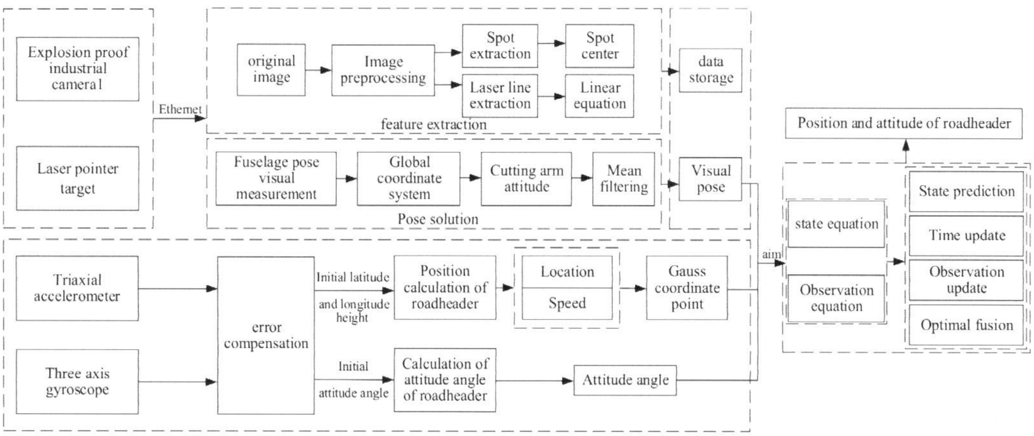

Figure 1. Vision and strap-down whip guide fusion of roadheader fuselage pose measurement technology scheme.

3. Vision Pose Measurement System for Roadheader

3.1. Coordinate Transformation

The schematic diagram of the visual measurement model of the body position and attitude of the roadheader, as shown in figure 2, the measurement coordinate system includes the roadheader fuselage coordinate system $O_oX_oY_oZ_o$, the rear camera coordinate system $O_{c2}X_{c2}Y_{c2}Z_{c2}$, and the laser measurement coordinate system $O_dX_dY_dZ_d$ and the roadway coordinate system $O_hX_hY_hZ_h$. The conversion matrix between the fuselage base coordinate system $O_oX_oY_oZ_o$ and the rear camera coordinate system $O_{c2}X_{c2}Y_{c2}Z_{c2}$ is set to be ${}^{c2}_{0}T$; The transformation matrix between the rear camera coordinate system $O_{c2}X_{c2}Y_{c2}Z_{c2}$ and the laser measurement coordinate system $O_dX_dY_dZ_d$ is ${}^{d}_{c2}T$; The transformation matrix between the laser measurement coordinate system $O_dX_dY_dZ_d$ and the roadway coordinate system $O_hX_hY_hZ_h$ is ${}^{h}_{d}T$; The pose transformation matrix between the roadheader base coordinate system $O_oX_oY_oZ_o$ and the roadway coordinate system $O_hX_hY_hZ_h$ is ${}^{h}_{0}T$.

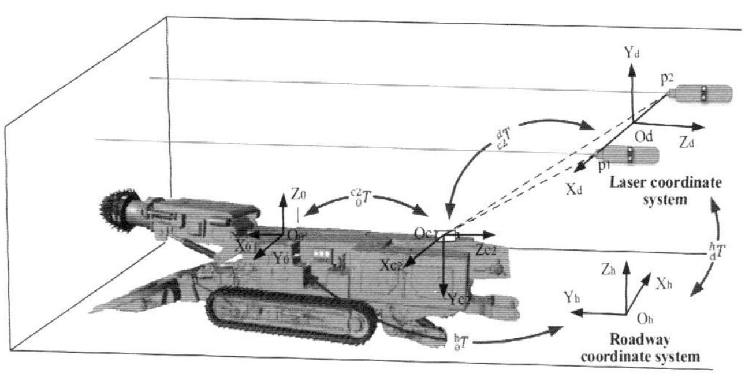

Figure 2. Schematic diagram of the visual measurement model of the body position and attitude of the roadheader.

3.2. Vision Measurement Model Based on Door-Shaped Structure

The image feature information of the laser pointer in the roadway includes two laser spots and two laser beam straight lines. The straight line features of the two laser beams and the straight line passing through the center of the two laser spots intersect at the center point of the two laser spots, which can form three coplanar straight lines. Considering that the monocular vision measurement results based on line features have higher accuracy, this paper uses the laser line features with known positional relationships to establish a vision measurement model based on the n-line perspective problem. The two laser pointers are parallel and fixed to ensure that the two laser beams are parallel. In the three straight lines formed by the laser features, the straight line passing through the center of the two laser spots is perpendicular to the two parallel laser beam lines, and the three straight lines form a gate-shaped structure. Combining the pinhole imaging model of the camera and the two-point three-line feature of the laser can form a three-line perspective geometric structure model based on the gate-shaped structure, as shown in figure 3.

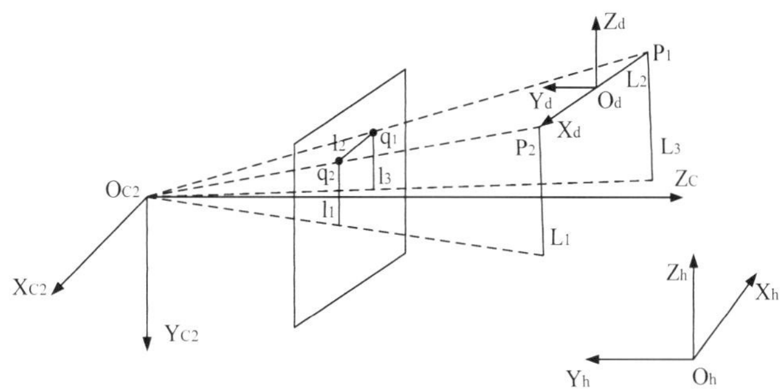

Figure 3. Schematic diagram of vision measurement based on "a gate with three lines".

In the figure, $O_{c2}X_{c2}Y_{c2}Z_{c2}$ is the rear camera coordinate system, L_1, L_2 and L_3 are three straight lines abstracted from the laser features, the straight lines L_1 and L_3 are the straight line features of the two laser beams of the laser pointer, and L_2 is the straight line passing through the center of the two light spots, L_1, L_2 and L_3 are in the same plane, where $L_1 \,//\, L_3$, $L_2 \perp L_1$, L_2 and L_1, L_3 intersect at points P_1 and P_2, and the distance between point P_1 and point P_2 is d. The origin O_d of the laser measurement coordinate system $O_dX_dY_dZ_d$ is located at the midpoint of P_1 and P_2. The straight line L_2 is the X_d axis, and the direction points to the point P_2. The Z_d axis is parallel to L_1 and L_3 through the O_d point, and the direction is opposite to the direction of the laser beam. The projections of the spatial straight lines L_1, L_2 and L_3 on the image coordinate system are l_1, l_2 and l_3, and the straight lines l_1 and l_3 intersect at points q_1 and q_2 in the image. The straight line equations of the three straight lines l_1, l_2 and l_3 in the image in the image plane can be determined according to the laser image information, regardless of the situation that the straight line degenerates into a point. In order to determine the transformation matrix ${}_{c2}^{d}T$ of the camera relative to the laser measurement coordinate system, it is necessary to calculate the position information of l_1, l_2 and l_3 in the camera space according to the position information of l_1, l_2 and l_3, and then obtain the rotation matrix ${}_{c2}^{d}R$ and the translation vector ${}_{c2}^{d}t$ between the camera coordinate system and the laser measurement coordinate system.

Journal of Physics: Conference Series **2363** (2022) 012014 doi:10.1088/1742-6596/2363/1/012014

4. SINS Position and Attitude Measurement System

4.1. SINS Positioning Coordinate System and Basic Principles
SINS usually consists of a three-axis gyroscope, a three-axis accelerometer and a three-axis magnetometer, and the solution method only uses the data of the accelerometer and the gyroscope. The three-axis gyroscope can obtain the instantaneous angular velocity of the carrier, and the three-axis accelerometer can obtain the instantaneous linear acceleration of the carrier. The position and attitude information of the carrier is updated using gyroscope data and accelerometer data. The basic principle of SINS pose calculation is shown in figure 4. After the angular velocity information output by the gyroscope is compensated by the cone error, the attitude transformation matrix is obtained according to the attitude update algorithm, and the attitude angle of the carrier relative to the navigation coordinate system is calculated. After the paddle error compensation is performed on the linear acceleration information output by the accelerometer, the coordinate transformation is performed to obtain the acceleration information in the navigation coordinate system. Due to the influence of gravity during the measurement process, it needs to be compensated before the calculation. Use the input initial velocity information to perform time integral operation on the compensated acceleration information to obtain the instantaneous velocity of the carrier at the current moment. Finally, using the initial position information, the obtained velocity can be integrated over time to obtain the position of the carrier at the current moment.

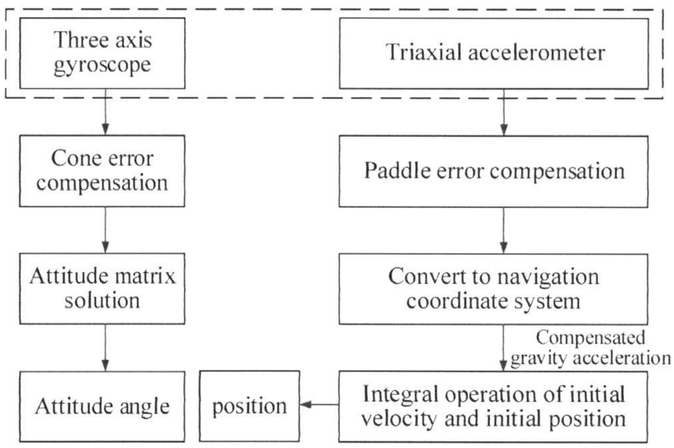

Figure 4. The basic principle of SINS pose calculation.

4.2. SINS Pose Calculation
In the actual tunnel excavation process in the coal mine, the cantilever roadheader moves slowly and swings frequently from side to side, so the rotation vector method is used to calculate the attitude.

Rotation Vector Differential Equation:

$$\dot{\phi} = \omega_{nb}^{n} + \frac{1}{2}\phi \times \omega_{nb}^{n} + \frac{1}{|\phi|^2}\left[1 - \frac{|\phi| sin|\phi|}{2(1 - cos|\phi|)}\right]\phi \times (\phi \times \omega_{nb}^{n}) \tag{1}$$

where ϕ is the equivalent rotation vector corresponding to the angular position change of the fuselage coordinate system from time t_k to time t_{k+1}, $|\phi|$ is the amplitude of the rotation vector, and ω_{nb}^{n} is the rotational velocity of the carrier from time t_k to time t_{k+1}. Because in the process of attitude calculation, the period of attitude update is short, and the rotation vector ϕ is very small, thus ignoring its high-order term, we can get:

$$\dot{\phi} = \omega_{nb}^{n} + \frac{1}{2}\phi \times \omega_{nb}^{n} + \frac{1}{12}\phi \times \left(\phi \times \omega_{nb}^{n}\right) \tag{2}$$

Using the above method to calculate ϕ, the attitude quaternion is updated:

$$\begin{aligned} q &= \left[\cos\frac{\phi}{2}, \frac{\phi}{|\phi|} sin\frac{\phi}{2}\right]^T \\ &= \left[\cos\frac{|\phi|}{2}, \frac{\phi_x}{|\phi|} sin\frac{|\phi|}{2}, \frac{\phi_y}{|\phi|} sin\frac{|\phi|}{2}, \frac{\phi_z}{|\phi|} sin\frac{|\phi|}{2}\right]^T \end{aligned} \tag{3}$$

The attitude calculation of the SINS is carried out by using the three-sub-sample algorithm of the rotation vector method. During the movement of the carrier, the SINS is used to obtain the motion parameters in the $[t_k, t_{k+1}]$ time period, and the obtained motion parameters are taken out three data at equal intervals to form a new continuous motion data, using a quadratic curve to fit the angular velocity of the carrier:

$$\omega_{nb}^{b}\left(t_k + \tau\right) = a + b\tau + c\tau^2 \tag{4}$$

$$\Delta\theta_i = \int_{\frac{i-1}{3}h}^{\frac{i}{3}h} \omega_{nb}^{b}\left(t_k + \tau\right) d\tau, \mathrm{i} = 1,2,3 \tag{5}$$

Simultaneous is obtained

$$\begin{aligned} \phi(h) = \Delta\theta_1 + \Delta\theta_2 + \Delta\theta_3 + \frac{33}{80}\Delta\theta_1 \times \Delta\theta_3 + \\ \frac{57}{80}\Delta\theta_2 \times \left(\Delta\theta_3 - \Delta\theta_2\right) \end{aligned} \tag{6}$$

During the movement of the carrier, the acceleration information on the three coordinate axes x, y, and z at time k output by the SINS is expressed as:

$$a_b(k) = \left(a_{bx}, a_{by}, a_{bz}\right)_k^T \tag{7}$$

According to the solution principle, it is first converted into the navigation coordinate system:

$$a_n(k) = C_b^n(k) a_b(k) \tag{8}$$

Among them, $C_b^n(k)$ represents the rotation matrix from the inertial coordinate system to the navigation coordinate system at time k, and the attitude is calculated and updated by the data of the gyroscope. In order to eliminate the influence of the earth's gravitational acceleration, it is necessary to compensate the gravitational acceleration component and then carry out the integral operation, and finally calculate the velocity and displacement information by integral calculation

$$a_g(k) = a_n(k) - g = C_b^n(k) a_b(k) - g \tag{9}$$

where $a_g(k)$ represents the acceleration value excluding the influence of acceleration.

First, the acceleration $a_g(k)$ is integrated to obtain the speed of the carrier in the navigation coordinate system

$$v_n(k+1) = v_n(k) + a_g(k)\Delta t \tag{10}$$

In the above equation, $\Delta t = t(k+1) - t(k)$, the position information can be obtained by integrating the above equation:

$$S(k+1)S(k) + v_g(k)\Delta t + \frac{1}{2}a_g(k)(\Delta t)^2 \tag{11}$$

The position information is decomposed in the three axes of the navigation coordinate system, and the pose information of the carrier at the current moment is calculated by this method. Combined with the initial pose information, the continuous motion trajectory can be obtained by iterative calculation in time sequence.

5. Fusion Algorithm Based on Extended Kalman Filter

5.1. Fusion Process of Binocular Vision and SINS

According to the visual pose measurement and the autonomous positioning of the SINS, the pose parameter information of the cantilever roadheader is obtained, when it merges, the two sets of positioning data must be aligned in time and space. Finally, the extended Kalman filter fusion strategy is used to achieve accurate motion estimation. The process of integration strategy is shown in figure 5.

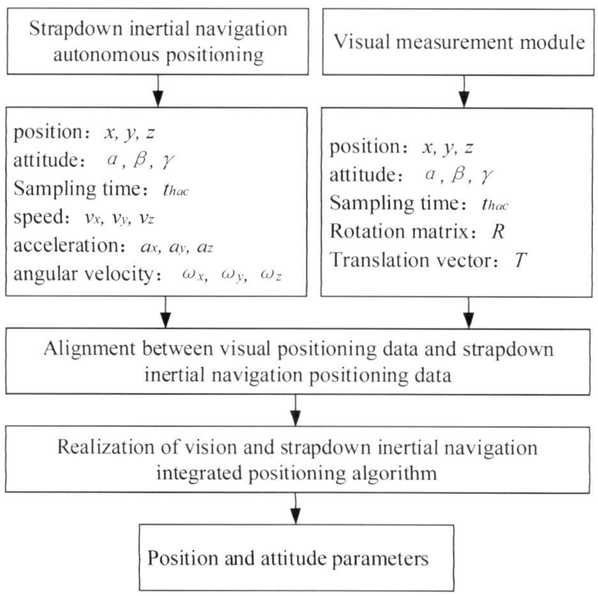

Figure 5. Extended Kalman filter fusion strategy.

5.2. Combination Positioning Data Alignment

Data fusion is the integration of a variety of sensor information to improve measurement accuracy and ensure measurement stability. The premise of data fusion is to achieve the alignment of multiple sets of data. Before the multi-sensor data are fused, it is a key step to achieve multi-sensor information fusion is to assign each sensor to a unified time benchmark. The sampling frequency of different sensors is different, which makes it difficult to fuse the measurement data without alignment. Therefore, the goal of time alignment is to unify the measurement data of multiple sensors for the same measured object to the same reference time. The frame rate of visual measurement is lower than the sampling frequency of SINS. Therefore, before the fusion of visual measurement data and SINS measurement data, all data must be aligned in time to ensure data synchronization. When only considering the difference of binocular vision measurement frequency and SINS sampling frequency, the visual sampling time with low sampling rate is taken as the reference time, the least square method is used to align the SINS data whose sampling rate is integer times of the visual sampling rate to the reference time.

Visual measurement is called sensor A and SINS is called sensor B. The sampling periods of A and B sensors are τ and T, respectively, $n=\tau/T$. If the latest update time of sensor A to the measured object is $(k-1)\tau$, and the next update time of the measured data is $k\tau=\left[(k-1)\tau+nT\right]$, that is, sensor A measures sensor B n times between two consecutive times of the measured object. We can use the least square method to align the measurement data of sensor B from time $(k-1)\tau$ to time $k\tau=\left[(k-1)\tau+nT\right]$, and fit it into a virtual value to replace the measured value of sensor B at $k\tau=\left[(k-1)\tau+nT\right]$. The fusion of this virtual measurement value and the measurement value of sensor A can eliminate the problem of deviation in the state estimation of the measured object caused by time synchronization.

We can use $Z_n=(z_1,z_2,...,z_n)^T$ to represent the n measurement sets of sensor B from $(k-1)$ to k. The starting point of measurement time is the same, and the measured value z_n and the measured value of sensor A are synchronized at k. If $U=(z,\dot{z})$ is used to represent the virtual measurement values and their derivatives of the fusion of the measured values $z_1,z_2,...,z_n$ of the sensor B from $(k-1)$ to k. The actual measurement value z_i of sensor B is:

$$z_i=z+(i-n)T\dot{z}+v_i \quad (i=1,2,3,\ldots,n) \tag{12}$$

v_i is a measurement noise, equation (12) transformed into a vector is:

$$Z_n=W_nUz+V_n \quad (i=1,2,3,\ldots,n) \tag{13}$$

Among them $V_n=(v_1,v_2,...,v_n)^T$, the mean is zero, and the variance is:

$$E\left[V_nV_n^T\right]=diag(\sigma_r,\sigma_r,\ldots,\sigma_r) \tag{14}$$

$$W_n=\begin{bmatrix}1 & 1 & \cdots & 1\\ (1-n)T & (2-n)T & \cdots & (n-n)T\end{bmatrix}^T \tag{15}$$

And σ_r is the measurement noise equation before fusion.

According to the least squares criterion:

$$J=V_n^TV_n=\left[Z_n-W_n\hat{U}\right]^T\left[Z_n-W_n\hat{U}\right] \tag{16}$$

To minimize J, the two sides of J seek deviation from $\hat{U}$ and make it equal to zero:

$$\frac{\partial J}{\partial\hat{U}}=-2\left(W_n^TZ_n-W_n^TW_n\hat{U}\right)=0 \tag{17}$$

The virtual measurements and noise variance of the fusion time of n measurements of sensor B are:

$$\hat{z}(k)=c_1\sum_{i=1}^{n}z_i+c_2\sum_{i=1}^{n}i\cdot z_i$$
$$Var\left[z(k)\right]=\frac{2(2n+1)}{n(n+1)}\sigma_r^2 \tag{18}$$

Among them, $c_1 = -\frac{2}{n}$, $c_2 = \frac{6}{n(n+1)}$.

5.3. Design of Fusion Algorithm Based on Extended Kalman Filter

Based on the extended Kalman filter fusion algorithm, the loose coupling method is used to fuse the data of binocular vision and SINS to achieve the combined Positioning of Cantilever Roadheader. The state variable of the composite system is expressed as a 15-dimensional column vector as equation (19):

$$x = \begin{bmatrix} P_k & \Phi_k & V_k & \delta_k & \varepsilon_k \end{bmatrix}^T \tag{19}$$

Among them, P_k, ϕ_k and V_k are the position, attitude and velocity calculated by SINS, respectively. The addition of δ_k, ε_k is the zero drift of accelerometer and gyroscope in SINS.

Position— $P_k = (X_k, Y_k, Z_k)^T$

Attitude— $\Phi_k = (\alpha_k, \beta_k, \gamma_k)^T$

Velocity— $V_k = (V_x, V_y, V_z)^T$

Accelerometer deviation— $\delta_k = (\delta_x, \ \delta_y, \ \delta_z)^T$

Gyroscope deviation— $\varepsilon_k = (\varepsilon_x, \ \varepsilon_y, \ \varepsilon_y)^T$

The process model is established based on the output data of the SINS. Specifically, the linear acceleration a_b and angular velocity ω_{nb}^b are taken as the input, and expressed as u_k :

$$u_k = \left(a_b, \ \omega_{nb}^b\right)^T \tag{20}$$

The actual measurement model considering noise is:

$$\begin{aligned} a_g &= C_b^n\left(a_b - \delta_b - \omega_a\right) \\ \omega &= \omega_{nb}^b - \varepsilon - \omega_\omega \end{aligned} \tag{21}$$

The state equation x_{k+1} and observation equation z_k of extended Kalman filter can be expressed as:

$$x_{k+1} = f\left(x_k, \ u_k, \ \omega_k\right) \tag{22}$$

$$z_k = h\left(x_k, \ v_k\right) \tag{23}$$

The nonlinear equation $f(x_k, u_k, \omega_k)$ can be obtained by the following formula:

$$f(x,u,\omega) = \begin{bmatrix} P_k + V_k\Delta t + \frac{1}{2}(\Delta t)^2 C_b^n\left[\left(a_b - \delta_b - \omega_a\right) - g\right] \\ \Phi_k + f_{Euler}(x,u,\omega) \\ V_k + \Delta t\left[C_b^n\left(a_b - \delta_b - \omega_a\right) - g\right] \\ \delta_k \\ \varepsilon_k \end{bmatrix} \tag{24}$$

Among them

$$C_b^n = \begin{bmatrix} cos\beta cos\alpha & cos\beta sin\alpha & -sin\beta \\ sin\gamma sin\beta cos\alpha - cos\gamma sin\alpha & cos\gamma cos\alpha + sinsin\beta sin\alpha & sin\gamma cos\beta \\ sin\gamma sin\alpha + cos\gamma sin\beta cos\alpha & cos\gamma sin\beta sin\alpha - sin\gamma cos\alpha & cos\gamma cos\beta \end{bmatrix}^T \tag{25}$$

$$f_{Euler}(x,u,\omega) = \begin{bmatrix} \left(\omega_{nby}^b sin\gamma + \omega_{nbz}^b \cos\gamma\right) tan\beta + \omega_{nbx}^b \\ \omega_{nby}^b cos\gamma - \omega_{nbz}^b sin\gamma \\ \left(\omega_{nby}^b sin\gamma + \omega_{nbz}^b \cos\gamma\right) sec\beta \end{bmatrix} \Delta t \tag{26}$$

Visual pose measurement system can measure the position and attitude of the target carrier, and construct the observation equation of extended Kalman filter with its output:

$$h(x_k, v_k) = \begin{bmatrix} \Phi_{c,k} \\ P_{c,k} \end{bmatrix} + v_c \tag{27}$$

In the above expression, v_c is the measurement noise of the visual pose measurement system.

By equations (24) and (27), the form of extended Kalman filter can be established:

$$A_k = \frac{\partial f}{\partial x}(x_k, u_k, 0) \tag{28}$$

Control input matrix:

$$B_k = \frac{\partial f}{\partial \omega}(x_k, u_k, 0) \tag{29}$$

Observation matrix:

$$H_k = \frac{\partial h}{\partial x}(x_k, 0) \tag{30}$$

Observation noise:

$$V_k = \frac{\partial h}{\partial v}(x_k, 0) \tag{31}$$

The above extended Kalman filter solution equation can be solved to realize the data fusion of visual pose measurement and SINS autonomous positioning.

6. Experiment

In order to verify the feasibility, measurement accuracy and stability of the combined positioning scheme of roadheader vision and SINS, we build a combined positioning test platform for experimental verification, as shown in figure 6. The main experimental equipment includes inspection robot, smoke maker, explosion-proof industrial camera, total station, prism, SINS and two laser pointers and so on. The explosion-proof camera and inertial navigation are fixed on different positions of the roadheader model, and use the total station to measure the initial pose of the roadheader body relative to the roadway. According to the pose measurement model of the roadheader body, the external parameters of the rear explosion-proof industrial camera are calibrated, and the two laser pointers are horizontally and parallel to form a laser target, fixed at a certain distance behind the fuselage, and the prism is fixed on the fuselage. Considering the difficulty of lateral movement of the roadheader in the test, in order to facilitate the system test, the movement of the roadheader is simulated by moving the camera and inertial navigation installed on the inspection robot. Explosion-proof camera is used to collect laser target image, and combined with SINS attitude output test. The inspection robot is moved to five positions, and each position is measured by the combined pose measurement system and the total station pose measurement system. The data are recorded and the pose measurement test results of the roadheader body are obtained, as shown in table 1. The combined pose measurement system is used to measure the pose of the roadheader body, and the total station measurement data is used as the real value. The combined positioning error of the roadheader body is obtained by comparing the measured pose with the actual pose, as shown in figure 7. And the combined positioning attitude angle error of the roadheader body, as shown in figure 8.

According to the measurement results of the roadheader body pose and the combined positioning pose error, it can be known that the measured error of the roadheader body position in the x-axis direction of the roadway coordinate system is within ±20 mm, the error in the y-axis direction is within ±35 mm, and the error in the z-axis direction is within ±10 mm, The yaw angle error is between -0.4° and 0.7°, the roll angle error is between -0.2° and 0.6°, and the pitch angle error is between -0.02° and 0.5°. According to the underground construction requirements, the accuracy of the visual measurement system meets the accuracy requirements of the section forming deviation in the range of -25 mm to 150 mm specified in the Coal Roadway Engineering Quality Acceptance Specification. The measurement test of the fuselage pose of the roadheader integrated with vision and SINS shows that this method can stably and accurately measure the fuselage pose of the roadheader in complex environments such as high dust, multiple stray light and low illumination underground.

Table 1. Test results of roadheader body pose measurement.

Project	Number	x/mm	y/mm	z/mm	yaw angle/°	roll angle/°	pitch angle/°
	1	-366	16236	1500	-7.21	-0.1	1.1
	2	-278	16180	1522	20.09	-0.5	-1.21
True value of fuselage pose	3	-158	16132	1520	25.14	-0.52	2.32
	4	-55	16192	1504	-14.98	1.41	0.1
	5	40	16214	1518	-21.4	0.3	1.9
	1	-346.8	16263.2	1525.1	-7.01	-0.23	1.5
	2	-293.1	16188.3	1536	19.39	-1.11	-1.67
Combined pose measurements	3	-177.5	16166.7	1530.7	24.62	-0.52	2.34
	4	-43.4	16168.2	1528.2	-14.61	1.31	-0.41
	5	60.1	16188.1	1527.5	-21.61	0.5	2.38

Figure 6. Combined positioning test verification platform.

Note: 1: Laser pointer; 2: Digital Total Station; 3: Computer; 4: Inspection Robot; 5: Prism; 6: Explosion-proof Industrial Camera; 7: Strapdown Inertial Navigation; 8: Smoke Maker.

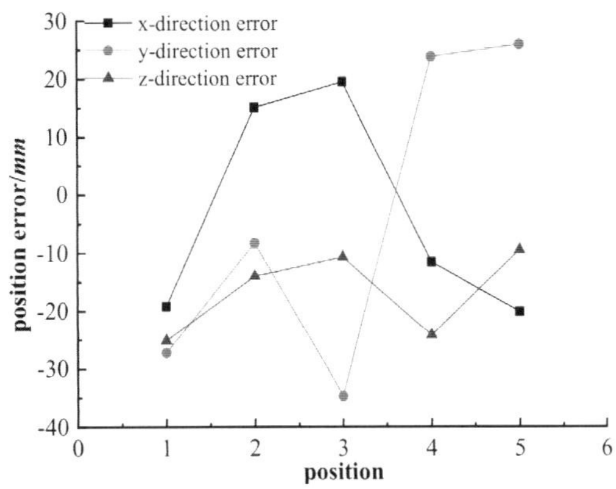

Figure 7. Combined positioning position error.

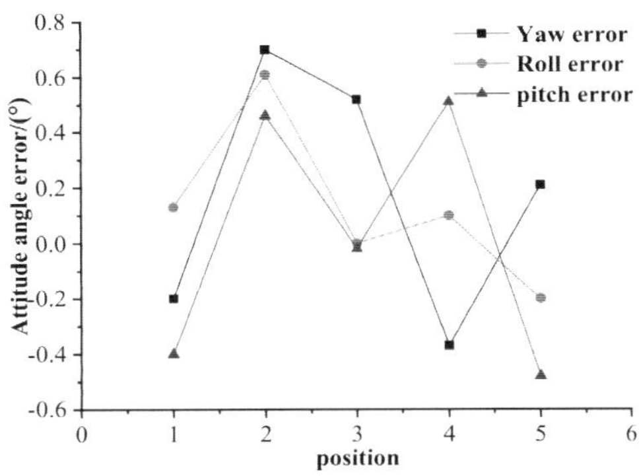

Figure 8. Combined positioning attitude angle error.

7. Conclusion

Aiming at the positioning problem of roadheader fuselage in low illumination and high dust environment in coal mine, this paper adopted visual pose measurement technology and SINS pose measurement technology to study a pose measurement technology of roadheader body based on fusion of vision and strapdown inertial navigation. A pose measurement technology of roadheader fuselage based on visual and SINS fusion is studied. Data alignment after acquiring visual and inertial measurement data. The extended Kalman filter method is adopted to fuse the data to realize the pose measurement of the roadheader fuselage with the fusion of vision and SINS, which improves the detection accuracy of the roadheader fuselage and realizes the precise positioning of underground equipment. The main conclusions of this paper are as follows:

1) The position and orientation measurement technology of the roadheader fuselage is combined with vision and SINS. The error of the measured position of the roadheader fuselage in the x axis direction of the roadway coordinate system is within ± 20 mm, the error of the y axis direction is within ± 35 mm, and the error of the z axis direction is within ± 10 mm.

2) The pose measurement technology of the roadheader fuselage based on the fusion of vision and SINS is adopted. The measured yaw angle error of the roadheader fuselage is between -0.4° and 0.7°, the roll angle error is between -0.2° and 0.6°, and the pitch angle error is between -0.02° and 0.5°.

References

[1] Ma H W, Wang S B, Mao Q H, et al. 2021 Key common technology of intelligent heading in coal mine roadway *Journal of China Coal Society* **46** (01) 310-320.

[2] Xie J and Wang F 2021 Discussion on key technologies of intelligent tunneling robot in coal mine *Industry and Mine Automation* **47** (S2) 39-42.
[3] Dan Q, Sun C J, Zheng H T, et al. 2020 Research on the key technology of robotization of roadheader *Coal Science and Technology* **48** (S2) 199-204.
[4] Hu X T, Zhu T, Su J M, et al. 2021 Key technology of intelligent drivage perception in coal mine roadway *Journal of China Coal Society* **46** (07) 2123-2135.
[5] Fu S C, Li Y M, Zong K, et al. 2017 Accuracy analysis of UWB pose detection system for roadheader *Chinese Journal of Scientific Instrument* **38** (08) 1978-1987.
[6] Zhao J X, Yang W J, Zhang X H, et al. 2021 Study on accurate positioning method and its visual measurement technology of cantilever roadheader underground *Coal Science and Technology* **49** (12) 192-201.
[7] Zhang X H, Zhao J X, Yang W J, et al. 2021 Vision-based navigation and directional heading control technologies of boom-type roadheader *Journal of China Coal Society* **46** (07) 1-11.
[8] Zhang C, Zhang X H, Du Y Y, et al. 2021 Measuring technique of cantilever roadheader position based on binocular stereo vision *Coal Science and Technology* **49** (11) 225-235.
[9] Zhang X G, Yang Y, Sun Y J, et al. 2020 Pose recognition method of roadheader based on multifrequency continuous-wave phase-difference ranging *Journal of China Coal Society*, **45** (06) 2056-2064.
[10] Tao Y F, Li R, Li J G, et al. 2017 Study on deflection displacement accuracy of roadheader by single station multi-point time-sharing measurement system based on iGPS *Coal Technology* **36** (02) 246-247.
[11] Hargrave C O, James C A and Ralston J C 2017 Infrastructure-based localisation of automated coal mining equipment *International Journal of Coal Science & Technology* **4** (3) 252-261.
[12] Zhang X H, Liu B X, Zhang C, et al. 2020 Roadheader positioning method combining total station and strapdown inertial navigation system *Industry and Mine Automation* **46** (09) 1-7.
[13] Hua H T, Ma H W, Mao Q H, et al. 2021 Pose measurement of intelligent roadheader based on strapdown inertial navigation and oil cylinder movement *Coal Engineering* **53** (11) 140-145.
[14] Xu A G, Yan K X, Gao S, et al. 2021 A monocular visual odometry/UWB combination indoor positioning metho *Journal of Navigation and Positioning* **9** (05) 82-88.
[15] Lu X M, Yan C Q and Yuan Z P 2020 Precisely positioning method for roadheaders and robotic roadheader system *Journal on Communications* **41** (02) 58-65.
[16] Mao Q H, Zhang X H, Ma H W, et al. 2018 Study on spatial position and posture monitoring system of boom-type roadheader based on multi sensor information *Coal Science and Technology* **46** (12) 41-47.

Text Independent Writer Identification Based on Pre-training Model and Feature Fusion

Zhenjiang Li[1*], Qianxue Zhang[1], Yiqun Wang[2]

[1] School of Cyberspace Security, Gansu University of Political Science and Law, Lanzhou 730000, Gansu, China

[2] School of Artificial Intelligence, Gansu University of Political Science and Law, Lanzhou 730000, Gansu, China

Email: *lzj6843@gsupl.edu.cn; 2904702518@qq.com; wyq6696@gsupl.edu.cn

Abstract. Handwriting is one of the natural biological characteristics of human beings. People have a high degree of acceptance of identity recognition technology based on handwriting. It has a wide application prospect in the fields of finance, government affairs, justice and so on. In this paper, firstly, the original document image is segmented into smaller regional samples by using the sliding window method. Secondly, multiple pre-training models are used to extract the sample features, and multiple features of the samples are fused. Finally, the euclidean distance is used to express the degree of difference between samples. Experiments show that this method has a high recognition rate and has a certain application value.

Keywords. Writer identification; off-line handwriting; feature fusion; pre-training.

1. Introduction

Handwriting is a very important means of personal identity authentication. Long before the information age, people used handwriting features to distinguish different people. Although information technology has been integrated into every corner of people's lives, handwritten characters still have their unique charm and are widely used in various formal occasions. Handwritten character recognition is an important research direction in the field of OCR.

According to different research objects, writer identification can be divided into signature identification, text related writer identification and text independent writer identification. Signature is a special kind of handwritten text with few contents but unique forms. Some signatures are also specially designed with exaggerated strokes and fonts, which are quite different from person's daily handwriting style. The research materials of text related and text independent writer identification are all texts written under normal circumstances. The difference lies in whether the content of the corpus used for comparison is the same. Obviously, the work of text independent writer identification is much more difficult than that of text related. This is because the abstract level of handwriting style features is higher than that of text content features. It is more difficult to extract them. However, in the real scene, it is difficult to obtain the content related text, so it is more practical to carry out the writer identification research on the content independent text.

At present, many scholars have carried out research on writer identification. Song et al. [1] proposed a contour feature based on path signature to extract the handwriting contour of the binary image, and then use the polygonization step to approximate each contour with fewer points. The way to define the path is to trace the polygon contour from any point to form the contour of the handwriting image, use the logarithmic path signature (LPS) to describe the geometric characteristics of the contour, and use the word bag construction method to represent the offline handwriting features after forming the logarithmic path signature features of the handwriting. Manhattan distance is used to measure the similarity between features to identify the handwriting writer. He et al. [2] constructed a residual recurrent neural network model using convolution neural network. The model consists of four convolution blocks and a global average pooling layer. The convolution neural network extracts global context information and local features. In this model, the output of the global average pooling layer is used as the global context feature of the handwriting, and the output of the third convolution block is the local feature map of the handwriting image. In order to input the local feature map into the residual cyclic extension network for feature classification, it is necessary to cut the local feature map. The residual of the local feature map after cutting is recycled to the neural network, and the context information is taken as the hidden state of the network to classify the features. Vincent [3] and others proposed an unsupervised handwriting recognition method, which combines the feature descriptor and convolution neural network to carry out the handwriting recognition task. The model takes the descriptor and pixel block of the handwriting image as the input of the convolution neural network. Firstly, the SIFT descriptor of the handwriting image is extracted by determining the key points, then the extracted descriptor and the cut pixel blocks are input into the 20-layer depth residual convolution neural network for training, and the trained neural network is used to extract the features of the query handwriting and the reference handwriting. By summarizing the existing research work, it can be found that, the core of text independent writer identification is feature extraction, while the difficulty lies in the separation of writer's style features and text structure features [4-9].

With the continuous development of the deep learning technology framework, scholars continue to put forward new backbone network models, such as VGG, ResNet, DenseNet, Inception, Xception, EfficientNet, etc. These models have been widely used in various fields. In the later stage, with the increasing complexity of the model, the number of model parameters increases exponentially, the hardware resources required for model training also expand, and the cost of a complete model training is high. In addition, the training process is also restricted by the quality and scale of sample data. Therefore, using the public released pre-training model to carry out transfer learning on its own data set is a more mainstream method.

Based on the pre-training model, this paper carries out the research on writer identification, fully explores the feature extraction potential of all kinds of existing backbone networks, and tries to integrate the features obtained by different models. The final recognition accuracy is high and has a certain application and promotion potential.

2. CNN and Pre-training Model

In the last decade, various AI applications have been fully implemented, such as face recognition, printed character recognition, etc. Among them, the development of convolutional neural network technology has contributed greatly. Since 2012, scholars have proposed new network model structures and achieved great influence every year, such as Alexnet, VGG, Resnet, Densenet, Senet, Xception, etc. When carrying out various application research, the above model structure is generally not changed greatly, but as a standard component, which is reorganized based on the characteristics of the application to meet its own application requirements. Therefore, the above model is usually called backbone network.

Alexnet won the first place in the 2012 ILSVRC competition with excellent results, which attracted the attention of all industries. Subsequent researchers continued to improve the CNN structure under the paradigm of Alexnet. With the increasing depth of the model and the explosion of parameter scale, only a few institutions can bear the cost of large-scale model training. Therefore, most application development is based on the backbone network model to carry out transfer learning. At first, a pre-trained backbone model is needed. Usually, the model will be trained to the best accuracy on large-scale general image data sets such as Image-net, and then the model parameters at this time will be trained on their own data sets as initial parameters. This form of model training is also called model fine-tuning. In the process of fine tuning, on the one hand, we need to consider whether all the parameters of the pre-training model participate in the training, on the other hand, we need to consider the size of the existing data set and the size and shape of the sample data.

3. Our Methods

3.1. Framework

It has been pointed out that one of the key points of writer identification is feature extraction. In this paper, the pre trained backbone network is used as the feature extractor. On this basis, the original whole document image is sliced, the dimension of the features of the sample slice is reduced by PCA method, and different types of backbone network features are fused for recognition. The overall framework of the paper is shown in figure 1:

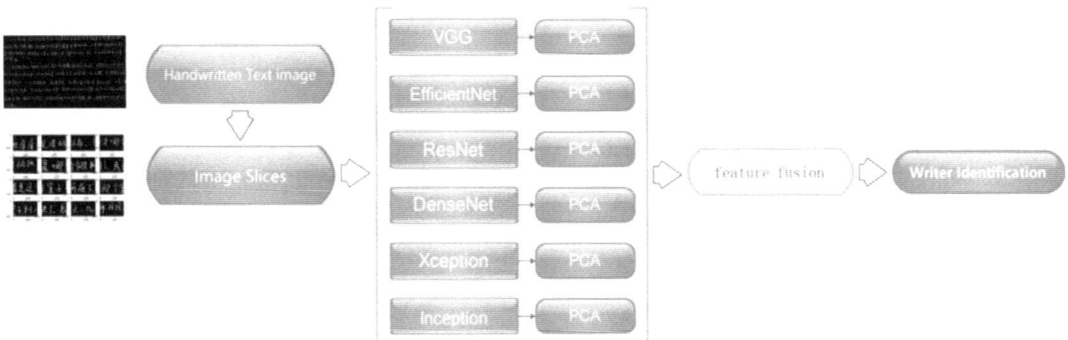

Figure 1. The frame diagram of our method.

3.2. Sample Slicing

In all kinds of writer identification task scenarios, it is usually difficult to obtain a large number of samples, and only a few or even single written text images can be used. Because the number is too small, it is impossible to directly use these samples for model training or recognition. On the one hand, the training process of deep learning model needs a large number of sample data to drive, and the lack of samples is easy to cause the model under fitting; On the other hand, the resolution of a single document image is usually large. Most of the pre-training models do not provide support for large-scale images, and a large amount of effective data will be lost after scaling them. Therefore, the most common method is to segment the document image into sub-images with less content, and describe the original image by statistics of the characteristics of the sub-images.

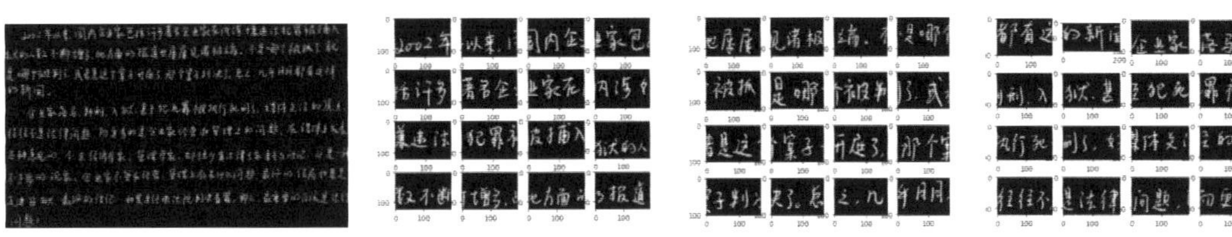

Figure 2. Handwritten text image sample slice.

Figure 2 is the slicing results of the sample in this paper, in which the first image is the handwritten text, and the subsequent images are the slicing results. In the segmentation process, we first divide the original document into several lines, then normalize all lines to the same line height while maintaining the aspect ratio, and finally use the sliding window with fixed width to get the slicing results.

3.3. Feature Extraction and Fusion
In this paper, the six backbone networks of inception V3, vgg19, resnet50v2, densenet121, xception and efficientNetB0 are used to extract the characteristics of slice samples. The above networks use the weight pre-trained on Image-net as the initial weight, remove the full connection layer of the original model, and add the maximum pooling layer after the last layer of convolution layer to convert the characteristic data into a one-dimensional matrix.

There are a large number of zero values in the original features. In order to describe the samples more compactly, the dimension of the original features of the image is reduced based on PCA method, and 90% of the principal components of the original features are retained in the dimension reduction. Different models have different emphasis on feature extraction capabilities. They can not only use the features obtained by a network alone, but also fuse the features obtained by multiple networks. When multiple models are used for feature extraction, all the features obtained are concatenated into one-dimensional vectors, and then the dimension is reduced by PCA method to complete the feature fusion.

3.4. Writer Identification
As the last step of writer identification is essentially a query task, which determines the category attribution of input samples by judging the distance between input sample features and existing sample features. In this paper, some samples are taken as registration information, and the sample center is determined by calculating its mean value.

Because the basic unit of feature extraction is not the original handwritten document image, but the slice of the image, it is necessary to use a certain number of slice classification results when judging the category. The smaller the number (Split_N) of slices required, the stronger the recognition ability of the model.

4. Experiment

4.1. Data
The work of this paper is based on the data set of casia-hwdb 2.0, which provides a total of 2092 handwritten document images of 419 writers. All documents have 20495 lines in total. The sample file is stored in dgrl format. Figure 3 shows some samples in the data set, in which the writers of each line of pictures are the same.

Figure 3. Samples in CASIA-HWDB 2.0.

When slicing the original whole document image, in order to ensure that the original text does not deform during normalization, we sliced it according to the ratio of width to height of 3:1 based on the line height of each line of text, and normalized it to the size of 360*180. According to this normalization method, a total of 172905 slice samples are obtained, and an average of 82.6 slice samples are obtained for each original document image. When classifying, select half of the samples of each writer as the category center, and use the other half of the samples for testing.

4.2. Evaluation
During the experiment, in order to verify the accuracy of the method proposed in this paper, we use the *Top-k* evaluation standard for the handwriting recognition task. If the target class in the *k* most similar classes, it is defined as a correct recognition, and *Top-k* is the percentage of all correct recognition times to the total recognition times.

4.3. Analysis
Figure 4 shows the *Top-1* accuracy of different model recognition obtained after feature extraction using single pre-training model. The models used in the figure 4 are resnet50v2, vgg19, inception V3, densenet121, xception and efficientnetbet0 respectively. The number of slices used for validation of each model is 10, 20 and 30 respectively.

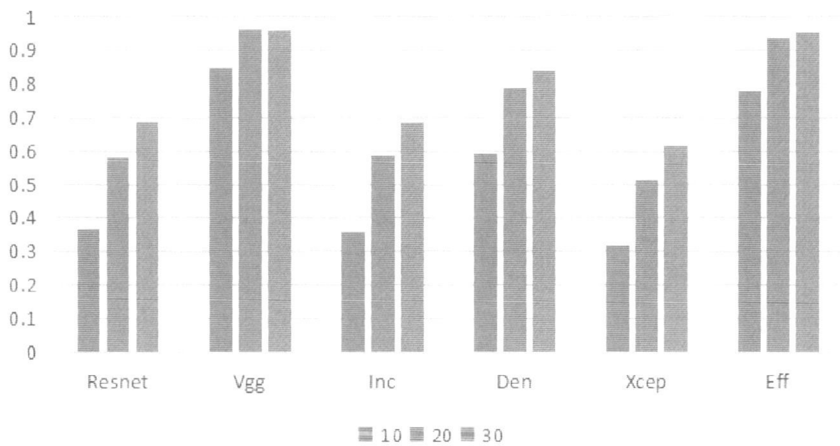

Figure 4. Training accuracy by single model.

It can be seen from the figure that the effect of vgg19 model is obviously better. When only 10 slices are used, the accuracy rate has exceeded 80%, which is higher than that of resnet50v2, inception V3, densenet121 and xception when 30 slices are used. The effect of efficientnetb0 model is second only to vgg19. Therefore, we chose the output features of vgg19 and efficientnetb0 for fusion.

Table 1 shows the top 1 accuracy of models vgg19, efficientnetb0 and their fusion features at different slice numbers.

Table 1. Top 1 accuracy of single model and fused model.

	10	20	30	40	50	60	70	80
Vgg	0.847	0.964	0.961	0.979	0.975	0.997	0.932	1.000
Eff	0.781	0.937	0.953	0.968	0.966	0.989	0.937	0.997
Fusion	0.873	0.973	0.965	0.982	0.979	0.998	0.937	1.000

As can be seen from the table, when the number of slices used continues to increase, the accuracy of all models continues to rise. Since the average number of slices obtained from a single handwritten document is 82.6, we set the maximum slice number to 80 when testing. At this point, it is equivalent to inputting a whole handwritten document for writer Identification, and the accuracy of the three models is close to 100%.

From the perspective of feature fusion effect, when the features obtained by vgg19 and efficientnetb0 are fused, their comprehensive performance is higher than that of the original two models in all different slice numbers, which also shows the effectiveness of the feature fusion method proposed in this paper.

Table 2 shows the top 1, top 3 and top 5 accuracy of fused features when the number of slices ranges from 10 to 80. It can be seen from the table that when only 10 slices are used, the writer can be effectively recognized, while when the whole document can be obtained conditionally, the recognition accuracy is close to 100%.

Table 2. Top 1, top 3 and top 5 accuracy of fused feature model.

	10	20	30	40	50	60	70	80
Top 1	0.873	0.973	0.965	0.982	0.979	0.998	0.937	1.000
Top 3	0.943	0.989	0.976	0.992	0.989	0.999	0.952	1.000
Top 5	0.961	0.993	0.979	0.993	0.991	0.999	0.958	1.000

During the experiment, we found that for all models, when the number of slices reached 70, the performance decreased significantly, which was different from the intuitive understanding and the trend before and after. Why there is such a singular point deserves further discussion.

5. Conclusion

In this paper, a writer identification method based on pre-training model and feature fusion is proposed. The experiment on CASIA-HWDB 2.0 Chinese handwritten dataset shows that this method has high accuracy, simple implementation and good application value. In the experiment, the slice of handwritten document image is used as the basic feature extraction unit, which is essentially a recognition method based on phrase granularity. In the future work, we can further reduce the identification granularity, reduce the information required for registration and enhance the interpretability of the model.

Acknowledgments

The research was funded by the National Natural Science Foundation of China, grant number 62166036. Science Foundation of Gansu Province, grant number 20JR10RA335. Project of Innovation Ability Improvement of Colleges in Gansu Province , grant number 2022B-152.

References

[1] Lai, S. and Jin, L. 2019 Offline writer identification based on the path signature feature. *ICDAR2019* pp 1137-1142.

[2] He, S. and Schomaker, L. 2021 GR-RNN: Global-context residual recurrent neural networks for writer identification *Pattern Recognition* **117** 107975.

[3] Christlein V, Gropp M, et al. 2017 Unsupervised feature learning for writer identification and writer retrieval *ICDAR2017* pp 991-997.

[4] Chahi A, El Merabet Y, et al. 2019 Off-line text-independent writer identification using local convex micro-structure patterns *Proc. of the New Challenges in Data Sciences: Acts of the Second Conf. of the Moroccan Classification Society* pp 1-5.

[5] Ayixiamu L, Yan Y, et al. 2021 Text-independent writer identification based on hybrid codebook and factor analysis *Acta Automatica Sinica* 2276-2284.

[6] Christlein V, Bernecker D, et al. 2015 Offline writer identification using convolutional neural network activation features *German Conf. on Pattern Recognition. Springer* pp 540-552.

[7] Nguyen H, Nguyen C, et al. 2019 Text-independent writer identification using convolutional neural network *Pattern Recognition Letters* **121** 104-112.
[8] He S, Schomaker L, et al. 2019 Deep adaptive learning for writer identification based on single handwritten word images *Pattern Recognition* **88** 64-74.
[9] He S, Schomaker L, et al. 2020 FragNet: Writer identification using deep fragment networks. *IEEE Transactions on Information Forensics and Security* **15** 3013-3022.

An Expression Recognition Model Based on Channel and Spatial Attention Fusion

Yunyun Cao

Department of Information Technology and Electromechanical Engineering, Shanghai Xing Jian College, Shanghai, China
Email: y.y.cao@163.com

Abstract. Expression recognition is an essential subfield in biometric recognition as well as a cutting-edge research area in computer vision. It has several uses in autonomous driving, human-computer interface, autism research, and other fields. This study offers an expression recognition model based on fusion attention that extracts expression features from the channel and spatial domains, respectively, then combines the two features. It is then trained on the FER2013 and ExpW datasets. Experiment findings suggest that the model may greatly enhance face emotion recognition accuracy.
Keywords. Channel attention; spatial attention; attention fusion; expression recognition.

1. Introduction

Human communication and information transfer rely heavily on expression. One of the most significant challenges in computer vision is facial expression recognition [1]. Traditional expression recognition is separated into two steps: first, manually construct features to extract the expression contained in the face picture, and then categorize the extracted expression features. Local binary patterns, histograms of oriented gradients, and Gabor features are examples of common manual features. These manual features are primarily used to extract the texture, shape, and other information in facial expressions, after which the features are sent to various classification models such as the K-neighbor algorithm, decision tree, and support vector machine, among others, to output the types of facial expressions.

Excellent convolutional neural network [2, 3] models, such as AlexNet, VGGNet [4], GoogleNet, ResNet [5], have evolved with the growth of deep learning and have achieved considerable success in the field of image categorization. With the introduction of face recognition expression datasets such as FER2013 and ExpW, an increasing number of academics are attempting to use convolutional neural networks [6] to facial expression recognition [7, 8]. Rose improves expression classification accuracy by combining the LeNet-5 network model with picture preprocessing technologies.

However, in real application situations, the acquired face photos frequently contain information other than facial expressions [9], such as occlusion, posture, and illumination. These interference effects will make expression recognition more challenging, resulting in retrieved expression features that frequently contain surplus information, lowering expression recognition accuracy. As a result, reducing the interference of extraneous information is critical to improving the accuracy of expression identification.

In this experiment, spatial attention is used to give a higher weight to the important areas in facial expressions, and the features on important channels are stressed through attention, the features are mixed to minimize the effects of redundancy on features, and the features are mixed to enhance the accuracy of expression recognition.

2. Related Work

2.1. Convolutional Neural Network
A convolution neural network [10] is a hierarchical network structure that performs convolution. Its most notable features are local perception, weight sharing, and multiple convolution cores, which result in fewer parameters, improved computing efficiency, and better feature expression ability. The heart of a convolution neural network is the convolution process, which may be characterized as two functions $f(t)$ and $g(t)$ that yield the third function, as given in the following formula.

$$f(t) * g(t) = \int_0^t f(u)g(t-u)du \quad (1)$$

VGGNet is the first suggested and implemented convolutional neural network model, focusing on the link between network depth and performance by employing a high number of convolutional kernels 3×3 rather than 7×7 convolutional kernels. GoogLeNet takes a bolder approach to network construction by utilizing the Inception module to boost network width and finally integrating the capabilities of many Inception modules.

2.2. Attention Mechanism
A typical module in deep learning is the attention mechanism. Its purpose as a resource allocation strategy is to direct limited computational resources toward more pressing challenges. Attention mechanisms are used in computer vision to focus on certain portions of an image, reducing the complexity of model image calculation and improving system performance. It may be separated into spatial attention and channel attention based on the distinct action modalities of attention. The former solely processes the task-related portions of the feature map in the spatial domain. The latter calculates each channel's attention weight independently and then aggregates all channels using linear weighting. The general algorithm of the attention mechanism is as follows:

$$Attention(Q, K, V) = soft\,max(\frac{QK^T}{(d_k)^{1/2}})V \quad (2)$$

2.3. Attention Based Expression Recognition
Expression recognition is a subset of face recognition that consists of three steps: face picture pre-processing, facial feature extraction, and expression categorization.
Most attention-based expression recognition algorithms, on the other hand, are primarily concerned with spatial domain information, neglecting the influence of distinct channel features on expression detection. This study gives a fusion attention technique to improve expression recognition accuracy by integrating spatial and channel attention.

3. Model Introductions

3.1. Channel Attention
After passing the feature map through several convolution cores in the convolution layer, each convolution core outputs a new feature map, thus N convolution cores produce the feature map of N channels. Each channel represents a distinct degree of information, some of which may be fine-grained data like texture and color. This study develops a channel attention method, which is illustrated in figure 1, to make the model pay greater attention to the information of key channels.

Assume that the feature map of the channel attention module input is $T \in R^{w \times h \times c}$, as shown in figure 1. The feature map is globally averaged and maximized, getting two feature vectors $1 \times 1 \times c$. The sigmoid function is then used to change the channel weight distribution from 0 to 1. The weight vector and feature map are then dot multiplied.

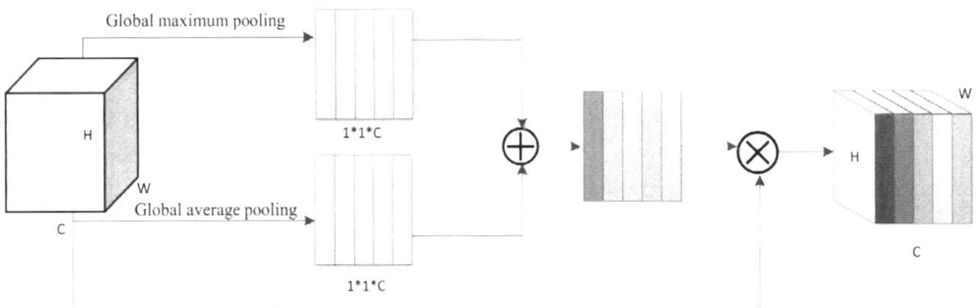

Figure 1. Channel attention model.

3.2. Spatial Attention

The spatial hierarchy information of the feature map is the focus of spatial attention. Collect the channels firstly, then do maximum and average pooling, convolute and link the pooled results, then decrease the spatial attention weight dimension to only one channel. Figure 2 depicts the principle:

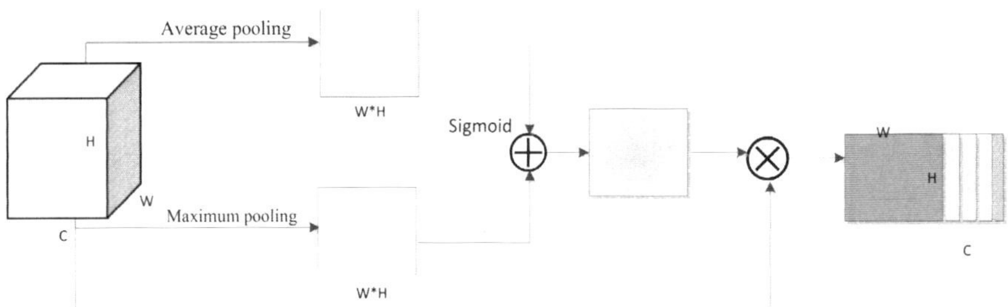

Figure 2. Spatial attention model.

Assume that the feature map of the channel attention module input is $T \in R^{w \times h \times c}$, as shown in figure 2. Firstly, global and average pooling are applied to the supplied feature map to generate the feature vector of $w \times h$, and then the sigmoid function is applied to the two feature connections to calculate the spatial attention weight. The spatial attention weight and feature map are then dot multiplied.

3.3. Fused Attention

VGG-19, GoogLeNet, and ResNet-152 are considered as backbone networks in this study. Following feature extraction, channel attention and spatial attention are combined to create a neural network model based on the fusion attention process. Analysing the link between different channels and assigning weights to them, generate spatial attention weights in the spatial domain to suppress or raise the relevance of certain regions. Figure 3 depicts the overall model's architecture:

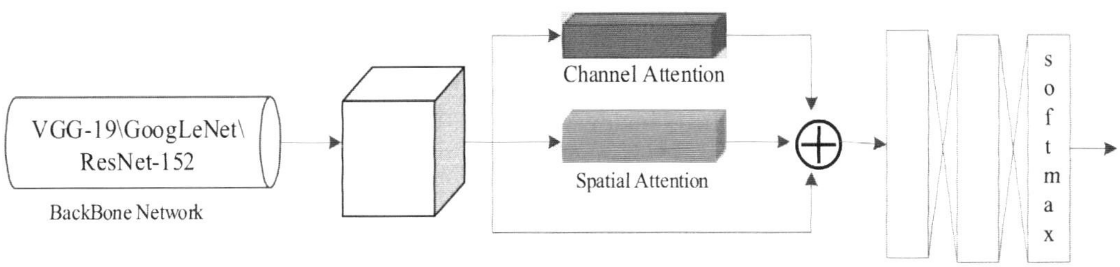

Figure 3. Expression recognition model based on fusion attention mechanism.

4. Experimental Design

4.1. Experimental Platform

At the moment, the most prominent deep learning development frameworks in the sector are PyTorch, TensorFlow and PaddlePaddle. TensorFlow and Keras are used in this research to create a face expression recognition model. TensorFlow is a deep learning framework developed by Google that was launched in November 2005. This framework is a computational graph programming approach that supports multiple CPUs / GPUs and mobile devices. It offers strong cross-platform capabilities and can handle a variety of network models. Keras is a Python-based advanced neural network API that facilitates convolutional neural network building and allows parallel processing across many CPUs and GPUs. All of the tests in this study were created with the aforementioned tools and ran on the CentOS 6.5 operating system whose hardware operating platform is shown in table 1.

Table 1. Hardware information of deep learning server.

Hardware	Detail information
Processor	Intel® Xeon® Gold 4214 12 core 2.2GHz
Graphics card	NVIDIA Geforce RTX 2080
Memory	512GB DDR4 ECC RDIMM
Hard disk	480GB SSD

4.2. Experimental Dataset

All experiments in this study use two datasets, FER2013 and ExpW. The FER2013 expression recognition data set contains 35886 expression images, including 28709 sets, 3590 public verification sets, and 3590 private verification sets, which are composed of 48×48 gray-scale image composition, a total of 7 expressions, corresponding to digital labels 0-6, and the specific expression categories and image distribution are as follows (table 2):

Table 2. FER2013 Expression category and data distribution.

Expression category	Training dataset	Testing dataset
angry	3995	491
hate	436	55
fear	4097	529
happy	7215	879
sad	4830	594
surprised	3171	416
normal	4965	626

Table 3. Expw Expression category and data distribution.

Expression category	Training dataset	Testing dataset
angry	2151	1435
hate	2318	1543
fear	631	422
happy	17545	11698
sad	6023	4016
surprised	4129	2753
normal	19585	13057

Expw is a picture of an expression that was retrieved from the Internet using the Google image retrieval feature. This dataset, like FER2013, has 7 expression photographs. In the experiment, the photos in the data are separated into training and test sets in a 6:4 ratio. This data set's expression categories and data distribution are as follows (table 3).

4.3. Analysis of Experimental Results

This study chooses the VGG-19, GoogLeNet, and ResNet-152 convolutional neural network models to validate the experimental results based on channel attention, spatial attention, and mixed attention, respectively. To improve the model training process, Adam optimizer is utilized. The training rounds is 20000, the initial learning rate is set to 0.001, the learning rate is attenuated to one eighth of the original for each 128 rounds of training, and the batch size is 128. The accuracy of the model on several data sets is displayed in tables 4 and 5.

Table 4. FER2013 Accuracy of expression recognition.

Network model	Channel attention	Spatial attention	Fused attention
VGG-19	85.6%	86.5%	91.2%
GoogLeNet	86.3%	87.6%	92.4%
ResNet-152	87.4%	88.2%	93.5%

Table 5. ExpW Accuracy of expression recognition.

Network model	Channel attention	Spatial attention	Fused attention
VGG-19	85.6%	87.2%	92.5%
GoogLeNet	87.3%	89.1%	94.7%
ResNet-152	89.4%	90.3%	96.3%

The test results reveal that channel attention only assigns different weights to each channel, making it easy to extract detailed information from different channels, whereas spatial attention focuses on task-related areas, suppresses interference from irrelevant areas, and slightly improves recognition accuracy. Fused attention combines channel attention and space, making full use of the rich information of the channel and spatial domain. This makes it simpler for neural networks to notice minor changes in expression, resulting in considerably enhanced identification accuracy.

5. Conclusion

Aiming at the existing challenges of facial expression recognition, this study first uses channel attention, calculates the attention weights of various channels of the feature map, and enhances the role of features on major channels in expression recognition. Then it uses spatial attention to make the model independently locate the area related to expression recognition, enhance the importance of the area, and finally integrates the two kinds of attention.

Reference

[1] Mollahosseini A., Chan D. Mahoor M. H., 2016, Going deeper in facial expression recognition using deep neural networks, 2016 IEEE Winter Conference on Applications of Computer Vision (WACV), pp. 1-10.
[2] Krizhevsky A, Sutskever I, Hinton GE. 2017, Imagenet Classification with Deep Convolutional Neural Networks. Communications of the ACM, 60(6):84-90.
[3] Szegedy C et al. 2015, Going deeper with convolutions, 2015 IEEE Conference on Computer Vision and Pattern Recognition (CVPR), pp. 1-9.

[4] He K M, Zhang X Y, Ren SH Q, Sun J. 2016, Deep Residual Learning for Image Recognition, Proceedings of the IEEE Conference on Computer Vision and Pattern Recognition (CVPR), pp. 770-778

[5] Huang G, Liu ZH, Laurens van der Maaten, Kilian Q. Weinberger,2017, Densely Connected Convolutional Networks, Proceedings of the IEEE Conference on Computer Vision and Pattern Recognition (CVPR), pp. 4700-4708.

[6] Hameed A A, Karlik B, Salman M S, 2016, Back-propagation algorithm with variable adaptive momentum, Knowledge-Based Systems, Volume 114, Pages 79-87.

[7] Tan L Zh, Rui D, 2020, Facial expression recognition based on convolution neural network and orthogonal neighbourhood preserving projection. International Journal of Autonomous and Adaptive Communications Systems,13(3), pp 246-259.

[8] Shen W, Li X, 2020, Facial expression recognition based on bidirectional gated recurrent units within deep residual network", International Journal of Intelligent Computing and Cybernetics, Vol.13 No.4, pp. 527-543.

[9] Liu M Y, Li SH X, Shan Sh G, Chen X L, 2015, AU-inspired Deep Networks for Facial Expression Feature Learning, Neurocomputing, Volume 159, Pages 126-136.

[10] Zhang W W, Murphey Y L, Wang T Y, Xu Q J, 2015, Driver yawning detection based on deep convolutional neural learning and robust nose tracking, International Joint Conference on Neural Networks (IJCNN), pp. 1-8.

Singular Value Decomposition (SVD) Co-training: A Semi-supervised Method for Object Detection

Siyang Fan[1], Rui Wang[1*], Jingwen Xu[1] and James Zhiqing Wen[2]

[1] School of Instrumentation and Optoelectronic Engineering, Beihang University, Beijing, China

[2] Engineering Research Center for Intelligent Robotics, Ji Hua Laboratory, Foshan, Guangdong, China

Email: fsy725@buaa.edu.cn, wangr@buaa.edu.cn, xujingwen9402@buaa.edu.cn, wenzq@jihualab.com

Abstract. Existing object detection algorithms based on Convolutional Neural Network (CNN) generally rely on supervised learning with a large amount of labeled samples, which is very time-consuming and laborious. To address the problem, we propose a semi-supervised learning method by using Singular Value Decomposition (SVD) co-training. Specifically, this paper uses two mutually independent features which are obtained by SVD to guide the co-training of two detectors in one CNN model, meanwhile an adaptive self-labeling strategy is used to ensure the quality of pseudo labels. To verify the efficacy of SVD co-training, we conducted experiments on GMU, AVD and self-made BHID datasets. The result shows that SVD co-training could achieve the accuracy comparable to full-supervised learning by using only 5% of labeled data.
Keywords. Computer vision; object detection; semi-supervised learning; self-labeling.

1. Introduction

Object detection based on Convolutional Neural Network (CNN) overcomes the limitations of traditional object detection and achieves more accurate results, however it requires a large amount of labeled samples, which takes a lot of labor and time to train a detection model. Therefore, an idea which use easily accessible unlabeled samples to train the model has emerged. Among them, the semi-supervised learning [1], which obtains high accuracy model by learning a small amount of labeled samples and a large amount of unlabeled samples, has attracted much attention.

Although semi-supervised learning has been widely studied and applied in image classification [2-5], there is relatively little work to apply it to object detection. The current semi-supervised object detection could be divided into two categories: one is based on consistency, and the other is based on pseudo-labeling. Consistency-based methods [6] use data enhancement and consistency regularization on unlabeled samples to improve the performance of the model, but its computational overhead is large. Pseudo-labeling based methods [7-9] leverages high accuracy pseudo labels from teacher model to train student model. Although the methods based on pseudo-labeling have reduced computational overhead compared to the consistency-based methods, they require training multiple models and still has a large decrease in accuracy compared to full-supervised learning.

Inspired by co-training [10] in semi-supervised learning, we propose a semi-supervised method for object detection. We use two mutually independent features obtained by Singular Value Decomposition (SVD) [11] to co-train two detectors in one CNN model, meanwhile an adaptive self-labeling strategy is used to ensure the quality of pseudo labels. We test the efficacy of SVD co-training on GMU [12],

AVD [13] and self-made BHID. The method we propose does not need to train additional CNN models, and could achieve the accuracy comparable to full-supervised learning only by using 5% of labeled data.

2. Related Work

2.1. Co-training

Co-training [10] is more widely used in semi-supervised learning. It proposes that if there are mutually independent views, the performance of the detectors could be improved by using unlabeled samples.

Two detectors are trained by using two labeled sample sets respectively, and used to label the images in the unlabeled sample set. The pseudo labels with high confidence are selected and added to each other's labeled sample set, and the above steps are repeated to finally obtain two high-performance detectors. Also these two detectors must be mutually independent.

2.2. Singular Value Decomposition

Singular Value Decomposition (SVD) is a matrix decomposition method [11], for an $m \times n$ matrix W can be decomposed as follows:

$$W = USV^T = U\begin{bmatrix} \sigma_1 & & \\ & \ddots & \\ & & \sigma_k \end{bmatrix}V^T \tag{1}$$

where S is an $m \times n$ diagonal matrix, and its diagonal elements are the non-negative singular values of W in order of decreasing magnitude. U and V are orthogonal matrices, and vectors in these matrices are mutually independent. Therefore, SVD could be one of the ways to satisfy the requirement for mutual independence of views in co-training.

3. Method

This paper combines SVD and co-training to propose a semi-supervised object detection method, and its training flow chart is shown in figure 1. First, training set ϕ_1 and ϕ_2 are used to train the network. The features are extracted through the convolutional layers and the fully-connected layers. Second, we perform SVD on the last fully-connected layer to obtain the SVD layer, separate two independent features, and use them to train independent detectors. Third, we use two detectors for self-labeling, filter out labels with high-confidence and add them to each other's training set. Repeating the above steps, a high-accuracy detection model could be obtained after multiple rounds of iterations finally.

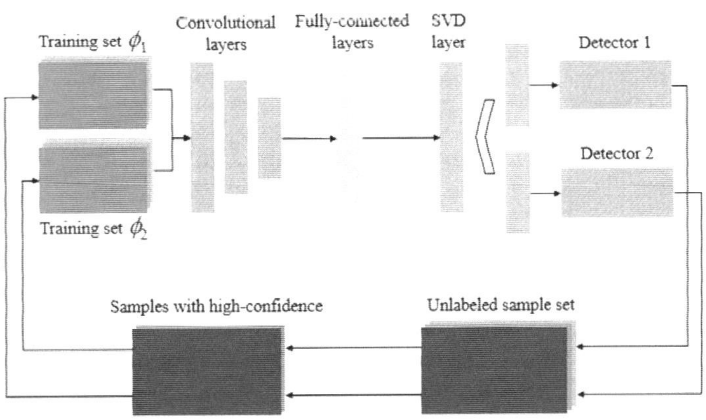

Figure 1. Flowchart of semi-supervised learning based on SVD co-training.

Since the SVD used in this paper is to decompose the weight of the fully-connected layer, we choose Faster RCNN [14] which contains the fully-connected layers and has excellent performance in object detection as the basic model for co-training.

3.1. Key Samples Selection
Co-training uses initial labeled samples to train the model, then label the unlabeled samples. In order to leverage all samples in co-training, the initial labeled samples should be diverse and informative to represent all samples, making the initial model has high generalization ability. In this paper, we refer to the initial labeled samples as key samples. First, the key samples should contain as many different kinds of scenes as possible to meet the diversity. Thus, we use color features and spatial features jointly to cluster the samples. The sample is divided into four sub-blocks, and the color features of each sub-block are extracted and concatenated to obtain the color-spatial features. According to the color-spatial features of each sample, hierarchical clustering is used to cluster them. We select the number of clusters with the largest silhouette coefficient, the number of clusters is 15 on GMU, 13 on AVD and 9 on BHID.

Then, we need to select the most informative samples in each cluster, the amount of information of key samples needs to be as close as possible to that of all samples. The information entropy [15] is used to measure the amount of information. The sample which contains more information has higher information entropy, as shown in figure 2. The samples in each cluster are sorted in order of decreasing entropy, and the first N samples are selected as the key samples.

Figure 2. Schematic diagram of information entropy from low to high.

3.2. Get Mutually Independent Detectors
To get mutually independent detectors, we use two independent features to train them, the whole process is divided into three steps. First, the training samples are used to train the model to convergence. However, the two detectors in figure 1 are highly correlated at this time and cannot be directly used for co-training.

Second, we decorrelate the features input to the two detectors. As shown in figure 1, the weight matrix W of the last fully-connected layer before the detectors is decomposed by SVD, and $W=USV^T$ is replaced with US to obtain SVD layer. The matrix US can be regarded as weighted orthogonal matrix according to equation (1), therefore the features output from SVD layer can still maintain the discriminative ability of the sample space after orthogonal transformation, and are mutually independent.

Third, we use the features obtained in second step to train the detectors again. In order not to destroy the independence of the two features, the SVD layer needs to be fixed, and only the parameters of the other layers, i.e., convolutional layers, fully-connected layers and the detectors in figure 1, are adjusted.

We use the Gram matrix [16] to measure the independence of SVD layer. As shown in figure 3, the values of diagonal elements in the Gram matrix are higher after iteration, and the values of non-diagonal elements are closer to zero, illustrating the independence of SVD layer are strengthened. Therefore, the independence of the features output from SVD layer are also strengthened after iteration.

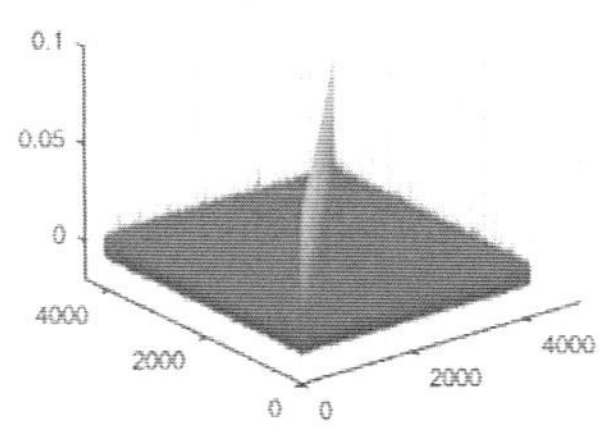

(a) The Gram matrix of SVD layer before iteration

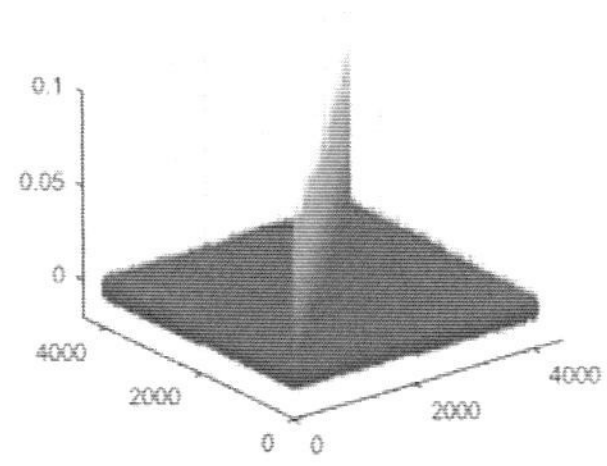

(b) The Gram matrix of SVD layer after iteration

Figure 3. The Gram matrix [16] of SVD layer before and after iteration.

3.3. The Adaptive Self-labeling Strategy

We score the self-labeling results according to the criteria of accuracy and recall, the scoring rule is expressed as:

$$score(x) = \sum_{i=1}^{n} I(accuracy1(w_i) > \delta_{acc}) \cdot I(accuracy2(w_i) > \delta_{acc}) \cdot I(iou(w_i) > \delta_{iou}) \quad (2)$$

where w_i is the object detected by both detector 1 and detector 2. $iou(w_i)$ is the intersection over union of the predicted frames of w_i in the two detectors. The high-confidence self-labeling results obtained by equation (2) can be regarded as both accurate and comprehensive.

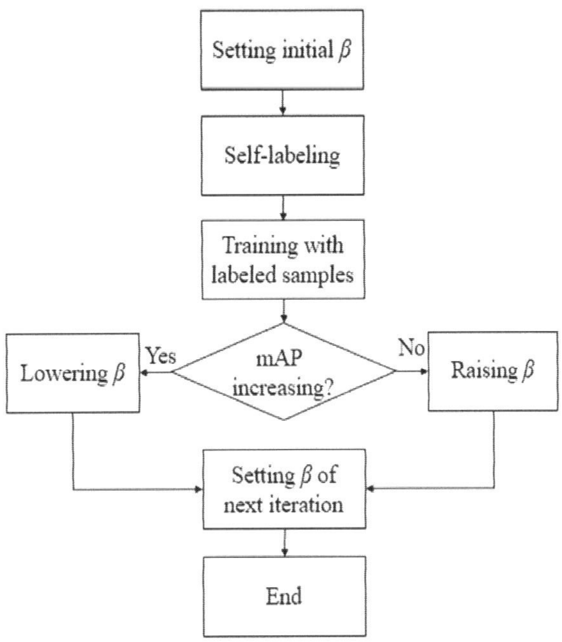

Figure 4. Flowchart of setting β.

Also we need to set a threshold to filter labeled samples and update the training set. A threshold too high will lead to a few updated labeled samples, and the accuracy of the detectors will not be improved. Otherwise, the quality of the updated labeled samples cannot be guaranteed, which could reduce the accuracy of the detectors. To contain the quality and quantity of labeled samples, the adaptive threshold setting method is expressed as:

$$\tau_k = \frac{\beta}{N_k}\sum_{i=1}^{N_k} score(x_i^k) \tag{3}$$

where N_k is the amount of samples contained in the kth cluster, and β is used to adjust the amount of labeled samples for each update.

We also propose a method used to set the β, as shown in figure 4. The accuracy of the detectors in initial training is lower, and the threshold needs to be high to filter out high-quality labeled samples. As the accuracy improves, the detectors have enough ability to perform high-quality self-labeling, so a larger amount of labeled samples are needed to prevent the detectors from overfitting.

4. Experiments

4.1. Datasets and Settings

In order to detect objects in unstructured environments, we use three datasets suitable for home, office and other scenes for evaluation, i.e., GMU [12], AVD [13] and self-made BHID. GMU dataset is the instance-level object detection dataset in the kitchen scene, contains 11 instance objects and 6728 images. AVD dataset contains 30 common instance objects, this paper selects 7262 images of 6 instance objects. BHID dataset is an instance-level dataset in the office scene, contains 7 instance objects and 1721 images. We select 3851 images as the training set and 2877 images as the test set on GMU. 5083 images are selected as the training set and 2179 images are selected as the test set on AVD. 1110 images are selected as the training set, and 611 images are selected as the test set on BHID. This paper uses the mean Average Precision (mAP) to evaluate the detection performance on all the 3 datasets.

4.2. Results and Analysis

We conduct experiments on GMU, AVD and BHID. The detection results are shown in table 1 and figure 5.

Table 1. mAP comparison of object detection models on different datasets.

Dataset	Amount of labeled samples	Method	mAP
GMU	199 (5% of total)	SVD co-training	79.3%
	3851 (total)	Full-supervised learning	81.3%
AVD	255 (5% of total)	SVD co-training	75.6%
	5083 (total)	Full-supervised learning	78.5%
BHID	55 (5% of total)	SVD co-training	76.1%
	1110 (total)	Full-supervised learning	78.4%

We select 5% of labeled samples from the training set as key samples for SVD co-training. The mAP on GMU is 75.6%, lower than the full-supervised learning by 2.0%, the mAP on AVD is 75.6%, lower than the full-supervised learning by 2.9%, and the mAP on BHID is 76.1%, lower than the full-supervised learning by 2.3%, which illustrates the effectiveness of SVD co-training. We use different amounts of key samples to perform SVD co-training on GMU, SVD co-training both achieves 79.3% mAP in the case of 5% and 10% of labeled samples, indicating the method we proposed can achieve detection accuracy comparable to full-supervised learning with 5% of labeled samples.

Figure 6 is the mAP variation curve of performing SVD co-training on different datasets. Since the amount of training samples of BHID is less than that of GMU, its mAP of full-supervised learning is lower, and the final iteration mAP is much lower than that of GMU. Although the amount of training samples of AVD is large, but the proportion of small objects is about twice that of GMU, which increases the difficulty of object detection, so the mAP decreasing is more compared with GMU. Therefore, SVD co-training performs better on the datasets with the moderate amount of training samples and the moderate object size. The same is that mAP of the two detectors in the iterative process is close and seesawing on different datasets, indicating that a supervision mechanism is formed between the two detectors, the two detectors can check and correct each other to avoid error accumulation.

Figure 5. Detection results on GMU, AVD and BHID datasets.

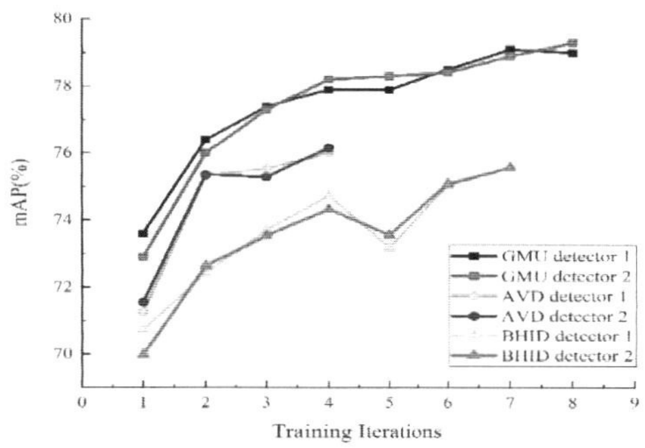

Figure 6. mAP variation curve of SVD co-training on different datasets.

4.3. Ablation Study

We also perform ablation study on SVD co-training, and analyse the influence on the detector performance of (1) whether to use SVD in co-training, and (2) whether to use the adaptive self-labeling strategy, as shown in table 2. We achieve 75.8% mAP by co-training without SVD, and mAP can be improved to 79.3% after adding SVD, indicating that independent detectors can enhance mutual ability to check and correct errors, compared with two detectors with strong correlation. 77.6% mAP is achieved by not using the adaptive self-labeling strategy, which shows that although the mAP of the detectors can increase with the expansion of the sample amount, the noise generated by the initial model with low mAP will have a negative impact on the final model.

Table 2. mAP comparison of various protocols.

Dataset	Amount of labeled data	SVD	Adaptive self-labeling	mAP
	199 (5% of total)	√		77.6%
GMU	199 (5% of total)		√	75.8%
	199 (5% of total)	√	√	79.3%

5. Conclusion

This paper proposes a semi-supervised object detection method using SVD and co-training. First, a small amount of key samples is manually labeled. Second, two independent detectors are obtained by SVD. Finally, an adaptive strategy is used to ensure the quality of pseudo labels. The results show that the high accuracy of object detection can be achieved by using SVD co-training with 5% of labeled samples, which significantly reduces the burden of manual labeling and provides convenience for detecting objects in an unstructured environment.

Acknowledgements
This work was supported by the National Natural Science Foundation of China (No. 61673039) and partly by Ji Hua Laboratory through Research on Multi-sensory Interactive Intelligent Nursing Robot in Complex Environment, the Science and Technology Research Programs in Foshan's Industrial Field in 2020 (No. 2020001006807).

References

[1] Van Engelen J E and Hoos H H 2020 A survey on semi-supervised learning *Machine Learning* **109** (2) 373-440.
[2] Sohn K, Berthelot D, Carlini N, Zhang Z, Zhang H, Raffel C A, Cubuk E D, Kurakin A, Li C L 2020 Fixmatch: Simplifying semi-supervised learning with consistency and confidence *arXiv preprint arXiv:2001.07685.*
[3] David B, Nicholas C, Ekin D C, Alex K, Kihyuk S, Han Z and Colin R 2019 ReMixMatch: Semi-supervised learning with distribution matching and augmentation anchoring *arXiv preprint arXiv:1911.09785.*
[4] Xu Z, Geng J, Jiang W, Zhang Z and Zeng Q 2021 Co-training generative adversarial networks for semi-supervised classification method *Optics and Precision Engineering* **29** (5) 1127.
[5] Huang H, Tang Y and Duan Y 2020 Feature extraction of hyperspectral image with semi-supervised multi-graph embedding *Optics and Precision Engineering* **28** (2) 443.
[6] Jeong J, Verma V, Hyun M, Kannala J and Kwak N 2021 Interpolation-based semi-supervised learning for object detection *Proc. of the IEEE/CVF Conf. on Computer Vision and Pattern Recognition (CVPR)* pp 11602-11611.
[7] Sohn K, Zhang Z, Li C L, Zhang H, Lee C Y and Pfister T 2020 A simple semi-supervised learning framework for object detection *arXiv preprint arXiv:2005.04757.*
[8] Zhou Q, Yu C, Wang Z, Qian Q and Li H 2021 Instant-teaching: An end-to-end semi-supervised object detection framework *Proceedings of the IEEE/CVF Inter. Conf. on Computer Vision (CVPR)* pp 4081-4090.
[9] Liu Y C, Ma C Y, He Z, Kuo C W, Chen K, Zhang P and Vajda P 2021 Unbiased teacher for semi-supervised object detection *arXiv preprint arXiv:2102.09480.*
[10] Blum A and Mitchell T 1998 Combining labeled and unlabeled data with co-training *Proc. of the Eleventh Annual Conf. on Computational Learning Theory* pp 92-100.
[11] Kalman D 2002 A singularly valuable decomposition: the SVD of a matrix *The College Mathematics Journal* **27** (1) 2-23.
[12] Georgakis G, Reza M A, Mousavian A, Le P H and Košecká J 2016 Multiview RGB-D dataset for object instance detection *Fourth Inter. Conf. on 3D Vision* pp 426-434.
[13] Ammirato P, Poirson P, Park E, Košecká J and Berg A C 2017 A dataset for developing and benchmarking active vision *IEEE Inter. Conf. on Robotics and Automation (ICRA)* pp 1378-1385.
[14] Shaoqing R, Kaiming H, Ross G, and Jian S 2017 Faster R-CNN: Towards real-time object detection with region proposal networks *IEEE Transactions on Pattern Analysis and Machine Intelligence* **39** (6) 1137-1149.
[15] Sergio V 1998 Fifty years of Shannon theory *IEEE Transactions on Information Theory* **44** (6) 2057-2078.
[16] Gatys L A, Ecker A S and Bethge M 2016 Image style transfer using convolutional neural networks *Proc. of the IEEE/CVF Inter. Conf. on Computer Vision (CVPR)* pp 2414-2423.

Construction of Entity and Entity Relationship Annotation Corpus for Test Identification

Bo Wang[1], Zixi Xu[1], Jianmei Huang[1], Teng Sun[1] and Xiaofeng Mu[2*]

[1] College of Information and Communication, National University of Defense Technology, Wuhan, Hubei, China
[2] Wuhan Children's Hospital (Wuhan Maternal and Child Healthcare Hospital), Tongji Medical College, Huazhong University of Science & Technology, Wuhan, Hubei, China
Email: 277825905@qq.com; 324138594@qq.com; 1078644258@qq.com; 1409320711@qq.com; *252187416@qq.com

Abstract. Currently, the test identification corpus faces the problems of small scale, poor consistency and poor standardization, which is difficult to meet the needs of intelligent applications such as knowledge mapping, question answering system and knowledge reasoning. Based on the entity labeling system for test identification developed by CCKS evaluation conference, this paper developed and improved the labeling specification. On this basis, the original text of test identification was obtained through news, Wikipedia and other channels. The entities in the original text were manually labeled and collated by the annotator, and the entity annotation corpus for test identification was constructed. The corpus constructed in this paper has a total of 86,000 words, with a total of 5980 entities and 4982 entity relations. The consistency of multi-round annotation is 86% and 85%, respectively.
Keywords. Test identification; entity; entity relations; corpus construction.

1. Overview

Test and identification refer to the test behavior of comprehensively assessing the subjects and making evaluation conclusions through standardized organizational forms and test activities. Due to the different objects, data acquisition forms and data acquisition purposes of test and appraisal, the data acquisition template structures of different test at different stages are different, and the data generally presents multi-source heterogeneous characteristics, which makes the data resource management of test and appraisal difficult [1]. At the same time, with the rapid development of communication, network, database and other information technology, the storage capacity of test and identification data is increasing, but the massive data is not effectively transformed into higher value knowledge units, which limits the use efficiency of test and identification data resources [2].

Named entity recognition (NER) and relation extraction (RE) are the key technologies of information extraction. NER identifies phrases with specific meaning or strong referentiality in unstructured text with entities as units, while RE aims to identify the correlation between entities [3, 4]. Information extraction technology can improve the knowledge density of unstructured texts and excavate rich experimental identification knowledge. In recent years, with the continuous development of deep learning technology, NER and RE are usually realized by deep neural network methods. Generally, the scale, consistency and standardization of corpus annotation will greatly affect the effect of NER and RE tasks performance. The construction of entity and entity relation annotation corpus is to mark the various entities and relations on unstructured text for computer reading [5, 6].

At present, many international evaluation conferences such as MUC, CoNLL, ACE, etc. have constructed entity tagging corpus, and formulated corresponding standards for entity types and tagging specifications. For example, the entity defined by MUC include the name of person, location, and organization, etc. While the entity defined by ACE include person, organization, geo-political entities, location, facility, weapon and vehicle etc. [7-9]. However, in the domain of test identification, the entities such as test elements, performance indicators, system composition, and task scenarios can better reflect the important conclusions and key technical indicators in the domain of test identification. The current labeling standards and norms for experimental identification are still in their infancy, and there is a shortage of entities and entity relationships corpus. At present, only 3398 entities in 400 test identification texts have been labeled by the Institute of System Engineering of the Academy of Military Sciences, and the corpus lacks the labeling of the relationship between entities, which is difficult to meet the needs of knowledge graph construction, question answering systems, knowledge reasoning and other fields.

Therefore, based on the entity annotation system of test identification developed by CCKS evaluation conference, this study explores and improves the corpus annotation specification, integrates the original text of test identification obtained from various channels such as news and Wikipedia, and manually annotates and collates the entities by using the annotation workers, and constructs the entity and entity relationship annotation corpus for test identification. The organization structure of this paper is as follows: Section 1 introduces the corpus annotation system and specification; Section 2 introduces the corpus construction method; Section 3 conducts data statistics and labeling consistency analysis on the corpus; Section 4 make a conclusion of this paper.

2. Annotation System

2.1. Entity Types

This study refers to the test identification corpus constructed by CCKS evaluation conference, and defines entity types as six categories: test elements, composition, performance indicators and task scenarios. On this basis, this study developed the definition criteria for the coverage of four types of entities, as shown in table 1.

Compared with the definition standard of entities in CCKS, the standards formulated in this study mainly include the following improvements:

(1) Reducing overlapping of entity definitions. By analyzing the entity definition standard in CCKS, we find that some definitions are overlaped, which makes it difficult for annotators to define entity types. For example, CCKS labeled condition of equipment as test elements and components of subjects as components, while some annotators may confuse the components of the equipment with the condition of the equipment, resulting in conceptual confusion between test elements and components entities ; in this study, six kinds of entity definition standards are formulated to remove the overlapping part of entity definition, which is helpful for annotators to understand the boundary conditions of annotation of various entities ,which improves the consistency rate of annotation.

(2) Reducing coverage of entity definitions. In CCKS, a class of entities usually contain multiple annotation objects, such as test elements annotates entities such as equipment name, test platform, test type, etc. There are obvious differences in semantics of these entities, and annotators tend to understand the annotation scope of test elements by literal meaning, and easily ignore the contents that need to be annotated such as test platform and test type. Therefore, we separate test platform and test type entities from the concept of test elements, thus reducing the complexity and superposition of entity definition.

(3) Unifying the description of similar concepts. In CCKS, there is a description confusion between some similar concepts. For example, object of test identification and subject appear in the definition of test elements and system composition respectively, which is not conducive to the unified understanding of the annotation system.

(4) Enhancing association between entity types name and definition. Annotators tend to associate the

definitions through the entity type name. In CCKS, the coverage range of all kinds of entity definition is obviously larger than the direct expression range of its name. For example, the definition of composition includes not only the subsystems and components of the subjects that are easy to associate, but also the related technologies used by the subjects. Therefore, we further standardize the definition of various entities, which significantly reduces the omissions caused by the lack of familiarity with labeling norms.

Table 1. Entity coverage and definition.

Entity type	Definition of CCKS	Definition of this article
Test elements	The objects of the test and identification work, such as the equipment listed as the assessment objectives, the conditions required to support the completion of the test and identification, and the condition of the equipment, etc.	Test and identify the subject of direct test and assessment, and the subject itself or the improved model
Test platform	Null	Platforms carrying test elements or the laboratories and test sites carrying out tests
Test types	Null	Types of test or identification of test elements
Components	Components of subjects, such as subsystems, components, techniques used, etc.	Software and hardware components or subsystems of test elements.
Performance indicators	The qualitative and quantitative description of test elements in terms of technology and use, such as weight, range, reliability, etc.	Quantitative description of test elements in technical performance, and the specific value or degree of technical indicators.
Task scenarios	The information involved in the actual utility and value of the test elements, such as personnel, confrontation objectives, system capabilities, etc.	The object that the test element performs the functions of enhancement, weakening and substitution, or the tasks performed and the resulting effectiveness.

2.2. Relationship Type

This article focuses on "experimental elements" and expands the relationship between "experimental elements" and other entity types. A total of 4 major types of relations are defined and 13 subtypes of relations are subdivided, as shown in table 2.

We define of six types of entities formulated in this study, test elements are the core concept of the annotation system, while test platform, test type, system composition, performance index and task scenario are all carried out around the concept of test elements, thus, establish entity relationship labeling system.

Table 2. Entity relationship.

Entity type 1	Relationship	Entity type 2	Example
Test elements	Attribute or description	Test elements	Translation, abbreviation, alias,etc.
Test elements	Test organization or site	Test platform	Test base, shooting range, etc.
	Installation platform		Vehicles, ships, aircraft, etc.
Test elements	Test	Test types	Capability test, prototype test, etc.
	Identification		Functional identification, etc.

Entity type 1	Relationship	Entity type 2	Example
Test elements	Functional unit	Components	Weapons, navigation and other functional modules
	Subsystem		Embedded software systems, etc.
Test elements	Performance profile	Performance indicators	Indexes of speed, height and precision, etc.
	Technical parameter		Values of performance indexes
Test elements	Object of utility	Task scenarios	Targets of attack, defence, enhancement, weakening
	Task, efficacy or application		Utility, such as destruction, weakening, enhancement, etc.
Components	Performance profile	Performance indicators	Indexes of speed, height and precision, etc.
	Technical parameter		Values of performance indexes
Components	Functional unit	Components	Communication, navigation and other functional modules
Components	Task, efficacy or application	Task scenarios	Utility, such as destruction, weakening, enhancement, etc.
Performance indicators	Numerical value	Performance indicators	Accurate values with unit, such as m / s, kg, etc.
	Degree		Qualitative description, such as, fast, light, large, etc.
	Attribute or description		Translation, abbreviation, alias, etc.
Task scenarios	Task, efficacy or application	Task scenarios	Utility, such as destruction, weakening, enhancement, etc.
	Object of utility		Targets of attack, defence, enhancement, weakening
	Attribute or description		Translation, abbreviation, alias, etc.
Task scenarios	Performance profile	Performance indicators	Indexes of speed, height and precision, etc.
Test platform	Attribute or description	Test platform	Translation, abbreviation, alias, etc.

3. Corpus Construction Process

Currently, there are three common modes for corpus annotation, namely, traditional annotation mode, crowdsourcing annotation mode and group annotation mode. This paper adopts the model of group labeling and expert review. The whole labeling process lasted for one month, with 2 experts and 6 undergraduates participating in the labeling work.

3.1. Original Data

The original data collected in this paper covers 11 types of equipment, such as missiles, rockets, satellites, combat vehicles, ships, submarines, aircrafts, radars, anti-missile systems, laser weapons and so on. A total of 600 pieces of test and appraisal news reports are involved. The distribution maps of news are shown in figure 1.

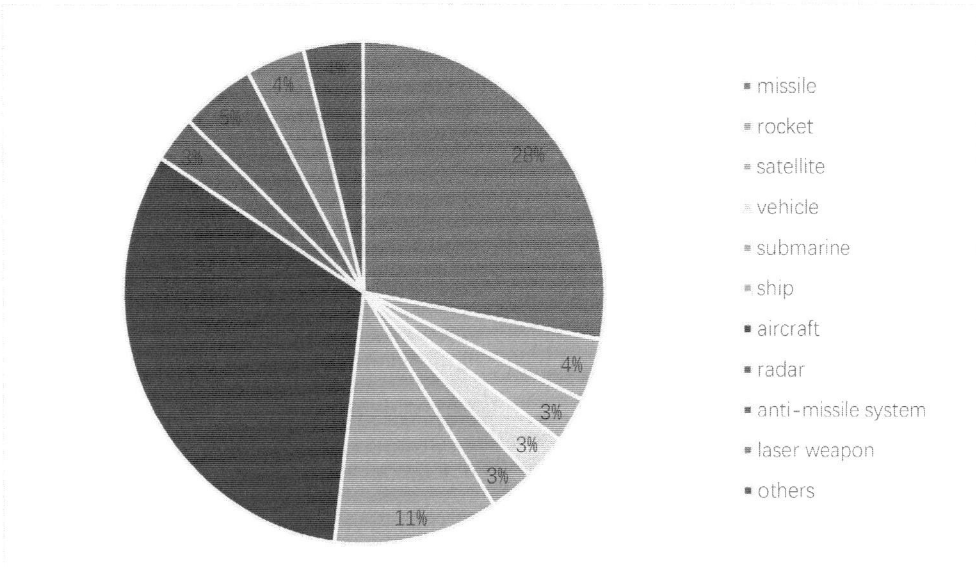

Figure 1. Original corpus distribution.

3.2. Annotation Process

In this paper, the annotation of corpus is divided into two stages. In the trial annotation stage, the annotation specification is formulated by repeatedly annotating and discussing strategies [10, 11]. In the formal annotation stage, the annotation consistency method is used to ensure the quality of annotation.

3.2.1. Trial Annotation. Firstly, two experts analyzed the characteristics of the original corpus text, established the classification method of each entity type, and formulated the preliminary annotation specification. Subsequently, in the multi-round test annotation, six annotators independently annotated the same 30 original corpus, calculated the annotation consistency rate and collected examples of annotation inconsistency. After referring to the opinions of experts, the annotation specification was updated, and then the new annotation specification was used to new 30 corpus until the annotation consistency of annotators was higher than 90%. The specific annotation process is shown in figure 2.

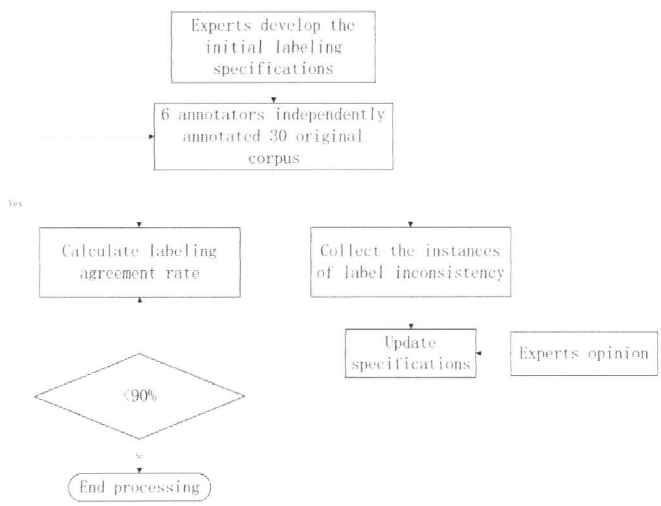

Figure 2. Trial annotation process.

3.2.2. Formal Annotation. Firstly, four annotators with high annotation quality in the trial annotation stage were selected and divided into A and B groups on average. Each group annotated 350 original corpus independently, of which 100 overlapping corpus were used for final consistency evaluation. When formal annotation is conducted, the two annotators in the group mark the same article independently, and the annotation inconsistent examples of the two annotators in the group are unified. Finally, the labeling consistency rate between groups was calculated after the labeling of both groups A and B was completed, and the specific labeling process is shown in figure 3.

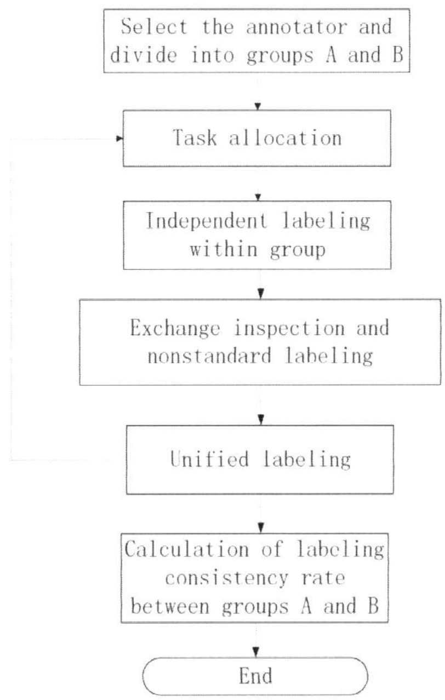

Figure 3. Formal annotation process.

4. Annotation Result

4.1. Corpus Quality Analysis
Iner-annotator agreement (IAA) is an indicator used to evaluate the degree of agreement of annotation results [12]. Based on IAA, the annotation quality and reliability of manually annotated corpus can be effectively evaluated [13]. Generally, IAA can be calculated by *F* value. When calculating the F value, a set of annotation results M1 is used as the standard answer, then the precision P, recall R and F value of the other set of annotation results can be calculated by equations (1)-(3).

$$P = \frac{Number\ of\ M1\ consistent\ with\ M2}{Number\ of\ M2} \tag{1}$$

$$R = \frac{Number\ of\ M1\ consistent\ with\ M2}{Number\ of\ M1} \tag{2}$$

$$F = \frac{2 \times P \times R}{P+R} \tag{3}$$

To determine the consistency of entity annotation, two conditions must be satisfied at the same time, one is that the beginning and ending positions of two entity strings are consistent, the other is that the labels of two entity types are the same; Determines the entity relation annotation consistency, requires the entity relation contains two entity annotation consistent, on this basis, also needs to guarantee the entity relation type identical. The consistency evaluation results of the corpus constructed in this paper after three rounds of trial annotation and formal annotation are shown in table 3.

Table 3. Iner-annotator agreement of entities and relations.

Rounds		Entities	Relations
	1st	0.69	0.66
Trial annotation	2nd	0.75	0.73
	3rd	0.90	0.89
Formal annotation		0.86	0.85

From table 3, it can be seen that the annotation consistency of different annotators gradually increases in the trial annotation stage, and the consistency of the formal annotation stage and the third round of pre-annotation is almost equal, indicating that the annotators understanding of the annotation specification tends to be consistent. It points out that when the annotation consistency reaches 0.8, it can be considered that the corpus we constructed is reliable.

In this paper, the number of named entities and relations constructed by final annotation are counted in tables 4 and 5. Compared with the existing Chinese and English test and identification corpus, the corpus annotation system constructed in this paper is the most complete and the annotation scale is the largest.

Table 4. Number of entities.

Entity types	Number
test elements	1055
test platform	459
test types	330
components	886
performance indicators	1547
task scenarios	1703

Table 5. Number of entity relations.

Entity 1	Relation	Entity 2	Number	
Test elements	Attribute or description	Test elements	401	639
Performance indicators		Performance indicators	70	
Task scenarios		Task scenarios	134	
Test platform		Test platform	34	
Test elements	Test organization or site	Test platform	174	174
	Installation platform		240	240
Test elements	Test Identification	Test types	300 -	300 -
Test elements	Functional unit	Components	587	676
Components		Components	89	
Test elements	Subsystem	Components	116	116
Test elements	Performance profile	Performance indicators	474	624
Components		Performance indicators	114	
Task scenarios		Performance indicators	36	

Entity 1	Relation	Entity 2	Number	
Test elements	Technical parameter	Performance indicators	185	238
Components		Performance indicators	53	
Test elements	Object of utility	Task scenarios	170	867
Task scenarios		Task scenarios	697	
Test elements	Task, efficacy or application	Task scenarios	509	649
Components		Task scenarios	43	
Task scenarios		Task scenarios	97	
Performance indicators	Numerical value	Performance indicators	398	398
	Degree		61	61

5. Conclusion

We have developed the largest, widest coverage, and most complete Chinese test identification corpus for entity and relation classification. The entities in the original text were manually labeled and collated, which has a total of 86,000 words, with a total of 5980 entities and 4980 entity relations. The corpus constructed in this paper implement the consistency of multi-round annotation is 86% and 85%. Based on this corpus, our follow-up work mainly includes two aspects. One is to study the automatic extraction method of entities and relationships, and the other is to explore the knowledge graph reasoning and question answering applications.

References

[1] Liu W, Wang X, Zhang Y and Liu Z 2017 An automatic-annotation method for emergency text corpus *Chin. Inf. Process* **2** (12).

[2] Huang Y Q, Xia J X, Yang X Z, et al. 2017 Big data on urban public safety *Geospatial Information* **1 5**(07) 1-5+9

[3] Jiang J F 2004 Research about the Pattern Acquisition for Free Text IE (Beijing: University of Chinese Academy of Sciences) Doctor's Thesis.

[4] Chieu H L and Ng H T 2002 A maximum entropy approach to information extraction from semi-structured and free text *AAAI/IAAI* pp 786-791.

[5] Zhang C, Hong S and Zhang P 2016 The research on event extraction of Chinese news based on subject elements *2016 IEEE/ACIS 15th Inter. Conf. on Computer and Information Science* pp 1-5.

[6] Lample G, Ballesteros M, Subramanian S, Kawakami K and Dyer C 2016 Neural architectures for named entity recognition *arXiv preprint arXiv*, 1603.01360.

[7] Chinchor N and Robinson P 1997 MUC-7 named entity task definition *Proc. of the 7th Conf. on Message Understanding* pp 1-21.

[8] Grishman R, Westbrook D and Meyersn A 2005 Nyu's English ace 2005 system description *ACE* **5**.

[9] Sang E F and De Meulder F 2003 Introduction to the CoNLL-2003 shared task: Language-independent named entity recognition *arXiv preprint* 0306050.

[10] Cornolti M, Ferragina P and Ciaramita M 2013 A framework for benchmarking entity-annotation systems *Proc. of the 22nd Inter. Conf. on World Wide Web* pp 249-260.

[11] Bontcheva K, Derczynski L and Roberts I 2017 Crowdsourcing named entity recognition and entity linking corpora *Handbook of Linguistic Annotation* pp 875-892.

[12] Hripcsak G and Rothschild A S 2005 Agreement, the f-measure, and reliability in information retrieval *Journal of the American Medical Informatics Association* **12** (3) 296-298.

[13] Ogren P, Savova G and Chute C 2008 Constructing evaluation corpora for automated clinical named entity recognition *Proc. of the Sixth Inter. Conf. on Language Resources and Evaluation.*

Domain-Adaptive Pre-training BERT Model for Test and Identification Domain NER Task

Bo Wang[1], Jianmei Huang[1], Ruicheng Yan[1], Yaofeng Su[1] and Xiaofeng Mu[2*]

[1] College of Information and Communication, National University of Defense Technology, Wuhan, Hubei, China
[2] Wuhan Children's Hospital (Wuhan Maternal and Child Healthcare Hospital), Tongji Medical College, Huazhong University of Science & Technology, Wuhan, Hubei, China
Email: 277825905@qq.com; 1078644258@qq.com; ruichengyan@126.com; suyf1982@163.com; *252187416@qq.com

Abstract. Name entity recognition (NER) is the foundation of a wide range of natural language processing (NLP) task in the domain of test identification. In this paper, we continue to train the pre-trained BERT model by unlabeled texts related to the domain of test identification, so as to inject domain knowledge into the pre-trained BERT model and realize the domain adaptation. The experiment results show that the proposed domain-adaptive pre-training method increases the F1 value by 1% compared with the baseline in the domain of test identification NER task.
Keywords. Test identification; NER; domain-adaptive.

1. Introduction

Due to different test objects and data acquisition forms, test identification data presents multi-source heterogeneous characteristics, which makes data resource management difficult; Besides, the test data shows the characteristics of big data, but the massive data is not efficiently stored and utilized, which limits the value of test data resources.

Name entity recognition (NER) is the foundation of a wide range of natural language processing (NLP) task in the domain of test identification. NER systems can help improving the value density of unstructured data, and plays a fundamental role in the application of question answering system [1, 2], knowledge graph [3, 4] and text summarization [5, 6], etc. in the domain of test identification.

In recent years, NER tasks are usually implemented based on Deep Neural Network (DNN), which can automatically learn complex features from data through nonlinear activation functions. BERT is a neural network model using transformer for bidirectional coding, it can effectively represent the semantic and syntactic properties of words or characters. The wide application of BERT significantly improves the performance of multiple downstream tasks including NER. However, due to the specificity of the test identification domain, a large number of proper nouns did not appear in the initial pre-training of BERT, which affected the performance of the BERT model for NER after fine-tuning.

Studies have shown that domain-adaptive pre-training is an effective method to inject domain knowledge into the pre-trained BERT model, which improves the effectiveness of NER in biomedical, computer, news and comment fields [7]. This paper uses domain-related raw texts to continue training the BERT model to improve the NER tasks performance in the domain of test identification, and constructs a set of method framework for domain-adaptive pre-training.

2. Domain-Adaptive Pretraining Methods

2.1. Pretraining

In recent years, most NER tasks include two stages. Firstly, a language model is pre-trained on large unlabeled corpora in the general domain, which generally has millions of parameters. Subsequently, the distributed word representation of pre-trained language model (PLM) is reused in NER tasks through a supervised fine-tuning phase.

The pre-training of BERT includes two tasks. One is to randomly mark one or more words, and let the encoder to predict the marked words according to the context. The second is let encoder to judge whether the input sentence conforms to the order in the original text, as shown in figure 1. The BERT model in task one learns word-level features by cloze task, and the BERT model in task two learns sentence-level features by judging the order relationship between sentences. After the training of a large number of unsupervised data in the general corpus, the BERT model has the ability to represent semantics in most domains.

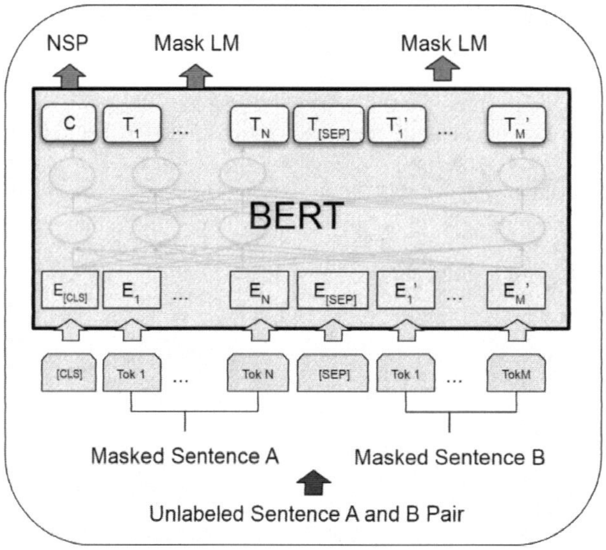

Figure 1. BERT pre-training method.

The Chinese pre-trained BERT model released by Google was trained on the Chinese Wikipedia text with a size of 0.4B. The Chinese Wikipedia covers multiple knowledge fields in a more formal language form, and ignores lists, tables and titles. The BERT model adopts GeLU hidden layer activation and AdamW optimizer during training, and contains 110 M parameters after training. BERT attains state-of-art performance on the Chinese NER task, making it the benchmark we choose.

Although Wikipedia training corpus covers a wide range, there are fewer informal language styles and insufficient coverage of proper nouns in special fields, which makes the performance of BERT-based NER systems in specific domains questionable. Therefore, this paper constructs a large corpus of specific domain texts, and continue training the pre-trained BERT model without supervision, so as to test whether the domain adaptability of the BERT model can be enhanced.

2.2. Domain-Adaptive Pretraining

Continuing pre-training using domain data to train the PLM, with the goal of gradually expanding the knowledge acquired and using it to improve the performance of domain NLP tasks [8, 9]. The domain-adaptive training method used in this paper is consistent with the Masked LM method in BERT, which randomly masks tokens in a certain percentage at the input end, and then predicts those masked tokens using pre-trained BERT model, as shown in figure 2.

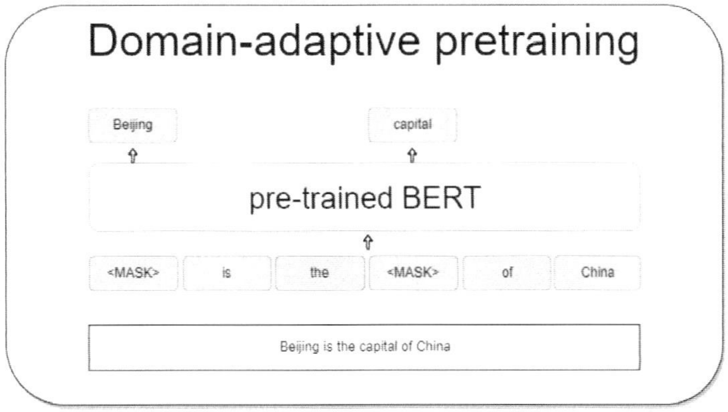

Figure 2. Domain-adaptive pretraining method.

Since the introduced [MASK] tag does not exist in the fine-tuning stage, that may cause mismatch between pre-training and fine-tuning stage. Related studies have shown that the mismatch may lead to catastrophic forgetting problems. To this end, the same strategy as BERT pre-training is adopted in this paper. First, 15% of tokens in the input sentence are randomly selected, these tokens were masked at 80% of the time, randomly converted into other words at 10% of the time, and remained unchanged at 10% of the time left.

2.3. Fine-Tuning

The Fine-tuning stage aims to modify the input and output of the model to supervised data for specific tasks by using the network structure and network parameters of the pre-trained BERT, and fine-tune all parameters from end to end. The advantage is that it can effectively use the generalization ability of deep neural network, and avoid the design of complex models and time-consuming training.

As shown in figure 3, in fine-tuning stage, the input and output are modified to entity-label text pairs in the domain of test identification by using the BERT model parameters after domain adaptation pre-training, and the model parameters are fine-tuned to transfer to the NER task.

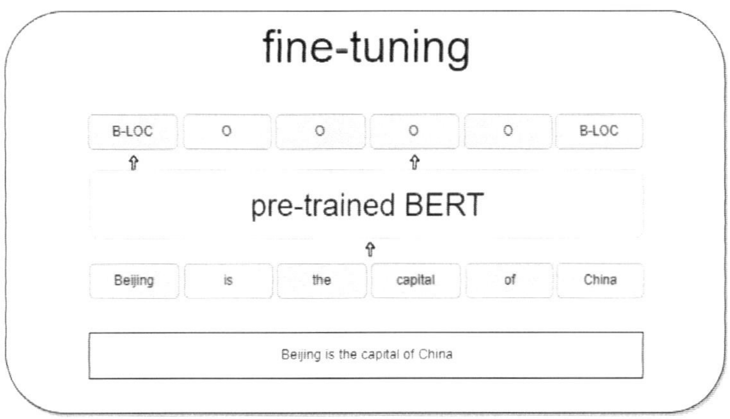

Figure 3. Fine-tuning method.

3. Experiment and Evaluation

In this section, the domain adaptation pre-training method is verified. The BERT-base-zh pre-trained model provided by Google is used as the baseline for domain adaptation pre-training, and then the pre-trained model is transfer to the NER task through fine-tuning.

3.1. Domain-Adaptive Pretraining Dataset
The description of weapons and equipment in Wikipedia includes a large number of descriptions of equipment performance and application scenarios, which are highly similar to the text description of test identification. Therefore, the domain-adaptive training data we use comes from weapon equipment related items in Chinese Wikipedia. Specifically, we choose more than 30 types of common equipment, such as aircraft, ships, missiles, radars, and remove the title of the article, leaving only the body part. The final domain-adaptive training dataset is shown in table 1.

Table 1. Domain-adaptive pretraining dataset.

Domain	Source	Number of items	Size	Tokens	Detail of items		
					Equipment types	maximum length	minimum length
Test identification	Chinese Wikipedia	5863	20.9M	21.94M	34	5064	156

3.2. Fine-Tuning Dataset
In this paper, the method of continuing pre-training is verified in the domain of test identification for two reasons. Firstly, there are a large number of domain-specific nouns in test identification domain. The recognition and classification of these words are challenging for the BERT model trained by general corpus. In theory, Continuing pretraining will show better performance in areas with stronger specificity. Secondly, CCKS evaluation conference organized corpus annotation in the domain of test identification. In this paper, we has also further improved the CCKS annotation system and annotation scale . The corpus for test identification domain is shown in table 2.

Table 2. Fine-tuning dataset.

Corpora	Task		Size	Number of labels					
	NER	Relation extraction		Test elements	Test platform	Test types	Components	Performance indicators	Task scenarios
CCKS	✓	×	61k	1188	-	-	694	854	662
ours	✓	✓	86k	1055	459	330	886	1547	1703

3.3. Evaluating Indicator
To evaluate our model, we use Precision (P), Recall (R), and $F1$ measure $(F1)$.

$$P = \frac{TP}{TP + FP} \tag{1}$$

$$R = \frac{TP}{TP + FN} \tag{2}$$

$$F1 = \frac{2*P*R}{P+R} \tag{3}$$

TP represents the number of entities correctly predicted, FP represents the number of entities that are wrongly predicted, FN represents the number of entities predicted as non-entities, and the P and R values can be balanced by F1 values.

3.4. Experiment Setting
We follow the BERT Masked LM settings in the domain adaptation pre-training process. Firstly, 15% tokens are randomly selected from the sentences in the domain adaptation dataset, and these tokens are

masked in 80% of the time. They are randomly converted into other words in 10% of the time, and remain unchanged in the remaining 10% of the time. After 60K steps training on two V100, we get the domain-adaptive BERT model after secondary pretraining. Then, the domain-adaptive pretrained BERT and BERT-base-zh are fine-tuned on the evaluation dataset after 1200 steps, and the performance between the two on NER tasks were compared.

3.5. Experiment Result

In figure 4, with the increase of fine-tuning steps, the F1 values of the domain-adaptive pre-training BERT model and BERT-base-zh first slowly increase and then stabilize on the evaluation dataset. Finally, the highest F1 value of the domain-adaptive pre-training BERT model is obtained at step 900, which is 60.41%, while the highest F1 value of the BERT-base-zh model is obtained at step 1000, which is 59.44%. It can be seen that our method improves the F1 value of BERT-base-zh model by 1%, the NER performance of the original BERT-base-zh model is improved by secondary pre-training on similar domain dataset.

In theory, our method should achieve better results in areas with more proper nouns, however, the experiment results show that our method has limited improvement compared with the baseline, which may be due to the following reasons : (1) In the stage of domain-adaptive pre-training, the dataset we used is only 22 M in size, while the original dataset used by BERT-base-zh in training is 0.4 B, the difference between the them is 18 times, which limits the BERT-base-zh model to learn the knowledge of test identification domain. (2) The original BERT-base-zh used Chinese Wikipedia dataset for pretraining, however, the data used in the domain-adaptive pre-training stage is the equipment-related items screened from the Chinese Wikipedia. So that the domain adaptation pre-training stage only strengthens some training texts, and does not introduce new domain knowledge.

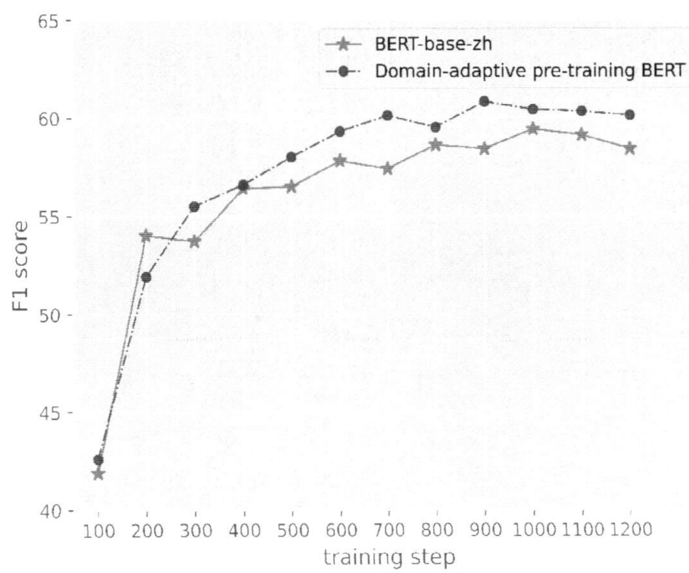

Figure 4. F1 values variation in NER task.

4. Conclusion

This paper constructs the framework of domain adaptation pre-training method. Firstly, we continue to train the pre-trained BERT model by unlabeled texts related to the domain of test identification, so as to inject domain knowledge into the pre-trained BERT model and realize the domain adaptation. On this basis, we transfer the pre-trained BERT model to the NER task by fine-tuning, and conduct comparative experiments on the evaluation dataset. The results show that our method can achieve certain domain knowledge fusion effect and improve the NER performance of the pre-trained BERT model in specific domain. In further research, we will try to study the influence of data similarity on the effect of domain adaptative pre-training.

References

[1] Mollá D, Van Zaanen M and Smith D 2006 Named entity recognition for question answering *Proc. of the Australasian Language Technology Workshop* pp 51-58.
[2] Lee C, Hwang Y G, Oh H J, Lim S, Heo J, Lee C H and Jang M G 2006 Fine-grained named entity recognition using conditional random fields for question answering *Asia Information Retrieval Symp.* pp 581-587.
[3] Abu-Salih B 2021 Domain-specific knowledge graphs: A survey *Journal of Network and Computer Applications* **185**,103076.
[4] Jia Y, Qi Y, Shang H, Jiang R and Li A 2018 A practical approach to constructing a knowledge graph for cybersecurity *Engineering* **4** (1) 53-60.
[5] Khademi M E and Fakhredanesh M 2020 Persian automatic text summarization based on named entity recognition *Iranian Journal of Science and Technology, Transactions of Electrical Engineering* 1-12.
[6] Hassel M 2004 Evaluation of automatic text summarization *Licentiate Thesis, Stockholm, Sweden*, 1-75.
[7] Gururangan S, Marasović A, Swayamdipta S, Lo K, Beltagy I, Downey D and Smith N A 2020 Don't stop pretraining: adapt language models to domains and tasks *arXiv preprint arXiv* 2004.10964.
[8] Devlin J, Chang M W, Lee K and Toutanova K 2018 Bert: Pre-training of deep bidirectional transformers for language understanding *arXiv preprint arXiv* 1810.04805.
[9] Chen Z and Liu B 2016 *Lifelong Machine Learning: Synthesis Lectures on Artificial Intelligence and Machine Learning* (San Rafael, CA, USA: Morgan and Claypool Publishers) pp 1-127.

Material Decomposition of Dual-Energy CT Based on CBAM Attention U-Net

Xinrui Zhang, Lei Li* and Bin Yan

Henan Key Laboratory of Imaging and Intelligent Processing PLA Strategic Support Force Information Engineering University, Zhengzhou, Henan, China
Email: 1196990264@qq.com; leehotline@163.com; ybspace@hotmail.com

Abstract. Spectral CT has the ability of quantitative material analysis by exploring the difference in the attenuation properties under different X-ray energies, and has become an important direction for the development of CT technology. However, due to the non-linearity and inconsistency of the observation, there is noise boosting in material decomposition and serious decrease of basis material images. In this paper, we design a U-Net convolutional neural network structure with double-entry and double-out to realize the effective material decomposition of high- and low-energy images. When designing the network, we introduce the attention mechanism, and add CBAM modules combining channel attention and spatial attention to improve the overall performance of the network. We also design an ablation experiment to test the module from a vertical perspective, and compare the material decomposition performance of each network from a horizontal perspective. The results show that the proposed Attention U-Net (AT-U-Net) has advantages in reducing noise boosting and improving the quality of basis material images.
Keywords. Spectral CT; material decomposition; convolutional neural network; U-Net; attention mechanism.

1. Introduction

Computed tomography (CT) is a detection technology that can achieve non-destructive reconstruction to obtain the internal structure of objects [1]. Compared with traditional CT, spectral CT can explore the difference of attenuation coefficients of substances under different X-ray spectra to extract rich material information, and have a better quantitative analysis ability of materials [2]. As one of the most common forms of spectral CT, dual-energy CT (DECT) [3] can deconstruct two different base materials with high and low energy spectra. However, classical spectral CT material decomposition algorithms may lead to noise boosting and the decline of image quality in the base material decomposition [4, 5].

Therefore, with the development of computer science and artificial intelligence, methods based on deep learning have good research and application prospects in the field of CT imaging [6]. By constructing an artificial neural network to achieve spectral CT material decomposition, spectral image features and hidden information can be effectively extracted, which is of great significance for suppressing the noise and improving quality of material decomposition. In 2018, Xu et al. used fully convolutional network (FCN) and Multiple perceptron (MLP) to solve the problems in base material decomposition, and verified that the FCN still has good performance under noise conditions [7]; in 2020, Zhang et al. designed a butterfly convolutional neural network (Butterfly Net), which has obvious advantages over the traditional FCN in material decomposition [2]. In 2021, Wang et al. adopted a U-Net architecture with double-entry and double-out, which further improved the effect of the material decomposition [4].

To further reduce the noise boosting and improve the quality of material decomposition images, this study construct a dual U-Net architecture based on attention mechanism. At present, information resources are constantly enriched, and a large amount of information is pouring into our production and life. However, only a part of important information attracts our attention. Based on this idea, people introduce the Attention Mechanism into convolutional neural network (CNN) [8]. In 2018, Sanghyun Woo et al. proposed Convolutional Block Attention Module (CBAM), which separates the spatial attention and the channel attention [9]. The CBAM attention model has been widely used in various scenarios due to its stable performance and lightweight structure.

Therefore, we use the U-Net based on CBAM attention mechanism to train and test on 1645 cranial cavity slice images, and evaluate the corresponding indicators of material decomposition images.

2. Methods

2.1. Construction of the Simulation Dataset

In experiments, we have obtained 1645 actual data of bone and tissue. In order to reduce costs and save resources, we use these data to reconstruct 1645 high- and low-energy images as our simulation data. In this way, we get high- and low-energy images as input of the network, as well as the ground truth of bone and tissue images, which lay a good foundation for subsequent training.

Before that, we need to understand the whole process of spectral projection and image reconstruction. As is shown in figure 1, when high and low energy X-rays pass through the object, the receiver forms a projected image by receiving the remaining photons after penetration.

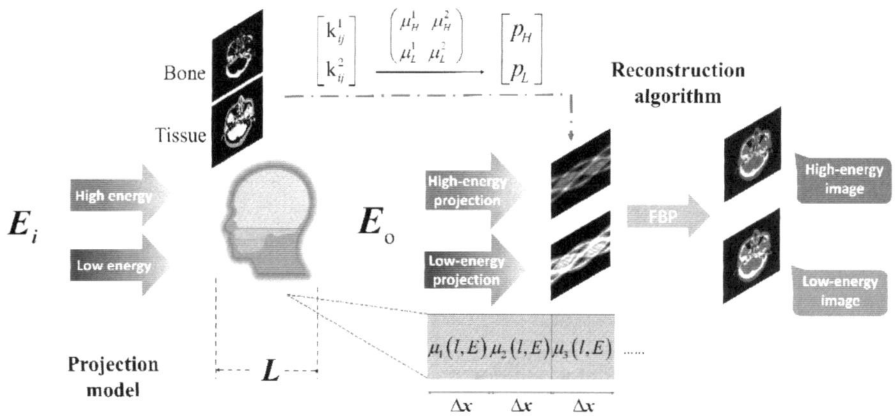

Figure 1. Acquisition of high- and low-energy images.

We assume that the intensity of the X-ray photon before getting into the object is E_i, and out of the object is E_o. For a nonhomogeneous object, it can be divided into innumerable homogeneous small pieces along the X-ray path; the length of which is Δx. The material attenuation coefficient of each small pieces is $\mu_k(l,E)$, $k=1,2,\ldots$, related to its location and the energy of X-ray photons. The relation of E_i and E_o obeys Beer-Lambert Law:

$$E_o = E_i \exp\left(-\sum_k \mu_k(l,E)\Delta x\right), \tag{1}$$

since the energy of X-ray photons is continuous and it presents a certain distribution in the object, we assume $S(E)$ as normalized energy distribution. When Δx is small enough, it can be expressed as:

$$E_o=E_i\int_0^{E_m} S(E)\exp\left[-\int_L \mu(l,E)\mathrm{d}l\right]\mathrm{d}E, \tag{2}$$

where E_m represents the maximum value of the energy distribution, and $\int_0^{E_m} S(E)\mathrm{d}E=1$. Therefore, we let $p=\int_L \mu(l,E)\mathrm{d}l$ and further obtain the following formula:

$$\begin{aligned} E_o&=E_i\exp\left[-\int_L \mu(l,E)\mathrm{d}l\right] \\ p&=-\ln\left(\frac{E_o}{E_i}\right), \end{aligned} \tag{3}$$

From equations (2) and (3), we can obtain:

$$p=-\ln\left(\int_0^{E_m} S(E)\exp\left[-\int_L \mu(l,E)\mathrm{d}l\right]\mathrm{d}E\right) \tag{4}$$

In practical applications, the above formula is usually converted into a simpler discretized model to facilitate numerical operations:

$$p=-\ln\left(\sum_{E=0}^{E_m} S(E)\exp(-\mathbf{W}\mu(l,E))\right), \tag{5}$$

where $\mathbf{W}$ is the projection matrix, which is a more convenient expression for the integration of the path and energy. This equation connects p with $\mu(l,E)$, which means the projection and the material information, respectively. And especially, $\mu(l,E)$ can be decomposed into two Attenuation coefficients of base materials:

$$\mu(l,E)=\mathrm{k}_1(l)\mu_1(E)+\mathrm{k}_2(l)\mu_2(E), \tag{6}$$

where $\mathrm{k}_1(l)$ and $\mathrm{k}_2(l)$ represent decomposition coefficients of bone and tissue; $\mu_1(E)$ and $\mu_2(E)$ represent attenuation coefficients of bone and tissue. We can look up NIST database to get $\mu_1(E)$ and $\mu_2(E)$.

In summary, from equations (5) and (6), combined with high- and low-energy spectral and position information, we can get the high- and low-energy projection from bone and tissue images:

$$p_{h,l}=-\ln\left(\sum_{E=0}^{E_m^{h,l}} S_{h,l}(E)\exp\left(-\mathbf{W}\left[\mathrm{k}_{ij}^1(l)\mu_{h,l}^1(E)+\mathrm{k}_{ij}^2(l)\mu_{h,l}^2(E)\right]\right)\right), \tag{7}$$

where h and l represent high- and low-energy, respectively; (i,j) represents the position of the pixel. Finally, we can use the FBP algorithm [10] to reconstruct the high- and low-energy images based on the $p_{h,l}$. The above process is the way we construct the simulation data set.

2.2. Dual U-Net Architecture

The dual U-Net architecture to achieve the decomposition of two different materials of bone and tissue in high- and low-energy CT images (MD-U-Net) is shown in figure 2. Each U-Net contains a contracting path and an expansion path [4]. The contracting path realizes the extraction of high- and low-energy image features, expansion path realizes the decomposition of base materials. Moreover, there is information interaction between the two paths and between the expansion paths of the two U-Nets to improve decomposition effect.

The contracting path includes 12 convolutional layers and 5 pooling layers [11]. The size of convolution kernel used in the convolutional layer is 3×3, and the kernel stride is 1. Every two convolutional layers is a group, and the number of filters is 32, 64, 128, 256, 512, and 1024, respectively, applying ReLu activation function. The size of kernel used in the pooling layer is 2×2, and the kernel stride is 2. We adopt the maximum pooling method; the expansion path includes 12 convolutional layers and 5 connection layers. The connection layer consists of three parts: a convolutional layer at a symmetrical position in the contracting path, an upsampling layer of this U-Net and the same position of the other. The number of convolution kernels in the expansion path is the reverse order in the contracting path. The size of kernel used in the upsampling layer is same as the pooling layer. We use interpolation upsampling to match the information in the pooling process.

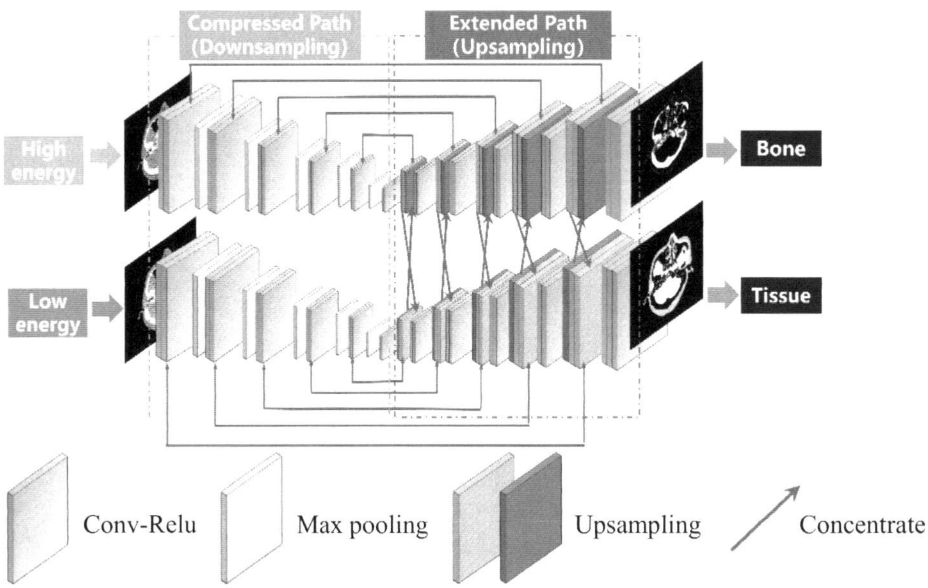

Figure 2. Material decomposition U-Net (MD-U-Net).

2.3. CBAM Attention Module

Figure 3 shows the implementation of the CBAM attention module. The block diagram on the left shows the implementation of the channel attention module. The channel attention module performs global max pooling and global average pooling [12] respectively on the input image $I_{inp} \in \mathbb{R}^{H*W*C}$ in the spatial dimension to obtain the feature map $F_{\max}$ and F_{avg}, they can be defined as:

$$F_{\max} = \max_{i,j=1}^{H,W} I_{inp}(i,j),$$
$$F_{avg} = \frac{1}{HW}\sum_{i,j=1}^{H,W} I_{inp}(i,j), \tag{8}$$

where $I_{inp}(i,j)$ represents the pixel value at the position (i,j) of the input image, $F_{\max}, F_{avg} \in \mathbb{R}^{1*1*C}$. Then, the two feature maps are respectively put into the shared MLP, and the down- and up-sampling process in the channel dimension is accomplished in the shared network. After that, the two channels of feature information are merged to activate through the function of sigmoid, and we can obtain feature map $F_C \in \mathbb{R}^{1*1*C}$. Finally, multiplying the output by the input image is the feature information of the channel attention $I_c \in \mathbb{R}^{H*W*C}$. The above process can be expressed by the following formula:

$$I_c = \sigma(\mathrm{MLP}(\mathrm{MeanPool}(I_i) + \mathrm{MaxPool}(I_i))) \circ I_i, \tag{9}$$

Next, we put the output from the channel attention module into the spatial attention module. The block diagram on the right shows the implementation of the spatial attention module. Firstly, the input feature map I_c is downsampled through the above two pooling methods sequentially, and we can obtain $F'_{\max}$ and F'_{avg} , $F'_{\max}, F'_{avg} \in \mathbb{R}^{H*W*1}$. Then, we connect $F'_{\max}$ and F'_{avg} in the channel dimension: $F_{concat} = [F'_{\max}, F'_{avg}] \in \mathbb{R}^{H*W*2}$. When the connection module passes through the convolutional layer with the size of 7×7 and the activation function, the channel dimension is reduced to 1. Finally, we multiply the output by the input feature map to get the feature information $I_s \in \mathbb{R}^{H*W*C}$ for spatial attention. The process of spatial attention can be expressed as:

$$I_s = \sigma(\mathrm{Conv}^{7\times7}([\mathrm{MeanPool}(I_c), \mathrm{MaxPool}(I_c)])) \circ I_c, \tag{10}$$

Since the two modules are cascaded, the output of the spatial attention module is the total output of the CBAM attention module.

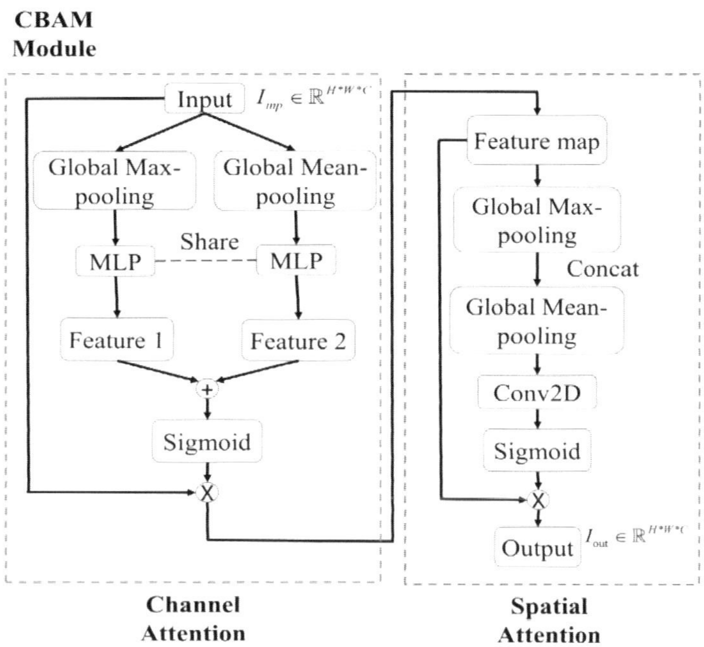

Figure 3. The implementation diagram of the CBAM attention module.

In the convolutional neural network, the introduction of the CBAM module can make the network pay attention to important feature map information and improve the overall performance of the network. At the same time, the CBAM module is also a plug-and-play lightweight module, so the load of the network is relatively small.

2.4. Ablation Experiment

Due to its good compatibility and stability, the CBAM attention module can extract the global information of the image well and improve the network performance in all directions. In the ablation experiment, we introduce CBAM attention mechanism to modify MD-U-Net. According to the different positions of the CBAM module, this paper proposes three options: one is to add the CBAM module in the symmetrical position of the contracting path and the expansion path. However, the experiment shows that, due to the cancellation effect of the CBAM modules in the symmetrical positions, the effect is not ideal. The second is to add the CBAM module in the contracting path. Although this option avoids the cancellation effect between modules, the CBAM module accelerates the loss of feature information due to the image feature compression in the pooling process; the third is to add the CBAM module in the expansion path, this solution can bring out the best of the attention module, and effectively improve the accuracy and quality of base material decomposition. Figure 4 shows the asymmetric attention U-Net structure of the third option. We conduct ablation experiments on the third asymmetric AT-U-Net. Each time a pair of CBAM attention modules are added to the upper and lower U-Net branches, a network architecture of up to six-layer modules can be obtained. We add modules at positions ① to ⑥ to form the type I to VI network respectively.

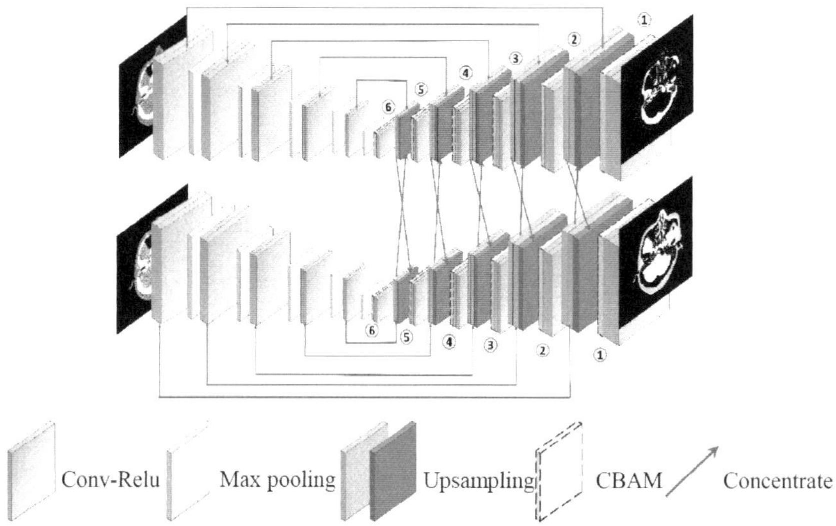

Figure 4. Dual U-Net architecture based on CBAM attention module.

3. Experiment

3.1. Data Set and Experimental Platform

The data set comes from cranial cavity slice images of 7 patients. The size of image is 512×512 pixels. We have 1645 decomposed bone and tissue images (with labels) each, as ground truth, high- and low-energy images each, as input data for the network. The high- and low-energy images at 80 and 120 kVp polychromatic spectrum are divided into a training set, a validation set and a test set of 1400, 140 and 105. Poisson noise is added with an initial total number of 1×10^4 photons to the high- and low-energy images. All experiments in this paper are performed on a ThinkStation P920 workstation with Intel Xeon Gold 6226R CPU 2.9 GHz and 64GB memory. We use a NVIDIA Quadro RTX 5000 GPU to train our model. This paper uses the deep learning framework based on Tensorflow to run the program in Pycharm.

3.2. Result Analysis

After the training process, the performance of the neural network is tested on a test set of 105 high- and low-energy images. After the test, it is necessary to compare the visual effect of images between the base material image output by the network and the ground truth, and further use three commonly used quantitative methods. Evaluation indicators: Peak Signal-to-Noise Ratio (PSNR), Structural Similarity (SSIM) and Root Mean Square Error (RMSE) are used to evaluate the decomposition quality of the base material. The overall performance indicators of the network are obtained by averaging the indicators of 105 test images. In the ablation experiment, with the increase of the number of CBAM modules, the trend of the three indicators is shown in figure 5:

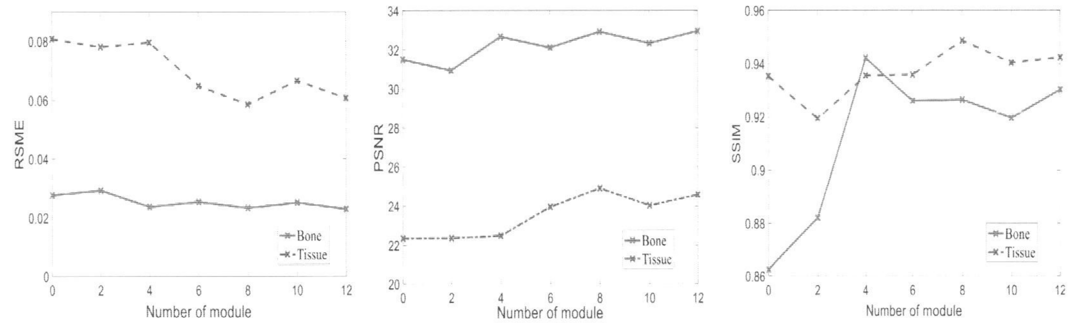

(a) (b) (c)

Figure 5. The trend of the three indicators.

In terms of indicators, the network performance has been greatly improved after adding the CBAM module. Among them, the type IV with 8 modules and the type VI with 12 modules have the best effect and the two have little difference in material decomposition performance. In terms of operating efficiency, since the number of modules added to the type IV is less, under the same number of iterations (76 epochs), the running time of type IV is shortened by nearly 2.5 hours compared with the type VI, as is shown in figure 6. And we can also see that the increase in the number of attention modules has little effect on the efficiency of the network from type I to V. It fully reflects the lightweight advantage of the CBAM attention module. All in all, the ablation experiments show that the type IV with 8 CBAM modules has the best performance compared to other networks in terms of parameters and operating efficiency.

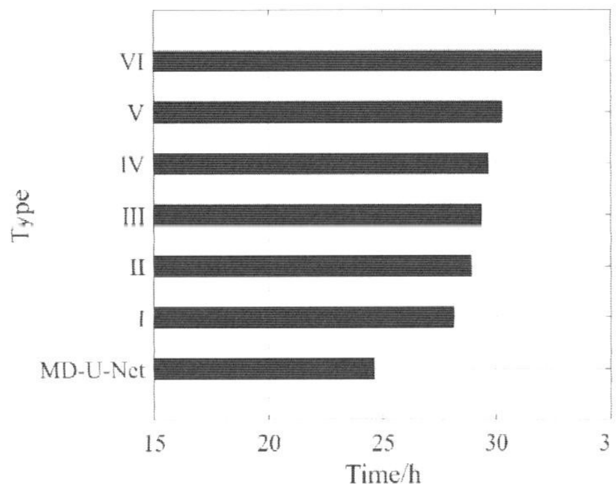

Type	Time (h)
MD-U-Net	24.60
I	28.12
II	28.88
III	29.34
IV	29.65
V	30.28
VI	32.02

Figure 6. The relationship between the running time and the number of CBAM modules.

We compare different types of networks horizontally, and select three network architectures: FCN, MD-U-Net, and the type IV AT-U-Net to evaluate network performance from two aspects: image effect and quantitative evaluation indicators. Figure 7 shows the image results of the base material decomposition for different. From top to bottom are decomposed bone material, the ROI of bone material, decomposed tissue material, the ROI of tissue material. From left to right are Ground truth, MD-U-Net, FCN and AT-U-Net, respectively. The bone and tissue image windows are [0, 0.05] and [0.015, 0.021]. We know that the type IV AT-U-Net introduces 4 pairs of attention modules based on MD-U-Net. Therefore, we can see from the region of interest (ROI) that, the bone and tissue images decomposed by AT-U-Net are better than MD-U-Net and FCN in terms of structure and texture.

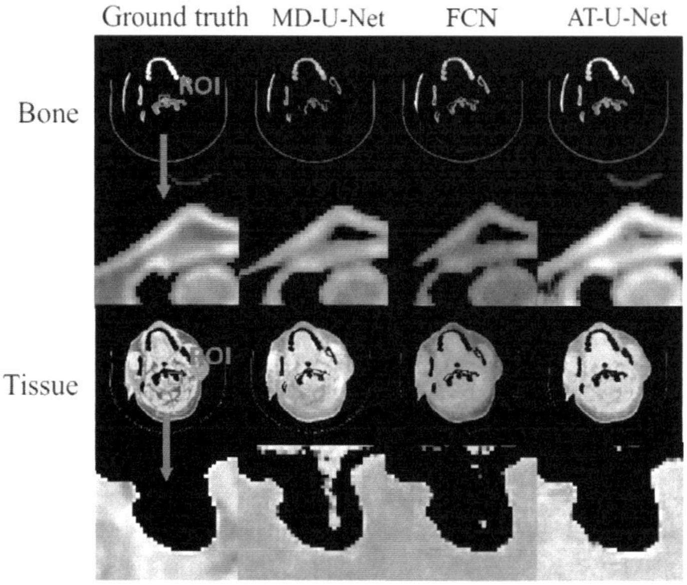

Figure 7. The material decomposition results of different networks.

Figure 8 shows the comparison of three networks' material decomposition images with ground truth from a qualitative view. Figures 8a and 8b show the pixel values of a certain section of the base material decomposition image, and the range of pixel position is [0, 512]. Whether it is bone or tissue, the results decomposed by AT-U-Net are the closest to the ground truth, followed by MD-U-Net, and finally FCN.

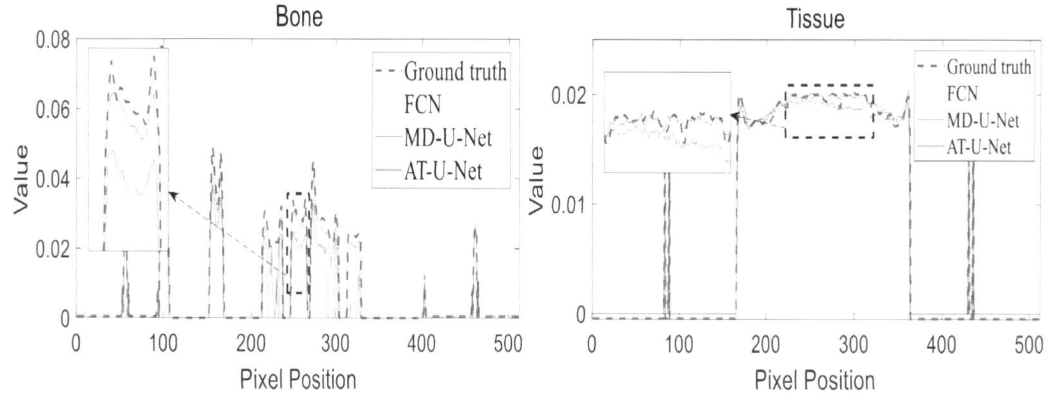

(a) (b)

Figure 8. Pixel values of decomposition results with different networks.

Table 1 lists three quantitative evaluation indicators for different kinds of networks. Compared with the traditional MD-U-Net, the RMSE of the bone and tissue images decomposed by the AT-U-Net is reduced by 0.0053 and 0.0123, the PSNR is increased by 1.66dB and 1.52dB, and the SSIM is more than 0.9. Regardless of the quantitative or qualitative results, the experiments of this paper show that the type IV AT-U-Net based on CBAM attention mechanism has significant advantages over the traditional network architecture.

Table 1. Quantitative evaluation indicators of different networks.

Index	Material	FCN	MD-U-Net	AT-U-Net
PSNR	Bone	31.0766	31.2662	32.9297
	Tissue	22.9691	23.3882	24.9077
RMSE	Bone	0.0293	0.0286	0.0233
	Tissue	0.0731	0.0709	0.0586
SSIM	Bone	0.8397	0.8777	0.9265
	Tissue	0.9319	0.9408	0.9487

4. Conclusion

In this paper, the method of deep learning is used to realize the decomposition of the base material by constructing a U-Net network architecture with two inputs and two outputs. In order to further improve the problems of noise boosting and image quality degradation in the decomposition of base materials, this paper introduces the CBAM attention mechanism based on the original MD-U-Net, and designs ablation experiments to verify the effectiveness of the attention module. Finally, we compare the improved network with other kinds of network architectures, and draw conclusions from both qualitative and quantitative perspectives: the AT-U-Net based on the CBAM attention mechanism proposed in this paper has significant advantages over other networks.

In the future, we will continue to optimize the network structure on the basis of asymmetric AT-U-Net to further improve the network performance. At the same time, we will continue to discuss the performance of other types of networks in the decomposition of base materials.

References

[1] Shi B 2010 *Research on Technology of Computed Tomography Based on GPU* (Huazhong University of Science & Technology) chapter 1 p 7.

[2] Zhang W 2020 *Research on Key Techniques for Spectral CT Reconstruction Based on Sparse Regularization and Deep Learning* (Strategic Support Force Information Engineering University) chapter 4 p 56.

[3] Bader M J, Haseke N, Johnson T, et al 2008 Dual energy CT (DECT) characterisation of urinary calculi: Initial in vitro and clinical experience *European Urology Supplements* **7** (3) 152.

[4] Wang Y, Feng Z, Yu X, Li L and Yan B 2021 Image-domain material decomposition for single-energy CT images using cascaded network *Conf. on Artificial Intelligence and Electromechanical Automation* pp 273-276.

[5] Frederick K, Joseph P M and Hilal S K 1980 Erratum: Noise considerations in dual energy CT scanning *Medical Physics* **7** (4) 388.

[6] Zhang W, Zhang H, Cai A, Wang L, Li L and Yan B 2017 A preliminary research on image-domain decomposition for dual-energy CT based on convolutional neural network *Proc. of 2017 IEEE Inter. Conf on Medical Imaging Physics and Engineering &the 8th National Annual Meeting of Medical Imaging Physics* pp 52-53.

[7] Xu Y, Yan B, Zhang J, et al. 2018 Image decomposition algorithm for dual-energy computed tomography via fully convolutional network *Computational and Mathematical Methods in Medicine* (4) 1-9.

[8] Zhang C, Zhu L and Yu L 2021 Review of attention mechanism in convolutional neural networks *Computer Engineering and Applications* **57**(20) 9.

[9] Woo S, Park J, Lee J Y, et al. 2018 CBAM: Convolutional block attention module *Proc. of the European Conf. on Computer Vision (ECCV)* pp 3-19.

[10] Liu L, Xia Q, Xu Z, Wang H, et al. 2019 Reconstruction of CT image and optimization of template by using FBP algorithm *Computer Era* **8** 16.

[11] Zhou F, Jin L and Dong J 2017 A review of convolutional neural network research *Chinese Journal of Computers* **40** (6) 23.
[12] Lin M, Chen Q and Yan S 2014 Network in network *CoRR abs/1312.4400.*

Pyramid Attention Based Early Forest Fire Detection Using UAV Imagery

Yongtuo Zhang[1], Sheng Chen[1], Weiyi Wang[2], Wei Zhang[1] and Li Zhang[1*]

[1] Tianjin College, University of Science and Technology Beijing, Tianjin, China
[2] Complex Systems Research Group, Faculty of Engineering, The University of Sydney, Darlington, New South Wales, Australia
Email: vic.ytzhang@outlook.com, chenshenguav@outlook.com, weiyi.wang@sydney.edu.au, kyoosan450@163.com, lee.lizhang@outlook.com

Abstract. Early forest fire detection using unmanned aerial vehicle (UAV) imagery is widely implemented in recent years. However, as the size of moving object in UAV imagery changes dramatically, the issues of focusing more informative areas of an image, large-scale variation as well as temporal consistency preservation have yet to be resolved due to the wide field of vision of UAVs in flight. To address these issues, in this work, we concerned the visual characteristics of early stage forest fire in UAV imagery and employ pyramid attention mechanism. We engineered a novel convolutional neural network (CNN) based on the state-of-the-art backbone by stacking our proposed PyrAtten blocks. We built a large dataset of early stage forest fire, including real world aerial photographs shot using our UAV prototype. Extensive evaluation of PyrAtten demonstrates its efficiency and effectiveness in detecting multi-scale objects of fire and smoke in UAV imagery, even if the object takes up a very small proportion of the whole image. Compared with existing similarly-sized networks, average detecting accuracy of PyrAtten-ResNet-50 reaches 97.5%, with negligible increase in computational overhead.
Keywords. Forest fire detection; UAV imagery; deep learning; pyramid attention mechanism.

1. Introduction

Forests always play a significant role in stabilising climate, regulating the water cycle, and providing habitat to thousands of life forms. Forests are also of vast economic benefits to human beings. However, in recent years, due to abnormal global climate change and continual human activities in forests, forest fires occur more frequently, which poses a huge threat to the earth's ecological environment and the safety of human life and property. As wood is inflammable, forest fire can spread incredibly quickly, and the latter it is found and controlled, the harder it is to turn it off. Hence, if the forest fire or its smoke is accurately and timely detected, localised and reported in the early stage of occurrence, it will help fire departments to take measures to minimise the spread of forest fire in time, so as to avoid causing disastrous consequences.

Conventionally, forest fires are detected basically via human patrol following fixed routes and observation from watchtowers located across the monitored area. However, human patrol is not only of low efficiency but also labour-intensive. Watchtowers have limited sight that can barely cover the whole area of interest. Although satellite remote sensing is capable of detecting large-scale forest fires, the cost of start-up and maintenance is pretty high, and the monitoring effect is often affected by weather, cloud thickness, orbital period etc., let alone sizes of early forest fires [1]. As flight control technology improves by leaps and bounds, unmanned aerial vehicles (UAVs) have been widely used in the field of forest fire detection due to flexibility and low cost. UAVs flying at a low altitude above the forest can detect forest fires, especially in their early stage, more efficiently and localise them more precisely [2]. Figure 1 shows the UAV prototype we assemble for the mission of early forest fire detection. As the embedded processor on it is computationally resource-restricted, the UAV in our system is used as the data collector, and transmits image streams to ground control station where forest fire and smoke detection works. Practically, a swarm of UAVs whose detection missions are planned in advance can sufficiently cover the task forest area searching for fire and smoke round the clock [1].

Figure 1. The UAV prototype for early forest fire detection.

Stimulated by the rocketing progress of deep learning in computer vision in the last decade, forest fire detection based on convolutional neural networks (CNNs) using UAV imagery is getting widespread as well. We therefore consider the characteristics of early stage forest fire and network complexity for low-cost UAV systems. In this work, we propose a novel UAV system, which utilise our proposed Pyramid Attention (PyrAtten) mechanism for early stage forest fire detection. We implement our pyramid attention blocks for channel-wise attention module and replace the 3×3 convolutions in the bottleneck blocks of the classic ResNet [3] with our PyrAtten modules. Extensive experiments demonstrate that our method achieves promising improvement with an acceptable computational cost increment for UAVs, remaining better detection accuracy and efficiency.

2. Related Work

2.1. Vision Based Fire Detection Using UAV

UAV based fire detection has been widely implemented in the world with the aid of computer vision technology. Hossain et al. [4] study and leverage features of forest fire and smoke extracted in colour space, setting a remarkable example of using feature engineering in computer vision to optimise forest fire detection algorithms. Sudhakar et al. [5] concern the false forest fire alert resulting from the characteristics of flying UAVs, considering the instability of both UAV and smoke recognition as challenges for detection algorithm researchers. CNN based forest fire detection using UAV has recently been widely deployed [6, 7], and the trade-off between high-performance detection and restricted computation resources of UAVs is taken into account as a practical issue to iron out.

2.2. Attention Mechanism

Attention mechanism has been proved to be efficient in exploiting discriminative areas of the image while mitigating the misalignment [8], making the model under certain context attending to the more informative parts adaptively.

SENet [9] extracts channel-wise attention by conducting two major operations on a feature map: "squeeze" it with global average pooling in channel dimension and then "excite" it to compute recalibrated feature map. CBAM [10] is another network module focusing on self-attention. Motivated by SENet, CBAM considers both channel attention and spatial attention which is computed by 2D convolutions with large kernel sizes, and finally combine them. ECA-Net [11] re-weights the feature map using a 1D convolution of kernel size *k* to locally capture inter-channel relations of each channel and its *k* neighbours. ECA-Net also alleviates the use of fully connected layer, making its implementation more light-weight. Coordinate attention [12] is designed to establish long-range dependencies for vision tasks and easy to be plugged into mobile networks, which makes it running efficiently on mobile devices without more notable computational cost. As early stage forest fire and smoke in UAV imagery are multi-scale and frequently change their sizes, we piggyback the idea of pyramid attention used in semantic segmentation [13] and propose a novel design of PyrAtten block, which can be easily plugged into existing network backbones and improve detection performance significantly with little model complexity increment.

3. Methodology

This section describes how our UAV system inspects the forest area of interest autonomously, and how the proposed PyrAtten works and is used on one of the state-of-the-art backbones.

3.1. Workflow of Forest Fire Inspection Using UAVs

Figure 2 shows the workflow of the task of fire inspection using UAVs in our system.

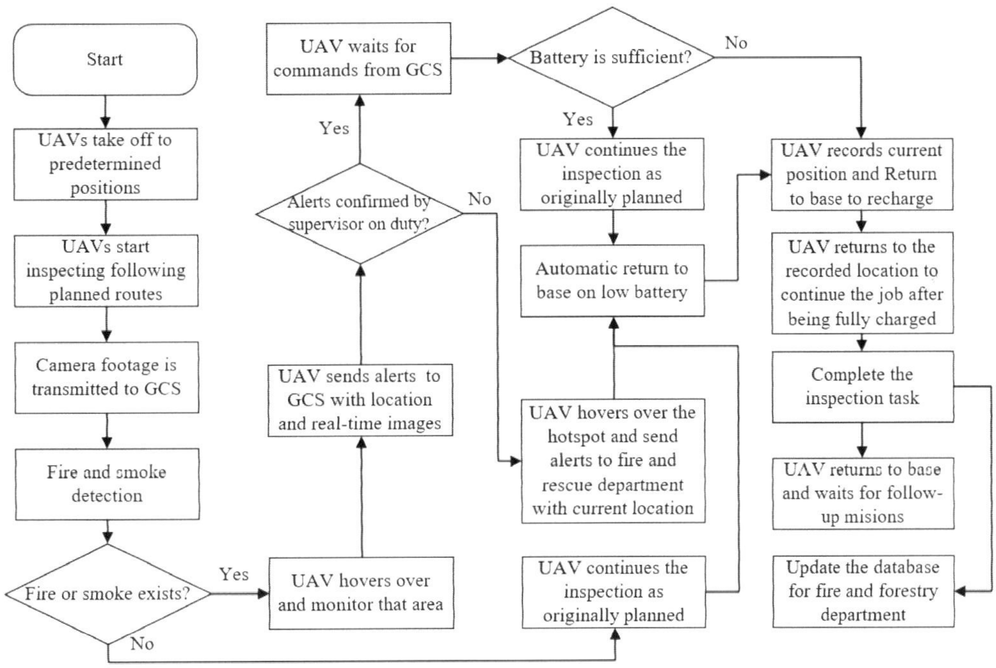

Figure 2. Workflow of fire inspection using UAVs in our system.

In our system, after mission planning on the ground control station, a UAV takes off from its automatic charging station for fire inspection, and can fly back to the charging station to get charged autonomously. The drone's thermal imager and visible-light camera capture images of the forest in real time. The data transmission module transmits the infrared image stream and the visible light image stream to the data processing module implemented on the ground control station, which uses the flame and smoke detector based on deep learning to detect flame and smoke. The inspection is conducted following the planned inspection task path until flame or smoke is detected. If an anomaly is found in the received infrared frame and is recognised as flames or smoke in a certain number of frames in the image stream that last for several seconds, a pop-up alert is continuously sent to the forest department personnel on duty until the manual confirmation is made. If it is artificially identified as a fire, the system automatically reports to the fire department that there is a fire alarm at the location of the drone, and sends a command for the drone to continuously hover around and monitor the fire location, so as to provide real-time pictures and geographic information for the firefighters to carry out rescue. The alarm will be cancelled if it is artificially determined that it is not a fire. If there is no manual confirmation for 5 minutes from the start of the alarm, the fire alarm will be reported directly to the fire department, and the drone will be to monitoring the location of the fire continuously. The data of the daily inspection of the UAV is compiled into a database of fire risk based on a geographic information system, which reflects the historical data and incidence of fires in the forest areas of interest on the monthly bases, which can provide a reference for forestry, fire protection and other departments to implement fire risk assessment and fire prevention. For very vast forest areas, the system can adopt UAV swarms. We use a mission planning software to divide the inspection area and assign them to UAVs, and the UAVs will conduct uninterrupted inspection of the allocated area according to the planned path.

3.2. Early Stage Fire Detection Motivated Design of Pyramid Attention

Pyramid attention mechanism is widely used in semantic segmentation and is proved effective in object detection. Empirically, early stage forest fire is commonly on different scales in UAV imagery, as the height at which a UAV flies varies with the changes in terrain to avoid collisions with trees or rocks. Performance of state-of-the-art CNN architectures nowadays is shown significantly improved, and thereby engineering a new backbone is not what we concern in this work. Therefore, piggybacking on previous high-performance deep neural networks, we investigate the focus on the structure of pyramid attention designated to suppress or emphasize feature maps for effective early stage forest fire detection.

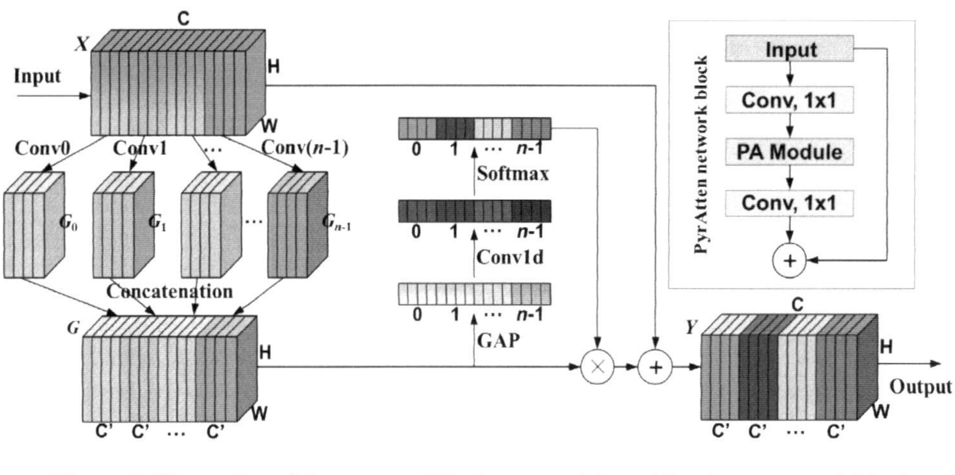

Figure 3. Illustration of the proposed PyrAtten module and PyrAtten network block.

We employ ResNet as the backbone of the network. As illustrated in figure 3, in our system, we engineer a novel PyrAtten module to generate pyramid attention with both channel and spatial information, and integrate them in ResNet as PyrAtten network blocks.

In a PyrAtten module, an input feature map $X \in R^{C \times W \times H}$ with C channels is first processed at n different scales in parallel to obtain it spatial information. Then a feature map pyramid can be calculated using group convolution method, from $Conv0$ to $Conv(n-1)$, of correspondingly n different kernels. G_i in the pyramid has the same number of channels, notated as C'. The group size of G_i is set to $2^{(k_i-1)/2}$, where k_i is the i-th kernel size, and G_i is therefore calculated by

$$G_i = \text{Conv}(i)(X) = \text{Conv}(k_i \times k_i, 2^{(k_i-1)/2})(X) \;\; i = 0, 1, \ldots, n-1 \tag{1}$$

where $G_i \in R^{C' \times W \times H}$. Using a concatenation operation in the channel dimension, the feature map G with multi-scale spatial information is then aggregated as

$$G = \text{Concat}([G_0, G_1, \ldots, G_{n-1}]) \tag{2}$$

where $G \in R^{C \times W \times H}$. Then we start to get the channel-wise information with a global average pooling, which reshape the feature map into $C' \times 1 \times 1$. A 1D convolution of an adaptively determined size is then used to make sure the interaction of neighbouring channels. We then apply a softmax function on it, thereby obtaining the attention vector, and an element-wise multiplication is performed to get the re-weighted version of G, which is later added by X to obtain the final output feature map $Y \in R^{C \times W \times H}$. Y is the emphasized tensor with both spatial and channel-wise information considered, and in our evaluation of the system, the network with PyrAtten network blocks is verified to be effective and efficient.

Table 1 shows the design of the proposed PyrAtten-ResNet-50, together with vanilla ResNet-50 and SE-ResNet-50. We engineer PyrAtten-ResNet-50 by stacking PyrAtten blocks like ResNet-50. Note that we elaborate kernel size with which group size of group convolution can be calculated.

Table 1. Network design of ResNet-50, SE-ResNet-50 and our proposed PyrAtten-ResNet-50.

Output size	ResNet-50	SE-ResNet-50	PyrAtten-ResNet-50
112×112	7×7, 64, stride 2		
56×56	3×3, max pool, stride 2		
56×56	$\begin{bmatrix} 1\times1, & 64 \\ 3\times3, & 64 \\ 1\times1, & 256 \end{bmatrix} \times 3$	$\begin{bmatrix} 1\times1, & 64 \\ 3\times3, & 64 \\ 1\times1, & 256 \end{bmatrix} \times 3$ w/ *fc*, [16, 256]	$\begin{bmatrix} 1\times1, & 64 \\ PyrAtten, & 64 \\ 1\times1, & 256 \end{bmatrix} \times 3$ w/ $k=3$
28×28	$\begin{bmatrix} 1\times1, & 128 \\ 3\times3, & 128 \\ 1\times1, & 512 \end{bmatrix} \times 4$	$\begin{bmatrix} 1\times1, & 128 \\ 3\times3, & 128 \\ 1\times1, & 512 \end{bmatrix} \times 4$ w/ *fc*, [32, 512]	$\begin{bmatrix} 1\times1, & 128 \\ PyrAtten, & 128 \\ 1\times1, & 512 \end{bmatrix} \times 4$ w/ $k=5$
14×14	$\begin{bmatrix} 1\times1, & 256 \\ 3\times3, & 256 \\ 1\times1, & 1024 \end{bmatrix} \times 6$	$\begin{bmatrix} 1\times1, & 256 \\ 3\times3, & 256 \\ 1\times1, & 1024 \end{bmatrix} \times 6$ w/ *fc*, [64, 1024]	$\begin{bmatrix} 1\times1, & 256 \\ PyrAtten, & 256 \\ 1\times1, & 1024 \end{bmatrix} \times 6$ w/ $k=7$
7×8	$\begin{bmatrix} 1\times1, & 512 \\ 3\times3, & 512 \\ 1\times1, & 2048 \end{bmatrix} \times 3$	$\begin{bmatrix} 1\times1, & 512 \\ 3\times3, & 512 \\ 1\times1, & 2048 \end{bmatrix} \times 3$ w/ *fc*, [128, 2048]	$\begin{bmatrix} 1\times1, & 512 \\ PyrAtten, & 512 \\ 1\times1, & 2048 \end{bmatrix} \times 3$ w/ $k=9$
1×1	GAP, 1000d-*fc*, softmax		

4. Experiments

In this section, our system is evaluated using a benchmark dataset of more than 8000 images of early stage forest fire and smoke. A UAV prototype we assembled is also used in real world experiments to collect images of forest fire and smoke. Fire and smoke objects are of different sizes in images as they are in the real world. All networks we test are implemented using PyTorch on a PC with one Intel(R) Xeon Silver 4208 CPU, one NVIDIA GeForce RTX 3070Ti GPU and 32GB memory.

Figure 4 gives some detecting results using PyrAtten integrated networks, in which from figures 4a to 4c, the size of the object is increasing while the image resolutions remain unchanged. Although the size of early stage fire and smoke in UAV imagery changes significantly, ResNet-50 with the PyrAtten blocks integrated is competent to extract multi-scale representation of feature maps layer by layer.

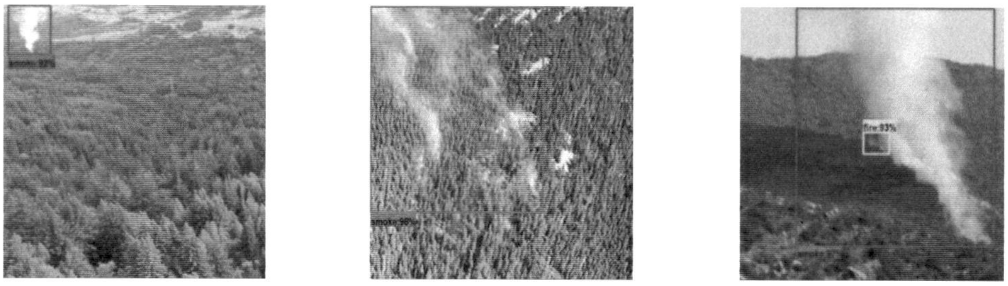

(a) (b) (c)

Figure 4. Detecting results of multi-scale smoke of early stage forest fire using PyrAtten blocks.

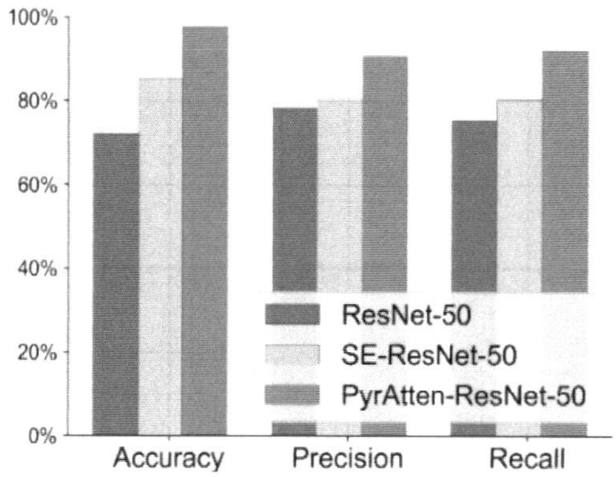

Figure 5. Comparison of accuracy, precision and recall for early forest fire detection in our dataset.

PyrAtten-ResNet-50 also outperforms in accuracy, precision and recall compared with ResNet-50 baseline and SE-ResNet-50 which consider only channel-wise attention, as shown in figure 5. The average accuracy for the early stage forest fire (smoke included in certain cases) detection of PyrAtten is 97.5%, 25.2% and 12.3% higher than that of ResNet and SE-ResNet, respectively. Besides, the number of SE-ResNet-50's parameters is 29.54M in our setting, 5.24% and 15.57% more than ResNet-50 and SE-ResNet-50, respectively, indicating that we manage to improve detecting performance eminently at the cost of minor increase in computational overhead.

5. Conclusion

The ability to focus on important areas of an image while discarding irrelevant ones is a hotspot worth studying, which can facilitate a wide range of applications s in the fields of deep learning based object detection. As early stage forest fire, especially the smoke caused by the fire, in UAV imagery is multi-scale and frequently changes their sizes. In this paper, we utilise pyramid attention combined with global context information to learn better feature representation and build long-range dependencies for the model. Taking into account the visual characteristics of early stage forest fire in UAV imagery, we engineer a novel CNN architecture equipped with our proposed PyrAtten blocks. Extensive evaluation of PyrAtten demonstrates its effectiveness and efficiency in detecting multi-scale objects of early stage forest fire and smoke in UAV imagery, even though the object takes up merely a small proportion of the whole image.

References

[1] Sherstjuk V, Zharikova M and Sokol I 2018 Forest fire monitoring system based on UAV team, remote sensing, and image processing *Proc. IEEE 2nd Inter. Conf. on Data Stream Mining Processing* pp 590-594.

[2] Hossain F A, Zhang Y M and Yuan C 2019 A survey on forest fire monitoring using unmanned aerial vehicles *Proc. 3rd Int. Symp. on Autonomous Systems* pp 484-489.
[3] He K, Zhang X, Ren S and Sun J 2016 Deep residual learning for image recognition *Proc. IEEE Conf. on Computer Vision and Pattern Recognition* pp 770-778.
[4] Hossain F A, Zhang Y M and Tonima M A 2020 Forest fire flame and smoke detection from UAV-captured images using fire-specific color features and multi-color space local binary pattern *J. Unmanned Veh. Syst.* **8** 285-309.
[5] Sudhakar S, Vijayakumar V, Kumar C S, Priya V, Ravi L and Subramaniyaswamy V 2020 Unmanned aerial vehicle (UAV) based forest fire detection and monitoring for reducing false alarms in forest-fires *Comput. Commun.* **149** 1-16.
[6] Zhao Y, Ma J, Li X and Zhang J 2018 Saliency detection and deep learning-based wildfire identification in UAV imagery *Sensors* (Basel) **18** (3) 712.
[7] Chen Y, Zhang Y, Xin J, et al. 2018 A UAV-based forest fire detection algorithm using convolutional neural network *Proc. IEEE 37th Chinese Control Conf.* pp 10305-10310.
[8] Chen G, Gu T, Lu J, Bao J A, Zhou J 2021 Person re-identification via attention pyramid *IEEE Trans. Image Process* **30** 7663-7676.
[9] Hu J, Shen L, Albanie S, Sun G and Wu E 2020 Squeeze-and-excitation networks *IEEE Trans Pattern Anal Mach Intell.* **42** (8) 2011-2023.
[10] Woo S, Park J, Lee J Y and Kweon I S 2018 CBAM: Convolutional block attention module *Proc. European Conf. on Computer Vision* pp 3-19.
[11] Wang Q, Wu B, Zhu P, Li P, Zuo W and Hu Q 2020 ECA-Net: Efficient channel attention for deep convolutional neural networks *IEEE/CVF Conf. on Computer Vision and Pattern Recognition* pp 11531-11539.
[12] Hou Q, Zhou D and Feng J 2021 Coordinate attention for efficient mobile network design *IEEE/CVF Conf. on Computer Vision and Pattern Recognition* pp 13708-13717.
[13] Sang H, Zhou Q and Zhao Y 2020 PCANet: Pyramid convolutional attention network for semantic segmentation *Image and Vision Computing* **103** 103997.

UAV Formation Transformation Based on Two-Stage Path Planning Method

Bingyu Song[1], Yahui Zuo[2*], Shan Huang[3], Xiaolu Liu[4] and Yingguo Chen[5]

College of Systems Engineering, National University of Defense Technology, Changsha, Hunan, China
Email: sbynudt@163.com; *1832538491@qq.com; 2549767334@qq.com; lxl_sunny@nudt.edu.cn; ygchen@nudt.edu.cn

Abstract. With the maturity of unmanned aerial vehicle (UAV) technology, the application of UAV in various fields is more in-depth. A method to quickly and accurately solve the flight paths of a large number of UAVs from one formation to another is crucial for application scenarios that require UAVs to perform formation transformation tasks. In this paper, the path planning of UAV swarm formation transformation is divided into two stages: rough planning and fine planning. In the rough planning stage, a 0-1 variable planning model is established to match the start and end points of the UAV flight path, and the model is solved using a Strengthen Elitist GA Algorithm to quickly obtain high-quality feasible solutions. In the fine planning stage, four models including behavioral mode, obstacle avoidance criterion, optimal waypoint selection and target exchange are established. On this basis, a heuristic fixed-step space search algorithm is designed and implemented to plan the path finely, and the path planning for the UAV formation transformation in two scenarios is finally completed.
Keywords. UAV formation transformation; 0-1 variable planning model; strengthen elitist GA Algorithm; heuristic fixed-step space search algorithm.

1. The Description of the Problem

1.1. Background

Unmanned aerial vehicles (UAVs) often need to quickly complete the task of formation transformation when participating in activities such as light shows, and the path planning between different formations is the key to formation transformation [1-3]. The difficulty of UAV swarm formation transformation task lies in the large number and collision constraints between UAVs [4]. In view of the above characteristics, this paper studies the formation transformation of UAVs in two scenarios: The UAV group first takes off on the ground and then forms a Chinese character pattern in the air; the UAV group transforms from the initial Chinese character pattern to the target Chinese character pattern in the air.

1.2. Analysis of UAV Terminal Position Acquisition

In order to complete the formation transformation task of UAVs, we first need to determine the space coordinates of the UAV group before and after the formation transformation. Then plan the flight route of UAVs, so that UAVs can quickly complete the task of formation change without collision (that is, the time to complete the formation change is very short) [5]. Therefore, we can divide the task into the following two stages: the matching of the starting point and the ending point of UAVs [6-8]; and the optimization of the collision-free path of UAVs.

1.3. Assumptions

- UAVs can take off directly in situ, without taking time to prepare for take-off.
- The flight speed of UAVs is always 3 m/s during flight.
- UAVs can only fly a fixed distance in a certain direction in a fixed time interval.
- UAVs can choose to wait in place for a time interval.
- At initial moment, UAVs are placed in a rectangle on the ground, and the distance between adjacent UAVs is about 3 m.

2. Rough Planning of UAV Path Based on Strengthen Elitist GA Algorithm

2.1. 0-1 Planning Model for UAV Endpoint Matching

Due to the upper limit of the flying height of UAVs and the need to reserve a certain space between UAVs, the position of the formation of the UAV group has certain spatial constraints. The spatial position constraints set in this paper are as follows:

$$z_i < 300 \tag{1}$$

$$z_i > 40 \tag{2}$$

$$\sqrt{(x_i - x_j)^2 + (y_i - y_j)^2 + (z_i - z_j)^2} < 1.5 \tag{3}$$

$$\max_k x_k + \min_l x_l = 0 \tag{4}$$

Equations (1) and (2) describe the spatial constraints, and UAVs should be distributed in the space above 40 m and below 300 m along the X axis.

Equation (3) describes the safe distance of UAVs, that is, the distance between UAVs should be greater than 1.5 meters.

Equation (4) makes the position data easy to process by centering the UAV on the X-axis.

Based on the spatial modeling of the preset formation position of the UAV swarm, this paper expresses the initial formation and target formation of the UAV swarm in the form of a set of spatial coordinates. The path planning of the UAV swarm formation transformation first needs to match the coordinates of the initial set with the coordinates of the target set, that is, to determine the starting point and end point of the UAV path.

In this paper, the space vector set of UAV position is set as the starting position vector set and the ending position vector set. Then, with the goal of completing the formation transformation of the UAV in the shortest time, a 0-1 planning model matching the starting point and the end point of the UAV is established:

$$\min z = \max\left(b_{ij} * d_{ij}\right) \tag{5}$$

Equation (5) is the objective function of the 0-1 planning model, where b_{ij} is a 0-1 variable, when the starting point i matches the end point j. Equation (5) indicates that the goal of the model is to minimize the farthest distance from the UAV to the end point.

$$\begin{cases} \sum_{i=1}^{n} b_{ij} = 1 & j = 1,\dots,n \\ \sum_{j=1}^{n} b_{ij} = 1 & i = 1,\dots,n \\ d_{ij} = \left| t_j - s_i \right| \end{cases} \tag{6}$$

Equations (6) represents the starting point of each UAV can only correspond to the end point of one UAV, and the solution of the Euclidean distance between the starting point and the ending point of the UAV.

2.2. Strengthen Elitist GA Algorithm

Due to the large number of UAVs, the complexity of solving the above-mentioned 0-1 planning model by conventional methods is relatively high. The use of genetic algorithm can reduce the complexity and achieve better results [9, 10].

2.2.1. GA Parameter Setting. In this example Coding can map the space of the target problem to the coding space of the algorithm. This paper uses permutation coding to connect the number of the target point with the starting point corresponding to the index. For example, a gene segment {4, 6, 8, 34} indicates that the starting point of 0, 1, 2, and 3 is the end point of 4, 6, 8, and 34 respectively [11].

The fitness reflects the degree of adaptation of a species to the living environment, which can guide the search improvement of the algorithm. If the objective function of equation (5) is directly used as the fitness function, the algorithm will easily fall into the local optimal solution because the fitness is only related to the maximum distance $D_{\max}$. The average distance D_{avg} can also be used as a part of the fitness, and the fitness function formula is as follows.

$$\xi(x) = \frac{1}{D_{\max} D_{avg}} \tag{7}$$

The tournament selection operator is used to select individuals. Two individuals with replacement are selected and the individual with better fitness is selected from them, and the process is repeated until the number of selections reaches the number of the original population [12].

After selecting an individual ζ_1 in the population according to the selection result, the random number comparison is used to determine whether to perform crossover operation. Then, a gene ζ_2 in the population is selected, and then crossover operation is performed. The crossover probability is set to 70% empirically, and the crossover mode is partially-matched crossover.

Mutations are made in selected individuals, and random numbers are required to determine the two gene values to be swapped with each other. In this paper, the reverse mutation operator is used and the mutation probability is set to 50% based on experience.

In order to make the genetic algorithm converge to the global optimal solution, the enhanced elite retention strategy is used to combine the parent population and the population with crossover mutation to form a population with a size of 2N, and then a new generation of population is selected from it.

The population size is set to 100. In order to approximate the global optimal solution as much as possible, the number of iterations of the algorithm is preset by running the program several times and observing the results. The complexity of scenario 1 contains 800 starting points and 800 ending points, which is higher than that of 800 starting points and 666 ending points of scene 2. Scenario one requires 20000 iterations and scenario two requires 8000 iterations.

2.2.2. Solution Result. The genetic algorithm is used to solve the two scenarios of UAV formation transformation respectively. The transformations of the optimal individual objective function value and the population average objective function value during the algorithm convergence process are shown in figures 1-4:

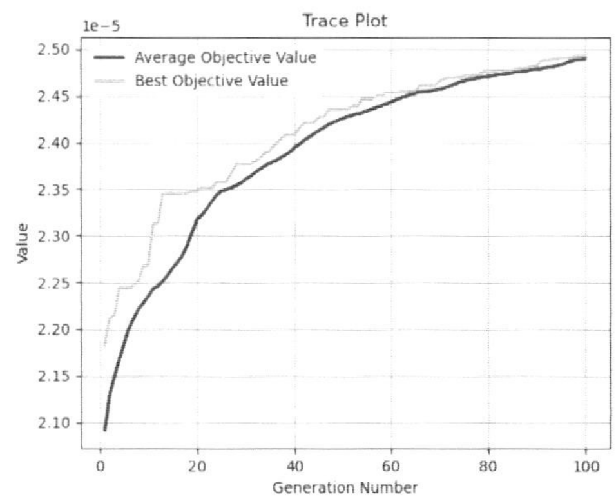

Figure 1. The local process of optimal value iteration for Scenario 1.

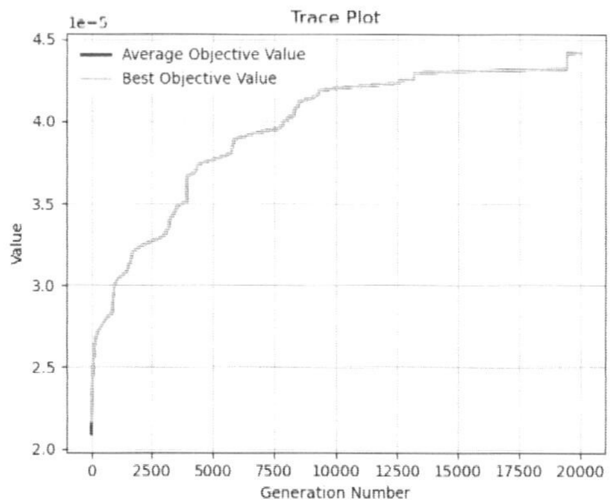

Figure 2. The global process of optimal value iteration for SCENARIO 1.

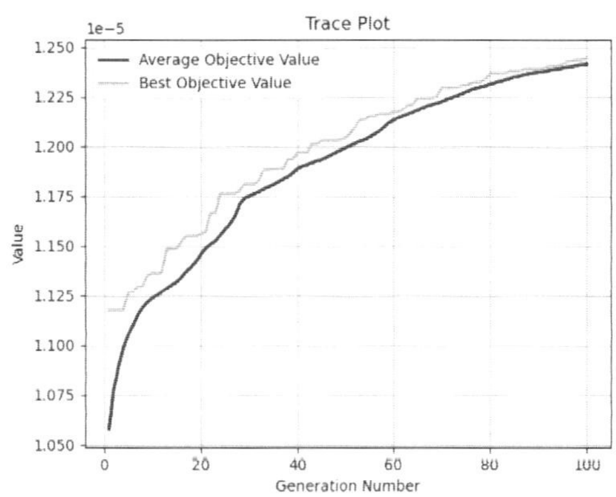

Figure 3. The local process of optimal value iteration for Scenario 2.

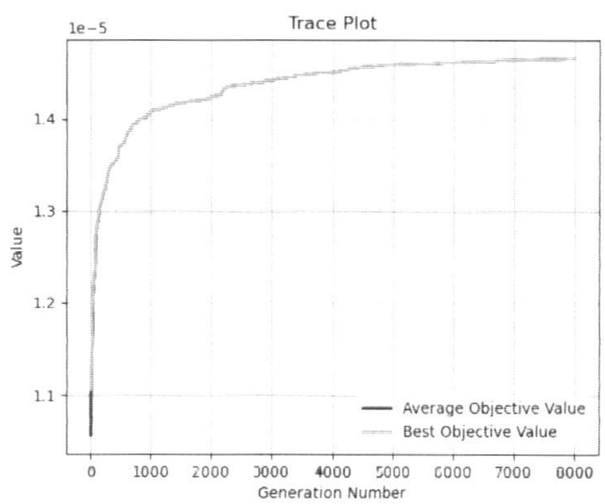

Figure 4. The global process of optimal value iteration for Scenario 2.

Since the path of the UAV swarm is planned in an offline manner, the time required for the algorithm to iterate for the two scenarios is 140 seconds and 48 seconds, respectively. From the fitness curve of the algorithm, the algorithm can achieve the purpose of improving the solution quality and converging to a satisfactory solution in the specified time in the application of the two scenarios. In scenario 1, the total distance required for the UAV swarm to complete the formation transformation is 93675.96 meters, and the longest distance of a single UAV is 193.22 meters. In the second scenario, the total distance required for the UAV swarm to complete the formation transformation is 126094.10 meters, and the longest distance of a single UAV is 360.19 meters. The result obtained by the genetic algorithm using the enhanced elite retention strategy is the matching of the starting point and the ending point of the UAV swarm, which cannot be used as the final planned route. In practice, the UAV will collide when flying in a straight line in the way of inverted points. In order to avoid the collision, it is necessary to further plan the path of the UAV in detail.

3. Path Fine Planning Based on Heuristic Fixed Step Space Search Algorithm

In order to obtain the coordinates of the UAV flight path without collision every five seconds, we need to further plan the unmanned route. We will model the UAV's behavioral pattern, obstacle situation, and behavioral strategy selection, and propose a fixed step-size directional space search method for UAV to complete the fine planning of the UAV's path, in which UAV can find the position of the next point in the directional space at any angle according to a fixed step size.

3.1. Unmanned Aerial Vehicle Behavioral Model

The step size of UAV is fixed as $step$, and the next step direction is any angle. Therefore, the next space position of the UAV is on the hemispherical surface with the current position point as center of and the fixed step as the radius.

Compared with some traditional path search methods that also discretize the UAV flight process, such as the A* algorithm [13, 14], the method designed in this paper has an obvious advantage in that the UAV can move at any angle in space, so it is closer to reality and can be found better feasible solutions.

The behavioral model of the UAV is established through the following two steps:

(1) The direction of a spherical curved surface is determined: The direction of the spherical surface is the line $\overrightarrow{pg}$ connecting the current position point p and the target point g.

(2) A spherical curved surface is established: The current UAV position o is taken as the coordinate origin and $step$ is taken as the radius to establish a spherical coordinate system, that is:

$$\begin{cases} r & ,r = step \\ \theta & ,\theta \in (0,\pi) \\ \psi & ,\psi \in (0,\pi) \end{cases} \tag{8}$$

The movement of the point can be described as the transformation of θ, ψ angle. The coordinates (x, y, z) of the point on the hemispherical surface can be expressed as:

$$\begin{cases} x = x_0 + r\sin\theta\cos\psi \\ y = y_0 + r\sin\theta\sin\psi \\ z = z_0 + r\cos\theta \end{cases} \tag{9}$$

The UAV can choose the next point on the hemispherical surface with the current point as the center and the fixed $step$ as the radius. When there is no feasible optimization path point, the central point of the sphere (x, y, z) is also included in the next feasible action.

3.2. Unmanned Aerial Vehicle Obstacle Avoidance Model

As shown in figure 5, the UAV may collide with other UAVs when searching the directional space with a fixed step size. That is, the ball with the UAV's current position p as the center and the fixed $step$ as the radius intersects with the ball with the position p' as the center and the safe distance as the radius. At this time, there are two situations in which UAVs may collide:

Scene 1: The position of the next flight path point of the UAV in P position is inside the ball with the UAV in P' position as the center and the safe distance d_{safe} as the radius.

Scene 2: The route of the UAV from the current position to the next flight path point crosses the safe distance range of another UAV as shown in figure 6.

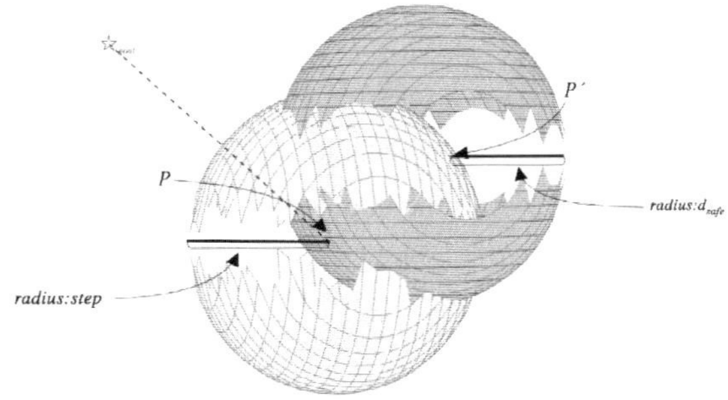

Figure 5. Schematic diagram of the possible collision of the UAV.

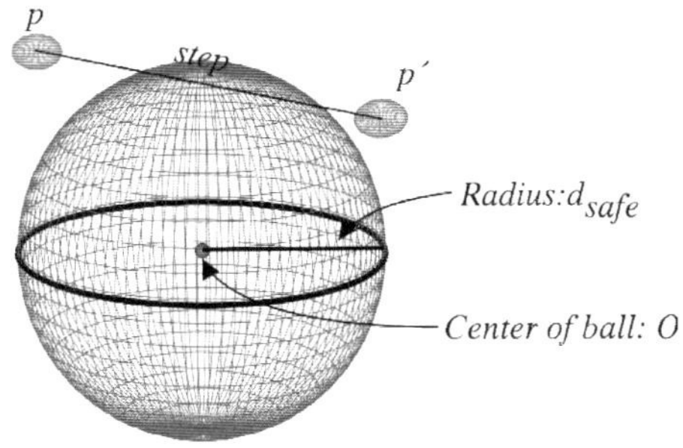

Figure 6. Schematic diagram of crossing other UAV's safe area.

For the first scenario, assuming that this is the choice of the k_{th} path point of the i UAV, we can use the following strategies to avoid the collision of UAVs:

$$P_{ik}\begin{cases} P_{ik} & ,\forall\rho\left(P_{ik},P_{jk}\right)>=d_{safe} \\ P_{ik}^{new} & ,\exists\rho\left(P_{ik},P_{jk}\right)<d_{safe} \\ P_{i(k-1)} & ,P_{ik}^{new}=\varnothing \end{cases} ,j\in\{0,1,2,...,i\} \tag{10}$$

In equation (10), $\rho\left(P_{ik},P_{jk}\right)$ represents the distance between two points of point P_{ik} and P_{jk}. The above formula indicates that the three behavioral options for the first collision scenario are to fly according to the original plan P_{ik}, re-plan the next path point P_{ik}^{new}, and wait in place for a time interval. For finding a new route point P_{ik}^{new} ,we will model and solve it in the following UAV route point selection model.

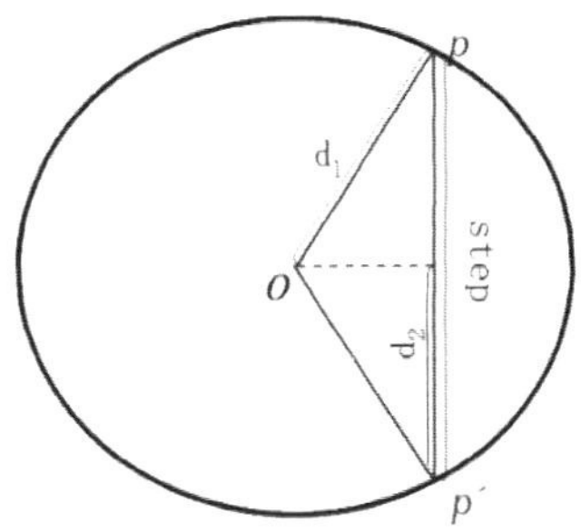

Figure 7. Two-dimensional schematic diagram of crossing other UAV's safe area.

The collision situation of the second type of UAV is simplified to a two-dimensional problem as shown in figure 7.

The UAV to be planned is in p position, its next path point is in p' position, and the distance between the two points is one step, that is $step$. The planned UAV is at point O, and its distance from point p and point p' is d_1. It is easy to know the situation depicted in figure 7, that is $|pp'| = step$, the flight path of one step of UAV is the closest to another UAV.

$$\begin{cases} (d_1 - d_2)^2 = d_{min}^2 \\ (d_1 - step + d_2)^2 = d_{min}^2 \end{cases} \tag{11}$$

In equation (11), d_{min} is the shortest distance in scenario 2. In this paper, the off-line route planning of the UAV is to plan the path of the UAV in advance, so the actual collision distance can be the radius length of the UAV. According to the actual situation, the radius of UAVs for lighting shows is generally less than 0.5 m, while the minimum safe distance of unmanned aerial vehicles is 1.5 m. We think that $d_{min} \geq \sqrt{3}d_{safe}$ can ensure that there will be no collision between UAVs.

3.3. Path Point Selection Model of UAV

The situation that the UAV encounters obstacles in flight can be divided into three situations: no obstacle, one UAV hindering, and multiple UAVs hindering.

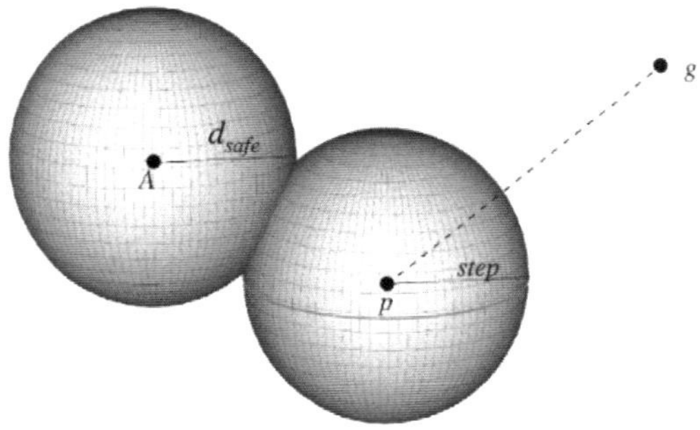

Figure 8. Barrier-free obstruction.

For the case where there are no other UAVs obstructed as shown in figure 8, due to the straight line between the two points, the shortest UAV's next path point should be the line connecting the point $p_{i(k-1)}$ and the point g_i with the sphere intersection. As the following formula:

$$p_{ik} = p_{i(k-1)} + step * \overrightarrow{p_{i(k-1)}g_i} \tag{12}$$

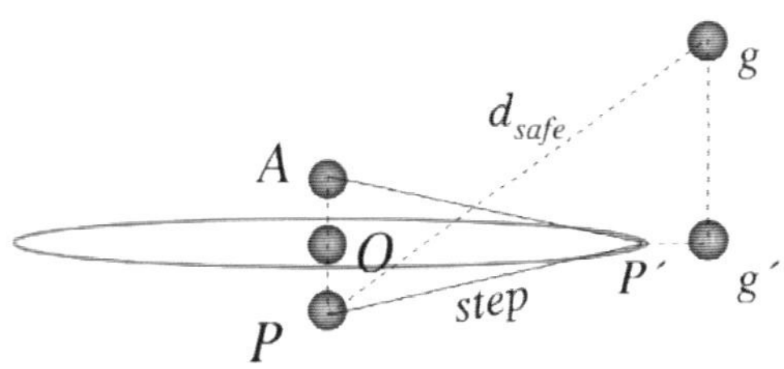

Figure 9. Single handicap situation.

For the second situation, because of the monotonic continuity of the distance from the point p', we know that the point should exist in the circle formed by the intersection line of two balls. For the convenience of discussion, we take out the circle formed by the intersection of two balls, as shown in figure 9.

Then the problem turns into finding a point from the circle, so that the point is closest to the target point. It is easy to know as the projection point g' from the g point to the plane where the circle is located, and the intersection point p' between the line connecting the center of the circle and the projection point and the circle is the desired point. The formulas are as follows:

$$\sqrt{step^2 - r^2} + \sqrt{d_{safe}^2 - r^2} = \left|\overrightarrow{pA}\right| \tag{13}$$

$$\vec{o} = \vec{p} + \sqrt{step^2 - r^2}\,\frac{\overrightarrow{pA}}{\left|\overrightarrow{pA}\right|} \tag{14}$$

$$\vec{p'} = \vec{o} + r * \frac{\left(\overrightarrow{pg} - \frac{\overrightarrow{pg} \bullet \overrightarrow{pA}}{\left|\overrightarrow{pA}\right|} * \frac{\overrightarrow{pA}}{\left|\overrightarrow{pA}\right|}\right)}{\left(\left|\overrightarrow{pg} - \frac{\overrightarrow{pg} \bullet \overrightarrow{pA}}{\left|\overrightarrow{pA}\right|} * \frac{\overrightarrow{pA}}{\left|\overrightarrow{pA}\right|}\right|\right)} \tag{15}$$

According to the above equations, the spatial coordinates of the next optimal path point in this scene can be obtained. For the case of being blocked by multiple UAVs at the same time, this paper only considers the case of being blocked by two UAVs at the same time as shown in figure 10.

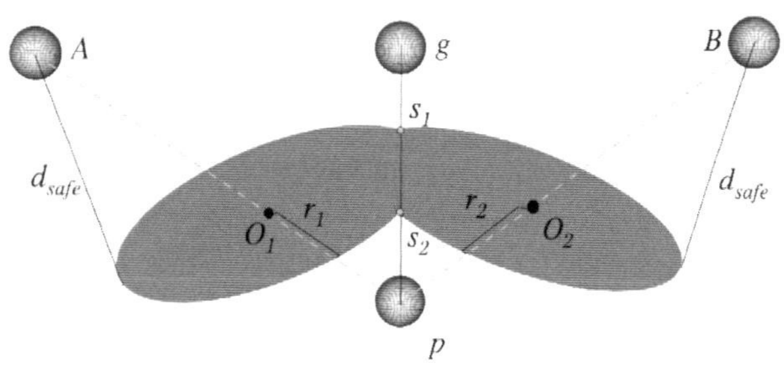

Figure 10. Blocked by two UAVs.

The point p is the current position of the UAV to be planned, the points A and B is the position of the other two UAVs, the point g is the target point, and the points s_1 and s_2 are the intersection of the two circles. Through the analysis of the second case, it can be seen that one of the two points s_1 and s_2 is optimal solution. The solution of the optimal solution is similar to the second case.

3.4. UAV Target Exchange Model

UAVs may encounter the situation shown in figure 11 in flight. Points P and A in the figure are the positions of two UAVs and points g_1 and g_2 represent the end points .

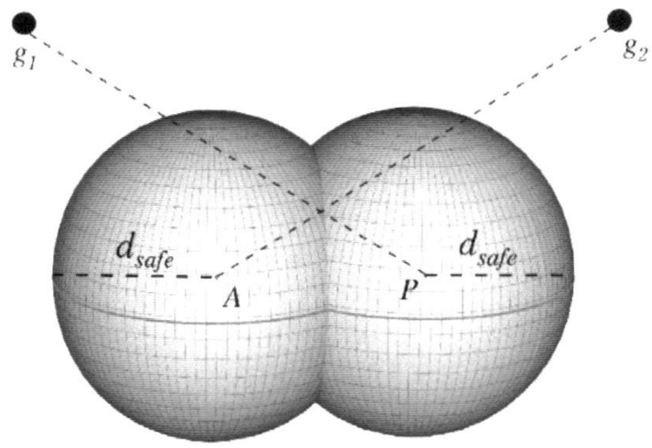

Figure 11. Circumstances suitable for swapping targets.

Obviously, if UAVs can swap each other's targets, better path can be obtained without detouring. There are two reasons for this situation: First, the solution method of the optimization algorithm used in the rough planning stage only obtains a satisfactory solution. The second is that the rough planning stage does not consider the situation of the UAV avoiding obstacles, so there will be different results. We designed the target exchange model of UAV, and dynamically adjusted the target of UAV to get a better route.

$$\begin{cases} g_i^{'} = g_j \\ g_j^{'} = g_i \end{cases} \quad \max\{\overrightarrow{p_i g_i}, \overrightarrow{p_j g_j}\} - step > \max\{\overrightarrow{p_i g_j}, \overrightarrow{p_j g_i}\} \tag{16}$$

In equation (16), $g_i^{'}$ represents the new destination after the UAV i exchanges target points. The condition of this formula represents the condition for the UAV i and the UAV j to exchange target points with each other. After the two UAVs exchange target points, the maximum distance between the two UAVs from the new end point is less than the distance from the old end point by one step.

3.5. Heuristic Fixed Step Space Search Algorithm and Solution Results

We designed a heuristic fixed step space search algorithm according to the established model. The algorithm can search the problem space from any angle in space and adopt the behavioral pattern of the model established above. The algorithm flow chart is shown in figure 12.

The matching result of the starting point and the ending point obtained in the rough planning stage is used as the input of the algorithm. The algorithm solves the UAV cluster path point coordinates, and the spatial scatter diagram of each path point is shown in figures 13-15.

The blue dots in the above figure represent the trajectory points of UAVs, and the endpoints are marked with red or black. It can be clearly seen that the UAV swarms spread out radially from the initial formation, and then reach the target point according to the originally planned path point. For the UAV swarm in Scenario 1, it took 65 seconds to complete the formation transformation task, while the UAV swarm in Scenario 2 took 121 seconds to complete the formation transformation task. Comparing with the roughest planned farthest path, it can be seen that the fixed-step algorithm can plan a better route while avoiding the collision of the UAV.

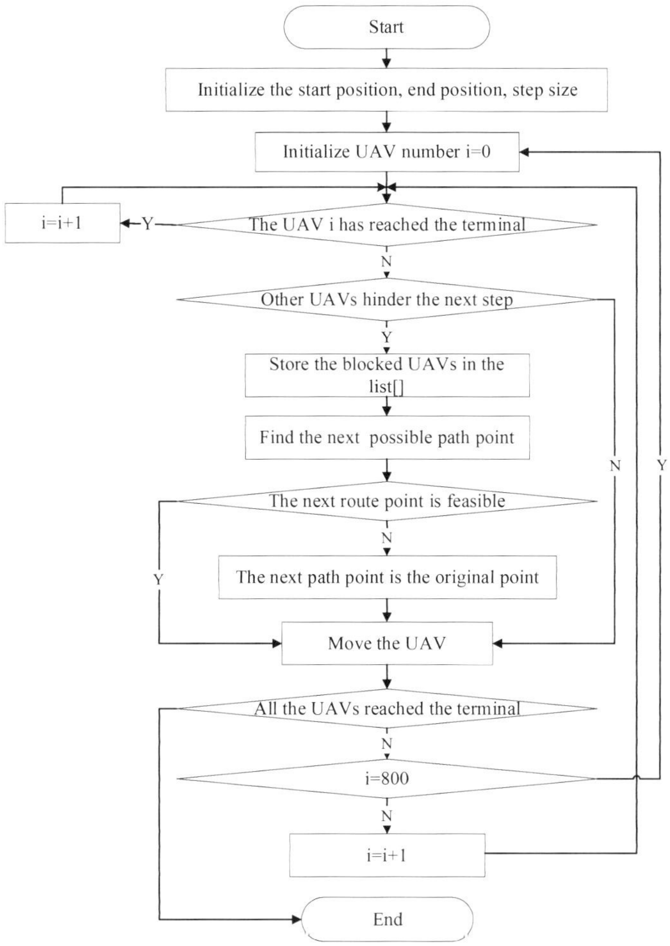

Figure 12. Heuristic fixed step space search algorithm flow chart.

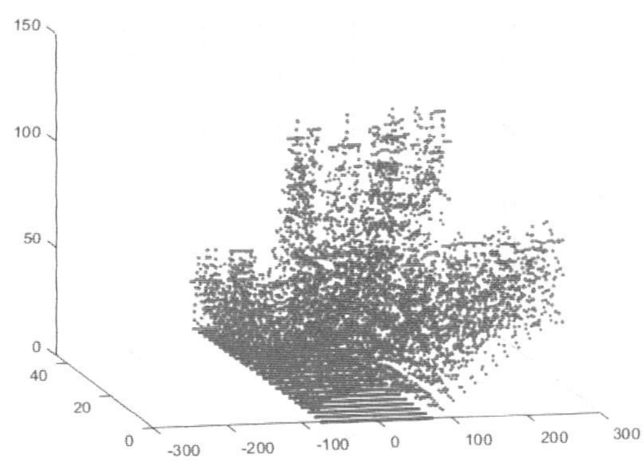

Figure 13. UAV swarm formation transformation path point in Scenario 1.

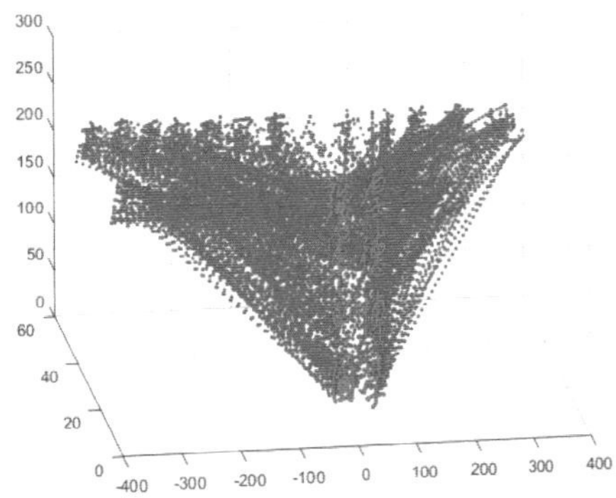

Figure 14. UAV swarm formation transformation path point in Scenario 2. (Front view)

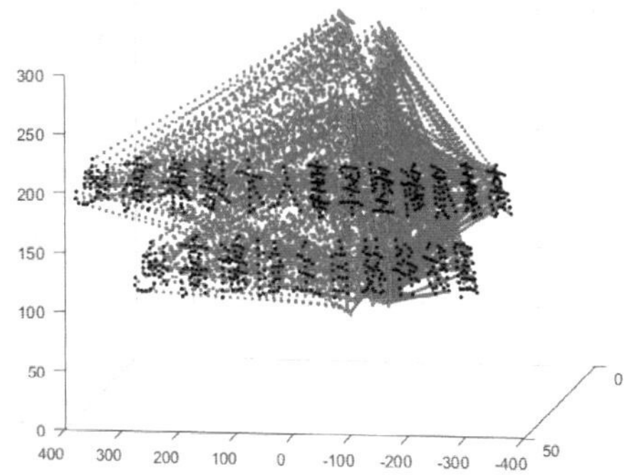

Figure 15. UAV swarm formation transformation path point in Scenario 2. (Back view)

4. Conclusion

In this paper, the model is built and the algorithm is solved by the single problem given in the title, and the results and conclusions for each problem are given.

For the path planning of UAV formation transformation, a 0-1 variable planning model is established and a genetic and particle swarm optimization algorithm is designed to preliminarily plan the path, which matches the starting point and the ending point. In the stage of fine planning, we set up four models to describe the behavioral patterns of unmanned aerial vehicles, including obstacle avoidance criterion, selection of optimal path points, and target exchange conditions of unmanned aerial vehicles, and designed a space search heuristic algorithm with fixed step size. In this paper, the formation transformation of UAV is divided into two stages: rough planning and fine planning. According to the characteristics of the algorithm, genetic algorithm and particle swarm optimization algorithm are combined, which greatly reduces the scale of the problem and can efficiently search the optimal flight path. In general, this paper has the following advantages:

- According to the characteristics of the problem, this paper divides the formation transformation of the UAV formation into two stages: rough planning and fine planning, which simplifies the problem and ensures the quality of the problem solving.
- In the rough planning stage, the genetic algorithm with enhanced strategy can be used to reduce the search space and improve the solution efficiency while ensuring the solution quality.
- In the fine planning stage, models such as UAV obstacle avoidance are established and a heuristic fixed step space search algorithm is designed to obtain a high-quality UAV formation transformation scheme.

The next improvement direction of this paper is as follows:

- In this paper, the unmanned aerial vehicle (UAV) does not consider the possibility of automatic obstacle avoidance mechanism in the path planning process, and only performs offline path planning for the UAV. How to endow UAV swarm intelligence with reinforcement learning is the direction of improvement.
- In this paper, the fixed step method is used for space search, which results in that some points in the space cannot be searched, which causes a certain waste of space search efficiency. How to further optimize the search algorithm is the direction of improvement.

References

[1] Zhao L C 2020 Research on Formation Transformation Method of Large Batch UAV (Yichang: Three Gorges University) Master's Thesis.

[2] Shenzhen Zhonglianxun Technology Co., Ltd. 2020 A UAV formation light show method CN201910978820.0.
[3] Zhu Z X and Yang X R 2021 Research on multi-UAV formation performance simulation based on Blender *22nd China System Simulation Technology and Application Annual Conf.*
[4] Wang F, Wu J and Zhou G Q 2015 Research on the development of multi rotor aircraft *Science and Technology Horizon* **13.**
[5] Wu L Y, Han W, Zhang Y and Xiong Y 2020 Formation transformation strategy of manned/UAV formation based on man-machine cooperation *Systems Engineering and Electronic Technology* **42** (02) 434-444.
[6] Gao Y 2020 Route Planning Algorithm for UAV Formation Transformation Based on Voronoi Diagram (Xi'an: Xidian University) Master's Thesis.
[7] Gu W, Tang J, Bai L and Lao S Y 2019 Time synergistic optimal efficiency model for formation transformation of multiple UAVs *Acta Aeronautica et Astronautica Sinica* **40** (6) 322599.
[8] Liu X, Zhou C P, Yu Q, Shao S and Ding M Y 2010 3D fast route planning based on hierarchical strategy *Journal of Astronautics* **31** (11) 2524-2529.
[9] Ghannami A 2021 Diversity and Initialization; Genetic Algorithm (Beijing: University of science and technology of China) Master's Thesis.
[10] Guo J, Zhao P and Ni J 2020 Fire station location planning model based on genetic algorithm *Journal of Computer Applications* **40** (S1) 41-44.
[11] Zhang C Q, Zheng J G and Qian J 2011 Comparison of genetic algorithm coding schemes *Computer Application Research* **28** (3) 819-822.
[12] Zhang C and Zhan Z H 2009 Comparisons of selection strategy in genetic algorithm *Computer Engineering and Design* **30** (23).
[13] Chen C 2020 Research and Application of Path Planning Based on Improved A-star Algorithm (Nanjing University) Master's Thesis.
[14] Taizhi L V and Zhao C X 2007 Path planning based on improved probabilistic cell decomposition *Computer Engineering* (21).

Analysis and Optimization of Planar Excitation Coil for EPS Torque Senor Based on GWO

Zhengyu Xie[1], Xiliang Chen[1,2*] and Yiwen Sun[1]

[1] School of Mechanical Engineering and Automation, Zhejiang Sci-Tech University, Hangzhou 310018, Zhejiang, China

[2] School of Aeronautical Engineering, Taizhou University, Taizhou 318000, Zhejiang, China

Email: 1067110849@qq.com; 110934982@qq.com; 837887616@qq.com

Abstract. Planar coil is usually used as a basic element in various research fields. This paper models a sensor equivalent circuit by taking the planar spiral excitation coil as an optimization object according to the electromagnetic coupling principle. The influence of coil electrical parameters on the coupling efficiency is analysed. The present result shows that the excitation coil should be small size, large inductance and high-quality factor. A two-dimensional planar spiral coil is modelled using ANSYS Maxwell software. The coil structure parameters with significant influence on the coil quality factor are chosen by the single factor experiment. The two-factor, five-level central composite deign principle is selected to establish a multivariate quadratic model by Response Surface Methodology (RSM). Then the paper implemented Grey Wolf Optimizer (GWO) to optimize response surface model, and obtained the optimal coil structure parameters. The conclusion was drawn that the inductance and quality factor of the optimized coil increase from 27.031 μH and 161.293 to 43.171 μH and 175.537, respectively. Compared to the control coil, the inductance is 1.597 times as the original coil and the quality factor is increased by 8.8%.

Keywords. Torque sensor; planar spiral excitation coil; response surface methodology; grey wolf optimizer; ANSYS maxwell.

1. Introduction

In electric power steering (EPS) system, torque sensors dynamically monitor the input torque of the steering shaft and input the signal to the ECU [1]. Due to space constraints, the excitation coil, as an important component of the torque sensor, should meet the design requirements of large inductance, small size, low-cost, high-quality factor, easy integration, and mass production. At present, the eddy current effect and the electromagnetic induction coupling principle are widely used in measurement systems [2, 3]. However, the excitation coil is less analysed and further research is needed. compared with the energized solenoid coil, The planar spiral coil has low mass, high volume efficiency, and good mechanical stability [4]. In addition, since planar coils can be printed on traditional circuit boards using flexible materials, they can be produced in high precision, low cost, repeatable, and predictable batches [5]. Planar coil units have been playing an irreplaceable role in wireless power transmission, non-destructive testing, displacement sensors, angular point sensor and planar transformers, etc. [6-9].

In this paper, a planar spiral coil is used as the excitation coil of the electromagnetic torque sensor, and a theoretical model of the sensor equivalent circuit is established according to the electromagnetic coupling principle to analyse the influence of the electrical parameters of the coil on the electromagnetic induction coupling efficiency. The optimal quadratic regression model of the coil quality factor is obtained by the single factor experiment and the central composite design, and the coil structure parameters by the GWO algorithm to improve coil quality factor.

2. Structure, Coil Inductive Coupling Efficiency and Coil Parameters

2.1. Structure

The structure of the electromagnetic torque sensor is shown in figure 1. The planar spiral excitation coil and signal processing unit are arranged on a circular printed circuit board (PCB), and the receiver coils are arranged on a flexible circuit board (FPC) with a forward and reverse wound 9-pair diamond structure.

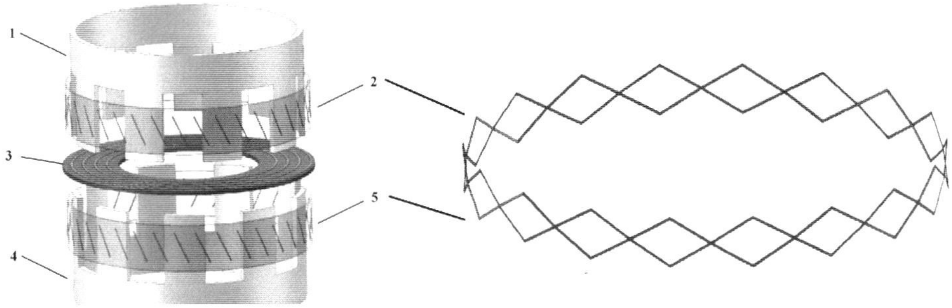

(a) (b)

Figure 1. Schematic of the sensor structure. (1): Rotor; (2): Receiving coil; (3): Planar spiral excitation coil; (4): Rotor; (5): Receiving coil.

2.2. Coil Inductive Coupling Efficiency

The inductively coupled equivalent circuit model is shown in figure 2a. The equivalent resistances of the excitation coil and the receiving coil are R_1 and R_2. Under the action of the high-frequency excitation alternating voltage V_{in}, current I_1 and high-frequency output alternating voltage V_{ece} at frequency f, the equivalent self-inductance generated by the current interaction between the adjacent conductors in the excitation coil and the receiving coil is $L_1$and L_2. In addition, due to the ring structure of the two coils, the equivalent capacitances generated are C_1 and C_2 Therefore, both coils can be considered as RLC high frequency oscillation circuits. Similarly, the rotor can also be considered as an RLC circuit due to the eddy current effect, and the equivalent resistance, inductance, and capacitance are R_3, L_3, and C_3, respectively. According to the mutual inductance principle, the mutual inductance coefficients between the excitation coil, the receiving coil and the rotor are M_{12}, M_{13} and M_{23}, denoted as M_{ab}; the coupling coefficients are k_{12}, k_{13} and k_{23}, denoted as k_{ab}. These equations are as follows.

$$M_{ab} = k_{ab}(L_a L_b)^{1/2} \tag{1}$$

$$\begin{cases} \dot{I}_1 R_1 + \dot{I}_1 j\omega L_1 - \dot{I}_3 j\omega M_{13} = V_{in} \\ \dot{I}_3 R_3 + \dot{I}_3 j\omega L_3 - \dot{I}_1 j\omega M_{13} = 0 \\ \dot{I}_1 j\omega M_{12} + \dot{I}_3 j\omega M_{23} = V_{ece} \end{cases} \tag{2}$$

When the circuit produces resonance, the power supply angular frequency is $\omega = (LC)^{1/2} = 2\pi f$, and the equivalent model circuit is resistive. Combining with equation (1), the excitation coil input power $P_{in}$and receiver coil output voltage V_{ece} can be obtained, as shown in equation (3).

$$P_{in} = \frac{V_{ece}^2}{R_1} = \left[\frac{V_{in}(R_3 j\omega M_{12} - \omega^2 M_{13} M_{23})}{[R_1 R_3 + (\omega M_{13})^2]}\right]^2 / R_1 = \frac{R_3 U_{in}^2}{R_1 R_3 + (\omega M_{13})^2} \tag{3}$$

M_{23} varies with the rotation of the rotor and, as shown in equation (3), the rotor angular displacement is output in the form of the receiving coil output voltage. Take a pair of diamond-shaped loops in the receiving coil to establish a differential measurement circuit, as in figure 2b, the output voltage V_{out} characterizes the direction and size of the rotor angular displacement, take the rotor is located in the position shown for analysis, set $x = R_1/L_1$, circuit output power P_{out}, coupling efficiency η as shown in equations (4) and (5).

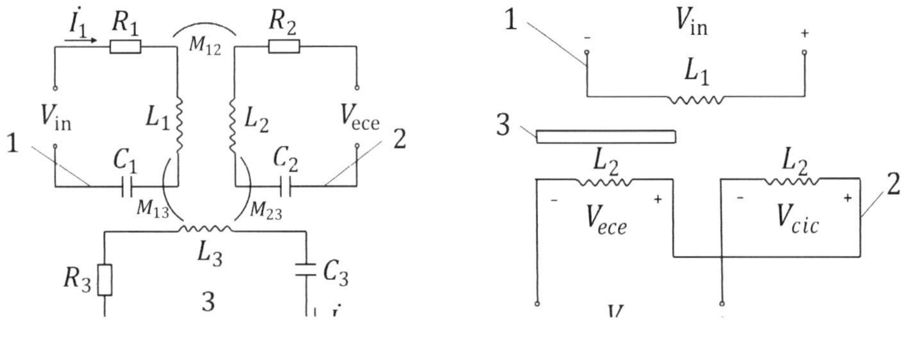

Figure 2. Equivalent circuit model of sensor. (1): Planar spiral excitation coil; (2): Receiving coil; (3): Rotor.

$$V_{out} = V_{ece} - V_{cic} = \frac{V_{in}\omega^2 M_{13} M_{23}}{[R_1 R_3 + (\omega M_{13})^2]} \tag{4}$$

$$\eta(x) = \frac{P_{out}}{P_{in}} = \frac{\left(\omega^2 k_{13} k_{23} L_3 \sqrt{L_2}\right)^2}{2R_2 R_3\left[R_3 x + \left(\omega k_{13}\sqrt{L_3}\right)^2\right]} \tag{5}$$

From equation (5), it can be obtained that increasing the inductance L_1 and reducing the resistance R_1 can make the coupling efficiency η increase, which is consistent with the conclusion in [10].

2.3. Coil parameters

The structural parameters of the planar spiral coil include the number of turns N, turn width w, turn spacing S, thickness h, inner radius r_{in} , outer radius r_{out}, and ratio of inner and outer diameter i, As shown in figure 3. Coil inductance L, resistance R, and coil quality factor Q are shown in equations (6)-(8).

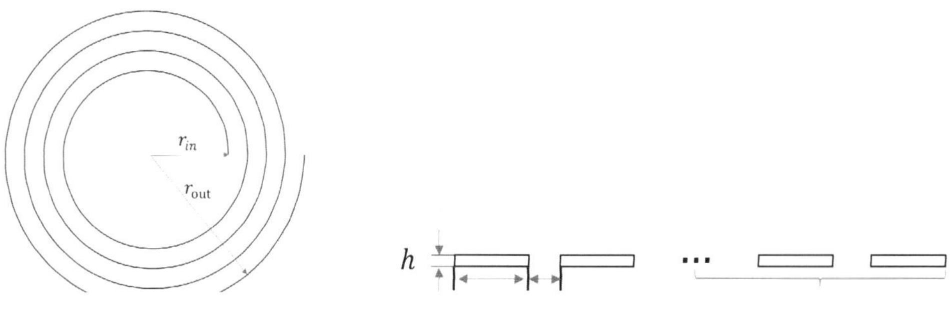

Figure 3. The model of planar spiral excitation coil structure.

$$i = \frac{r_{in}}{r_{out}} = \frac{r_{in}}{r_{in}+N(w+s)/2} \tag{6}$$

$$L = \frac{\mu_0 N^2 (r_{in}+r_{out})}{2} ln\left[2.46\frac{(1+i)}{(1-i)}\right] + 0.2\frac{(1-i)}{(1+i)} \tag{7}$$

$$Q = \frac{\omega L}{R} = \frac{\mu_0 \sigma \pi f w h r_{out}}{w+s}(1-i)ln\left[2.46(\frac{1+i}{1-i}) + 0.2(\frac{1-i}{1+i})\right] \tag{8}$$

where, μ_0 is the air permeability, $4\pi \times 10^{-7}H/m$; σ is the reciprocal of the resistivity ρ; l is the total length of the coil, m; When w is a constant value, it can be seen from equation (8) that the main influencing factor of the coil Q value is the coil structure parameters, so the influence of each structure parameter on the electrical parameters should be considered comprehensively in the coil design, combined with the conclusion of Section 2.2, when the larger the coil inductance and the smaller the resistance, the larger the coil Q value and the higher the coupling efficiency, so the coil design should meet the large inductance, small resistance and high quality factor.

3. Simulation and Grey Wolf Optimizer

3.1. Simulation of Coil

Due to the axisymmetric structure of the coil, a two-dimensional planar spiral coil model is established in the Eddy current solution mode of ANSYS Maxwell. The excitation current is set at 1000 mA with a frequency of 2 MHz, and the adaptive mesh is used to consider the skin effect. In the single factor experiment, the influence of each coil structure parameter on quality factor were analyzed, some structural parameters are selected for central composite design, and a multivariate quadratic response surface model is established.

3.1.1. Single Factor Experiment. According to equation (8), the single factor experiment selects the ratio of inner and outer diameters i, thickness h, turn width w, and turns spacing s as influencing factors. The relationship between coil structure parameter and quality factor is shown in figure 4.

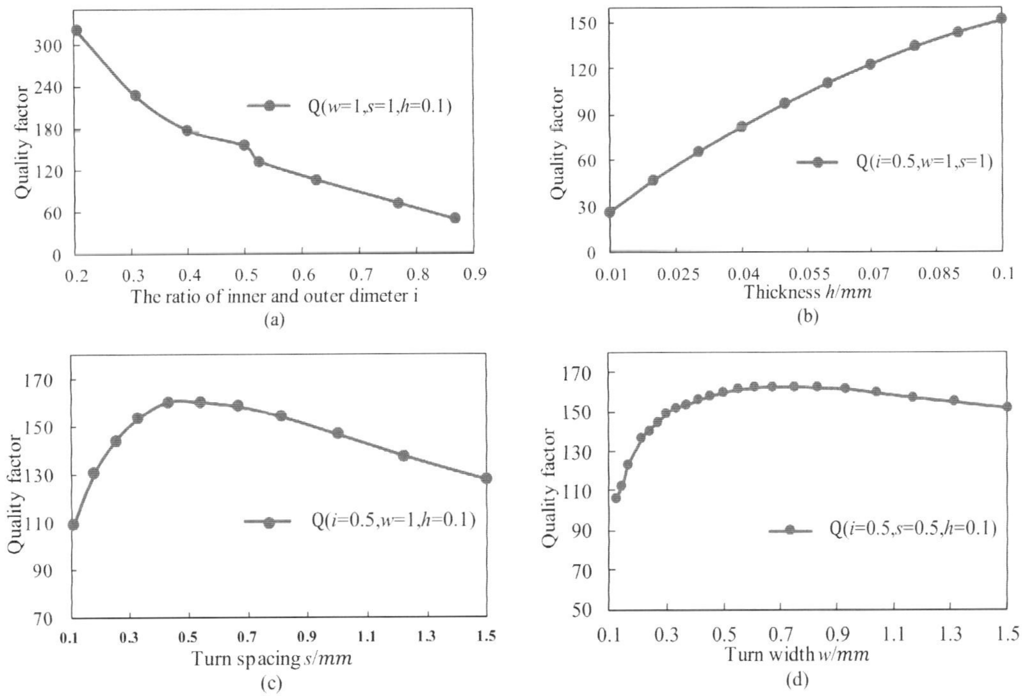

Figure 4. The result of single factor experiment.

In particular, considering the design requirement of small sensor size, the ratio of inner and outer diameter i should be taken as 0.5 according to the conclusion in Reference [11]. According to the single factor experimental results, the optimal structural parameters are $i = 0.5$, $h = 0.1mm$, $s = 0.5mm$, $w = 0.6\ mm$, and $r_{in} = 20\ mm$, numbered C-2. From figures 4c and 4d, it can be seen that the turn spacing and turn width have a significant nonlinear relationship with the coil quality factor, so these two structural parameters are selected as the influence factors in response surface design.

3.1.2. Response Surface Modelling. In the RSM, the independent variables are associated with target response y as in equation (9). Two factors, turn spacing $s(\mathrm{x}_1)$ and turn width $w(\mathrm{x}_2)$, which have a significant impact on the coil quality factor are selected by using the central combination circumscribed design principle for the two-factor, five-level response surface modelling.

The experimental design is shown in table 1, and the experimental results are shown in table 2. The multivariate quadratic model obtained by Minitab software is shown in equation (10), and the variance regression analysis is shown in table 3.

$$y = f(x_1, x_2, x_3, \dots, x_n) + \varepsilon \tag{9}$$

$$y = 142 + 88.4x_1 + 75.7x_2 - 296.7{x_1}^2 - 144.6{x_2}^2 + 210x_1x_2 \tag{10}$$

where: $x_1, x_2, x_3, \dots, x_n$ are independent variables, ε is the error of the target response y.

Table 1. Variables and levels in the central composite circumscribed design.

Variable	Level				
	-1.414	-1	0	1	1.414
Turn spacing (mm)	0.1670	0.2304	0.3835	0.5366	0.6000
Turn width (mm)	0.3000	0.3659	0.5250	0.6841	0.7500

Table 2. The result of the central composite circumscribed design.

Serial number	Turn spacing (mm)	Turn width (mm)	Quality factor
1	0.3835	0.5250	172.071
2	0.3835	0.5250	175.072
3	0.2304	0.6841	164.419
4	0.2304	0.3659	172.450
5	0.5366	0.3659	152.925
6	0.3835	0.5250	176.071
7	0.5366	0.6841	165.362
8	0.1670	0.5250	166.600
9	0.6000	0.5250	154.732
10	0.3835	0.3000	166.794
11	0.3835	0.7500	167.712

As can be seen from the variance analysis in table 3, the linear and quadratic terms of the response surface model are extremely significant, indicating that the influence of each factor on the response values is not a simple linear relationship. The interaction term is extremely significant and shows that the interaction between the factors is good and the whole response surface is composed based on the interaction between the factors. In addition, the total regression term is extremely significant, which proves that the regression equation fits well, and indicates that the experiment has a high reliability and accuracy. Finally, the lack of fit is extremely insignificant, which indicates that the error of the experiment is narrow.

Table 3. Variance analysis for each item of the multivariate quadratic model.

Variance source	Degree of freedom	Adj SS	Adj MS	*F* value	*P* value	Significance
Linear term	2	160.408	80.204	31.22	0.004	**
Quadratic term	1	236.549	118.274	46.04	0.002	**
Interaction term	1	104.738	104.738	40.77	0.003	**
Total regression	6	550.502	91.75	35.71	0.002	**
Lack of fit	2	1.612	0.806	0.19	0.843	
Error	4	10.276	2.569			

Note: *: Significant (*P* value< 0.05); **: Extremely significant (*P* value< 0.01).

3.2. Grey Wolf Optimizer

GWO is an optimization algorithm based on the hunting action of the gray wolf population, with adaptive regulation and information feedback mechanism with convergence factor. The global search ability is more balanced with the local search ability, The principle of the algorithm is shown in equations (11)-(14) [12]. The algorithm is used to analyze the response surface model. The resulting parameter space is shown in figure 5a, the target space is shown in figure 5b.

$$X = (X^1, X^2, \dots, X^n),\ X_p = ({X_p}^1, {X_p}^2, \dots, {X_p}^D),\ X_i = ({X_i}^1, {X_i}^2, \dots, {X_p}^D) \tag{11}$$

$${X_{ip}}^d = {X_p}^d - A \cdot \left|C \cdot {X_p}^d - {X_i}^d\right| \tag{12}$$

$$A = rand(-a, a),\ C = rand\{0,2\} \tag{13}$$

$${X_i}^{t+1} = \frac{{X_{i\alpha}}^t + {X_{i\beta}}^t + {X_{i\delta}}^t}{3} \tag{14}$$

where, X is the location of all individual gray wolf populations, X_p is the location of target gray wolf, X_i is the current gray wolf location. t is the number of generations of the current iteration, α is the fittest solution, β and δ are the second and third best solutions. a is a uniform random number, and the value from $-a$ to a, a is a constant with an initial value of 2, which decreases linearly from 2 to 0 as the algorithm iterates. C is a random number with a value of 0 or 2.

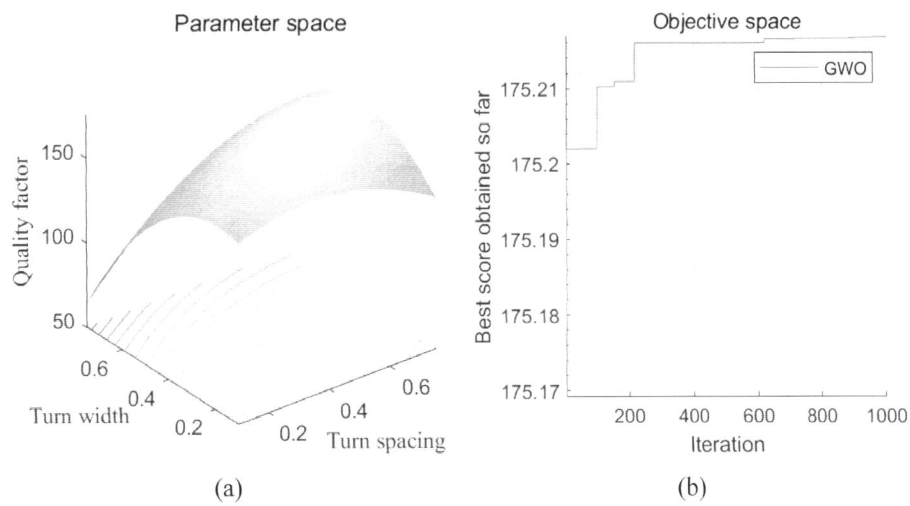

Figure 5. The parameter space and the objective space of RSM based on GWO.

Setting the number of populations as 50 and the maximum number of iterations as 1000, the optimal structure of the coil with $i = 0.5$, $h = 0.1\ mm$, $s = 0.3251\ mm$, $w = 0.4978\ mm$, *and* $r_{in} = 20\ mm$, numbered as C-1, was obtained from GWO, and the response surface model was optimized for solution y=175.2168. In order to verify the accuracy of the regression model and the superiority of the optimized coil structure, the coil C-2 was selected for the control experiment, and simulation results are shown in table 4. It can be seen that the actual simulation results of the coil C-1 quality factor and the prediction results of response surface model are 175.537 and 175.2168, respectively, with minimal errors, which further proves the reliability and accuracy of the response surface model. The inductance and quality factor of optimized coil C-1 increased from 27.031 μH and 161.293 to 43.171 μH and 175.537, respectively. Compared with the control coil, the inductance is 1.597 times as the original coil and the quality factor is increased by 8.8%.

Table 4. The electrical parameters of the coil model of the control coil.

Number	w/mm	s/mm	N	$L/\mu H$	R/ohm	Q
C-1	0.4978	0.3251	24	43.171	3.091	175.537
C-2	0.6	0.5	18	27.031	2.106	161.293

4. Conclusion

In this paper, a planar spiral coil is used as the excitation coil for the electromagnetic torque sensor. According to the principle of electromagnetic induction coupling, the model of the sensor equivalent circuit model is established, and analyse the influence of coil electrical parameters on the coupling efficiency. It is concluded that the coupling efficiency of sensor mutual inductance coil can be improved by increasing the quality factor, and the coil should meet the design requirements of small size, large inductance and high-quality factor.

Taking the coil quality factor as the optimization target, the single-factor experimental results show that the coil turn width and turn spacing have a significant effect on the quality factor, and the above structural parameters are selected for a two-factor, five-level central composite circumscribed design to obtain a multivariate quadratic response surface model. Finally, the GWO algorithm is used to optimize the response surface model, and the optimal structural parameters are $i = 0.5$, $h = 0.1\ mm$, $s = 0.32514\ mm$, $w = 0.49782\ mm$, and $r_{in} = 20\ mm$, which improved the inductance of the coil to 1.597 times as the original coil and improved the quality factor by 8.8% compared with the control coil, which satisfied the coil design requirements. This study has a certain research reference value.

References

[1] Jin L Q, Song C X and Hu C J 2009 Driving force power steering for the electric vehicles with motorized wheels *2009 IEEE Vehicle Power and Propulsion Conf.* pp 1518-1524.
[2] Zhang C, Li Z, Chen J, et al. 2021 Design and research of a novel non-contact vertical inductive torque sensor *Measurement* **177** (5) 109252.
[3] Zhao H 2015 A differential induction torque sensor and its finite element analysis *Journal of Sensing Technology* **28** (5) 654-659.
[4] Anil K, George B and Mukhopadhyay S C 2020 Technologies and applications of angle sensors: A review *IEEE Sensors Journal* 7195-7206.
[5] Faria A, Marques L, Ferreira C, et al. 2021 A fast and precise tool for multi-layer planar coil self-inductance calculation *Sensors* **21** (14) 4864.
[6] Zou X and He X 2020 Mutual inductance of coils for wireless power transfer *2020 Cross Strait Radio Science & Wireless Technology Conf.* (CS-RSWTC) pp 1-3.
[7] Anil Kumar A S and George B 2020 A noncontact angle sensor based on eddy current technique *IEEE Transactions on Instrumentation and Measurement* **69** (4) 1275-1283.

[8] Reddy B P, Mrali A and Shaga G 2017 Low cost planar coil structure for inductive sensors to measure absolute angular position *2017 2nd International Conference on Frontiers of Sensors Technologies* (ICFST) pp 14-18.
[9] Li B, Li Q and Lee F 2018 High-frequency PCB winding transformer with integrated inductors for a bi-directional resonant converter *IEEE Transactions on Power Electronics* pp 6123-6135.
[10] Zhang Z J, Zheng L F and Yang R 2021 Optimal design of limited-size wireless power transmission coil *Science, Technology and Engineering* **21** (9) 7.
[11] Wang H B 2015 Design Strategies of Eddy Current Sensors with Sub-nanometer Accurancy (Hefei: University of Science and Technology of China) Doctor's Thesis.
[12] Mirjalili S, Mirjalili S M and Lewis A 2014 Grey wolf optimizer *Advances in Engineering Software* **69** (3) 46-61.

Recognition of Power Quality Disturbance Based on RCNN

Rui Fan[1], Huipeng Li[1], Heju Xiao[2], Hong Wang[2] and Linhai Qi[2]

[1] State Grid Shanxi Electric Power Company Electric Power Research Institute, Taiyuan, Shanxi, China

[2] School of Control and Computer Engineering, North China Electric Power University, Beijing, China

Email: 531927670@qq.com; huipeng_li@163.com; xhj18811365829@163.com; neiep_wh@163.com; qilinhai@ncepu.edu.cn

Abstract. With the development of the smart grid, the problem of power quality disturbance is becoming more and more serious. To improve the classification ability of power quality disturbance signals, this paper designs a recurrent convolutional neural network model to realize the classification and identification of power quality disturbance signals. The model first uses the bidirectional long short-term memory neural network module to encode the disturbance signal. Then it uses the convolutional layer to perform spatial feature extraction and enforces the disturbance classification through the fully connected layer. The model is verified and analyzed by simulation data, which shows that the model has both good generalization ability and higher accuracy.

Keywords. Deep learning; power quality; disturbance classification; recurrent convolutional neural network.

1. Introduction

A new power system with new energy as the main body has gradually begun to be constructed, and various new high-precision power electronic instruments have been continuously put into use in the new power system. The power grid is becoming increasingly intelligent, and the power system environment is constantly changing and upgrading [1]. This further increases the complexity of the grid environment and the instability of the system. With the adjustment of the energy structure, new energy technologies play an increasingly important role, such as various distributed photovoltaic solar power generation, wind power generation, and so on. However, due to the instability of new energy power generation, the power quality is affected to a certain extent, resulting in a series of power quality problems such as voltage and frequency [2]. It is of great significance to effectively solve the negative impact of power quality disturbances, improve the governance ability for power quality problems, and enhance the ability to identify disturbance types accurately [3].

At present, researchers have proposed a variety of solutions for the identification of power quality disturbance types, mainly including two kinds. One is the traditional method of feature extraction based on artificial experience, which includes primarily two parts [4]. Usually, firstly, a feature extraction algorithm [5-9] is used to convert the disturbance signal mathematically. Corresponding features represent the different disturbance signals and construct the feature matrix. Then, the machine learning classification algorithm [10-12] is used for the extracted signal features to realize the classification and identification of disturbance signals. This method based on manual feature extraction has a complicated calculation process and difficult parameter selection. At the same time, due to many interference factors

 2363 (2022) 012024 doi:10.1088/1742-6596/2363/1/012024

in the actual environment, it is difficult to extract a complete feature expression disturbance signal, which affects the recognition accuracy in practical engineering applications. The other is to use deep learning algorithms based on data-driven [13], end-to-end training methods to optimize network parameters and complete the automatic extraction of disturbance signal features [14-16]. Deep learning algorithms are data-driven and have strong generalization capabilities. It can automatically and efficiently realize feature extraction, effectively solving the shortcomings of manual feature design in traditional methods. Reference [17] extracts disturbance signal features by training an autoencoder in an unsupervised manner and then implements feature classification through Convolutional Neural Networks (CNN). However, due to the step-by-step training method, the accumulated errors in different stages are difficult to eliminate. Reference [18] uses incomplete S-transform on the disturbance signal for feature extraction and constructs a three-layer deep feedforward network as a classifier to realize disturbance signal recognition. However, the feature extraction process adopts traditional methods, which are too complicated to take advantage of the data-driven advantages of deep learning. Reference [19] uses the principle of recursive graph to convert the one-dimensional disturbance signal into a two-dimensional recursive image, then realizes the classification of the disturbance signal through a recurrent neural network. However, the generation of the recursive graph trajectory image needs to select an appropriate threshold. Otherwise, it is difficult to present the characteristics of the disturbance signal.

This paper proposes an identification method of power quality disturbance signals using the Recurrent Convolutional Neural Network (RCNN). First, it uses a Bi-directional Long-Short Term Memory (Bi-LSTM) network to encode the disturbance signal and learn the time-series features of the disturbance signal. The bidirectional encoding method can also better capture bidirectional dependencies and reduce information loss. Then, spatial feature extraction is performed on the output hidden state matrix through CNN to realize the classification and identification of disturbance signals. At the same time, dropout regularization is used to improve the anti-interference ability and generalization of the model. Through experimental comparison and analysis, the model achieves not only high-precision identification of power quality disturbance signals but also has good performance in different noise environments, with good anti-noise ability and generalization.

2. Power Quality Disturbance Signal Analysis

The power quality disturbance phenomenon not only appears in the form of a single disturbance signal, but also often appears in the form of multiple disturbance signals. This paper lists the mathematical models and parameter selection ranges of standard power quality disturbance signals according to Reference [20]. These include Normal Signal (P1), Voltage Swell (P2), Voltage Sag (P3), Voltage Interruption (P4), Harmonic (P5), Oscillation Transient (P6), Pulse Transient (P7), Voltage Fluctuation (P8). The simulation platform simulates the above eight signals for model training and experimental verification. Figure 1 is a simulation effect diagram.

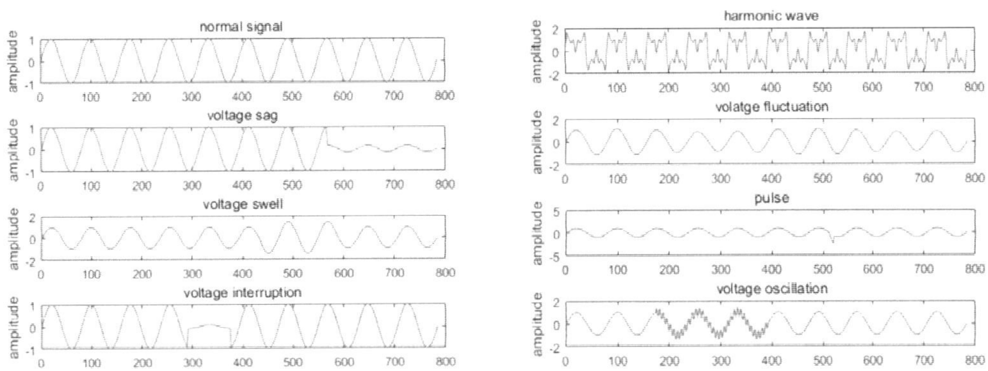

Figure 1. Disturbance signal simulation.

The system sets the sampling frequency of the simulated signal as 3200Hz, and samples ten cycles, with 640 sampling points in total. At the same time, the simulation signal is normalized to make the model focus on the disturbance trajectory and avoid the influence caused by the amplitude factor.

3. Recurrent Convolutional Neural Network

Figure 2 shows the structure of the RCNN model, and the RCNN model mainly consists of an input layer, a Bi-LSTM layer, a CNN layer, and an output layer. First, the disturbance signal is preprocessed at the input layer to perform dimension transformation. Then, the Bi-LSTM layer is used to realize the time-series coding of the disturbance signal and learn the time-series correlation of the disturbance signal. The Bi-LSTM layer is trained with a bidirectional structure to improve the coding effect and reduce the loss of information. The encoded bidirectional latent vectors are spliced, and the hidden state matrix is output for the classification and recognition of the CNN layer. This paper uses a multi-layer convolutional neural network in the CNN layer to extract features from the hidden state matrix encoded by the Bi-LSTM layer. The recognition ability of the network is enhanced through the fully connected layer. Finally, the Softmax layer is added to different output categories of disturbance signals to realize the feature classification of disturbance signals.

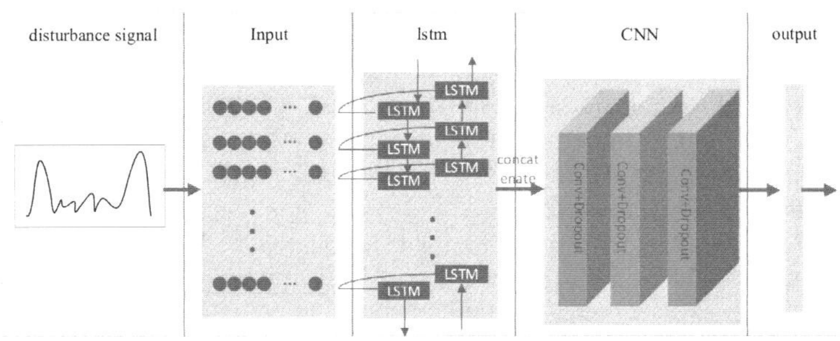

Figure 2. Recurrent convolutional neural network model.

3.1. Input Layer Modeling

In this paper, the length of the disturbance signal is 640, which is obtained from the simulation platform. First, the signal is preprocessed and periodically intercepted. The input matrix of the perturbation signal dimension is transformed to 32*20 as the input of the LSTM layer.

3.2. LSTM Layer Modeling

The input matrix is temporally encoded at the LSTM layer to extract the temporal features in the perturbed signal. It consists of a bidirectional chained structure of LSTM cells. It can not only encode time series signals, but also extract time series features. Moreover, the unique gate structure in LSTM can effectively solve the trouble of long-range dependencies and eliminate the impact of gradient explosion and gradient disappearance during training. Figure 3 shows the structure of the LSTM unit

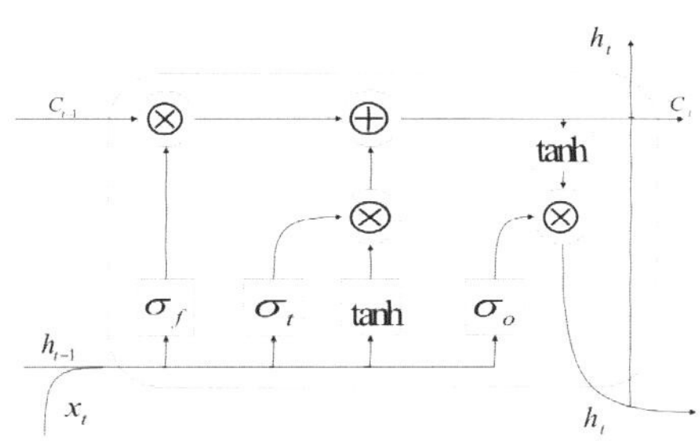

Figure 3. LSTM Unit structure.

The LSTM unit contains a variety of gate structures, and the preservation of historical information and long-term state is achieved through the gate control mechanism of these gates [21]. The sigmoid function and the dot product operation realize the gating mechanism. Its general expression is:

$$g(x)=\sigma(Wx+b) \tag{1}$$

In the formula, σ is the sigmoid function, which can nonlinearly activate the input value and map it to the 0-1 interval, indicating the degree of information retention.

Among them, the discarding degree of the information in the forget gate f_t control unit C_{t-1}

$$f_t=\sigma(W_f \bullet [h_{t-1};x_t]+b_f) \tag{2}$$

where f_t represents the forget gate, h_{t-1} represents the output value of the previous unit, x_t represents the current input value, and W_f and b_f represent the training weights and biases.

The input gate i_t controls how much information in the candidate cell state $\tilde{C}_t$ can be retained. Determine how much information can be added to cell C and update the cell.

$$i_t=\sigma(W_i \bullet [h_{t-1};x_t]+b_i) \tag{3}$$

$$\tilde{C}_t=\tanh(W_c \bullet [h_{t-1};x_t]+b_c) \tag{4}$$

$$C_t=f_t \odot C_{t-1}+i_t \odot \tilde{C}_t \tag{5}$$

where i_t represents the input gate, W_i is the weights of input gate, and b_i is biases of input gate. $\tilde{C}_t$ represents the candidate unit state, and W_c and b_c are weights and biases of it. C_t represents the state of the unit, $\tilde{C}_t$ and C_t are controlled by forget gate and input gate.

The output gate o_t decides how much information can be output from C_t, and the expression is as follows:

$$o_t=\sigma(W_o \bullet [h_{t-1};x_t]+b_o) \tag{6}$$

where o_t is the output gate, and W_o and b_o are the weights and biases of the output gate.

The coding effect of the LSTM layer will directly influence the overall recognition result of the model, in which the network structure and the number of neurons play a decisive role. To improve the coding effect of the disturbance signal in the LSTM layer and find the best network parameters, this paper has verified through many experiments, using the single-layer network structure and setting the number of neurons to 16. At this time, the increase in the number of network layers and hidden neurons cannot further improve the recognition rate of the model but will prolong the training time of the model and cause overfitting. A Bi-LSTM layer bi-directionally encodes the perturbed signal to prevent bias loss due to unidirectional encoding. The forward LSTM in the Bi-LSTM layer can extract the forward dependency in the perturbed signal, and the backward LSTM is used to extract the backward dependency in the perturbed signal. In this way, the bidirectional dependence information in the disturbance signal can be guaranteed, and the model coding effect can be improved. The encoded bidirectional latent vector is combined as the encoded output. The next step of classification and identification continues.

The network structure of the LSTM layer is shown in figure 4, where x_i is the input data, which is encoded by the LSTM neural unit. $\overrightarrow{h_i}$ and $\overleftarrow{h_i}$ are encoded by forwarding LSTM and backward LSTM, respectively. The hidden state h_i can be obtained by splicing the forward hidden vector $\overrightarrow{h_i}$ and the backward hidden vector $\overleftarrow{h_i}$, and the hidden state matrix can be obtained by splicing the hidden vectors of each time step. Afterward, the classification and recognition of the disturbance signal are realized through the convolutional neural network.

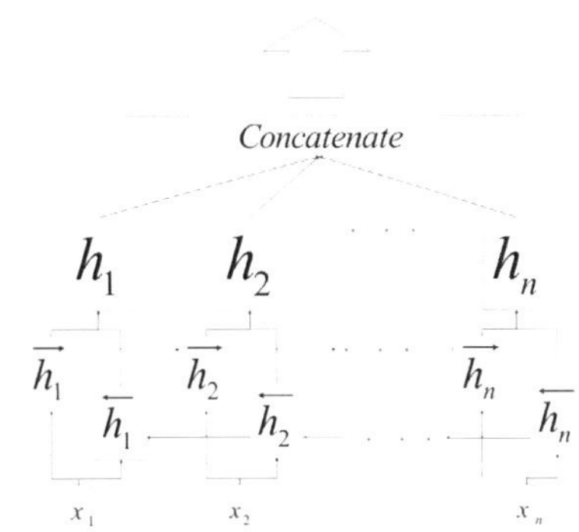

Figure 4. Bi-LSTM layer structure.

3.3. CNN Layer Modeling

The CNN reduces the amount of computation in the model training process through local connections and weight sharing, including convolutional layers, pooling layers, and fully connected layers [22]. The convolution layer is composed of several convolution kernels, and the input information is convolved by sliding the convolution kernel to extract the spatial features in the disturbance signal. The convolutional layer usually follows the pooling layer to extract the signal features twice, which reduces the sensitivity of the convolutional layer to the position.

This paper adopts a multi-layer convolutional layer in the CNN layer to extract the spatial features in the hidden state matrix. Setting the convolution stride to 2 in the convolution layer can encode the spatial correlation of the hidden state matrix and compress the spatial information. After three layers of convolution, the identification of the disturbance signal is finally realized. Considering the output dimension of the Bi-LSTM layer and the number of perturbed signal classifications, figure 5 shows the structure of the CNN layer.

The LSTM layer obtains the hidden state matrix by encoding the perturbed signal, and then three convolutional layers can extract the state matrix's spatial features. Before the convolution calculation, fill the features with the same padding, and figure 5 shows the output size after each convolution. Furthermore, the model uses the Dropout operation to prevent overfitting and improve network performance, set its ratio to 0.2. The input in the CNN layer structure is the 32*32 hidden state matrix output by the Bi-LSTM layer. After three feature extraction layers, the structure obtains 64 channels of size 4*4 as the feature map. Finally, it identifies the disturbance signal type.

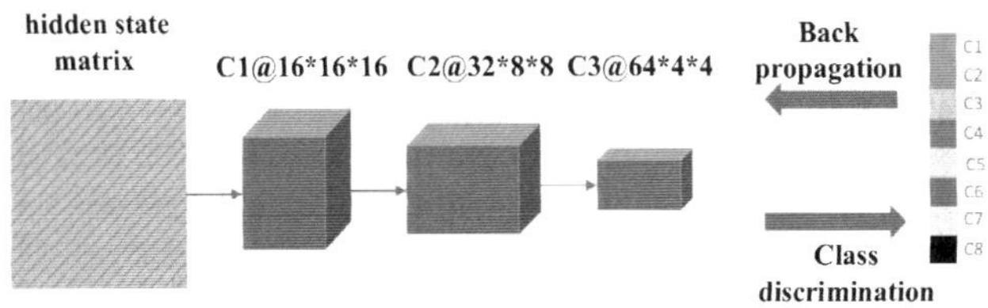

Figure 5. CNN layer structure.

After each convolutional layer, the activation function uses ReLu. Then, the output of the convolutional layer uses batchnorm before activation. Meanwhile, the model optimizer chooses Adam, and the learning rate is set to 0.005. The loss is passed back to the model through backpropagation to train the model. And the optimization of the model is driven by minimizing the loss value.

4. Experimental Design and Result Analysis

According to table 1, this paper simulates the data set required for the experiment through the simulation platform. The fundamental frequency of the disturbance signal is set to 50Hz, the sampling frequency is 3200Hz, and the sampling time is 0.2 s (10 cycles of sampling). The simulated signal data with a data dimension of 640 can be obtained. There are 2000 samples of each disturbance signal in the training set, with a total of 16000 pieces of data. The same method simulates the data in the validation set and the test set, both with 4000 pieces of data.

Table 1. Mathematical model and parameters of Power quality disturbance signal.

Disturbance type	Mathematical model	Parameter value range
P2	$x(t)=[1+\partial(u(t-t_1)-u(t-t_2))]\sin(\upsilon_0 t)$	$0.1\le\partial\le0.8, T\le t_2-t_1\le9T$
P3	$x(t)=[1-\partial(u(t-t_1)-u(t-t_2))]\sin(\upsilon_0 t)$	$0.1\le\partial\le0.9, T\le t_2-t_1\le9T$
P4	$x(t)=[1-\partial(u(t-t_1)-u(t-t_2))]\sin(\upsilon_0 t)$	$0.9\le\partial\le1, T\le t_2-t_1\le9T$
P5	$x(t)=\partial_1\sin(\upsilon_0 t)+\partial_3\sin(3\upsilon_0 t)$ $+\partial_5\sin(5\upsilon_0 t)+\partial_7\sin(7\upsilon_0 t)$	$0.05\le\partial_3,\partial_5,\partial_7\le0.15,\sum(\partial_i)^2=1$
P6	$x(t)=\sin(\upsilon_0 t)+\partial\exp(-(t-t_1)/\tau)$ $\sin(2\pi f(t-t_1))u(t-t_1)$	$8ms\le\tau\le40ms,$ $0.1\le\partial\le0.8300Hz\le f\le900Hz$
P7	$x(t)=\sin(\upsilon_0 t)+\partial(u(t-t_1)-u(t-t_2))$	$1\le\partial\le3, 1ms\le t_2-t_1\le3ms$
P8	$x(t)=[1+\partial\sin(u\upsilon_0 t)\sin(\upsilon_0 t)]$	$0.1\le\partial\le0.2, 0.1\le u\le0.5$

First, this paper trains the RCNN model using the simulated dataset and visualizes the accuracy-loss curve to analyze the model performance. Then, it classifies and identifies the disturbance signal by the RCNN model in different noise environments to investigate the anti-noise ability of the model. Finally, by comparing the method used in this paper with the current commonly used techniques, the model's performance is further analyzed in depth.

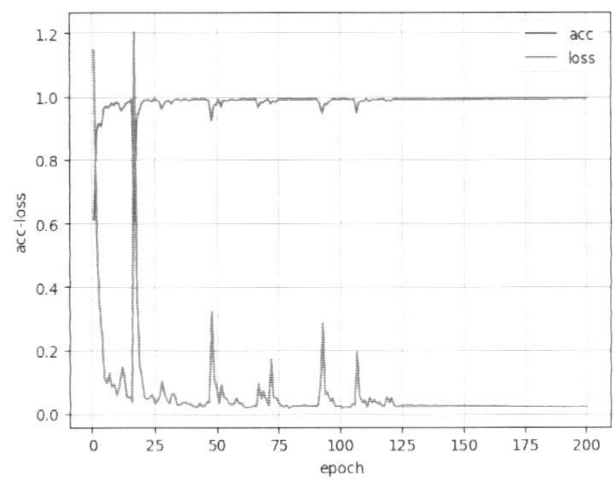

Figure 6. Accuracy-loss curve.

The training set and validation set are input into the RCNN model for training, and in each training batch, the loss value and accuracy of the model after training are recorded. Plot the accuracy-loss curve as shown in figure 6.

The RCNN model, combining the advantages of the traditional CNN and LSTM models, can fully and effectively extract the spatiotemporal features of the disturbance signal. After the model starts training, its training error decreases rapidly, and the effect is noticeable. After training 50 times, the accuracy rate has maintained a high level. Behind about 120 iterations, the model gradually stabilized, and the accuracy rate was stable at around 99.7%. Finally, the model obtains an accurate classification of disturbance signals.

To further explore the classification performance of the RCNN model, a confusion matrix is drawn for the classification results of the model through the data in the test set. Among them, there are 4000 pieces of data in the test set, 500 pieces of each type.

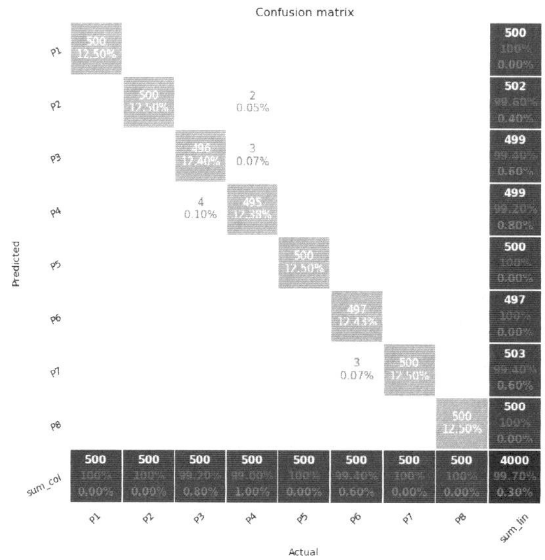

Figure 7. Confusion matrix.

Figure 7 is the result of the confusion matrix. The horizontal axis in the figure represents the actual category of the sample, and the vertical axis represents the predicted category of the sample. Taking the P3 category as an example, 496 pieces of data were correctly classified, and four pieces were wrongly classified into the P4 category. The rest of the categories are classified as described above. The sum_col row represents the recall rate of each category, that is, the proportion of all samples that are positive classes that are classified as positive class samples. The sum_lin column represents the precision rate, that is, the ratio of all samples classified as positive classes that are positive classes. The data grid in the lower right corner shows the overall prediction accuracy of the RCNN model of 99.70%. Green in the figure represents the number of correct classifications of the perturbed signal, and red represents the misclassification. The RCNN model has a high discrimination rate for each category. The misclassification mainly occurs in the two types of disturbance signals, P3 and P4, which may be confused due to the incomplete feature extraction of the disturbance signal by the model.

Table 2. The effect of noise on the classification effect of the model.

Disturbance type	ACC			
	No noise	50dB	40dB	30dB
P1	1.000	1.000	1.000	1.000
P2	1.000	1.000	1.000	0.968
P3	0.992	0.988	0.968	0.932
P4	0.990	0.982	0.952	0.914
P5	1.000	1.000	1.000	1.000
P6	0.994	0.992	0.970	0.958
P7	1.000	1.000	1.000	1.000
P8	1.000	1.000	1.000	1.000

Average	0.997	0.995	0.986	0.971

The RCNN model was trained using data sets superimposed with noises of different signal-to-noise ratios to verify the anti-interference ability of the model. In table 2, ACC represents the recognition accuracy of disturbance signals. According to the experimental results, even in a high noise environment, the recognition ability of the model still maintains a high level. When the signal-to-noise ratio is 30dB, the accuracy of the RCNN model remains above 96%. It shows that the model has good accuracy under different signal-to-noise ratios and has a certain anti-noise ability.

To further illustrate the model's superiority, this paper designs a variety of methods as a comparative experiment. It mainly includes five methods: RCNN (the method in this paper), CNN, LSTM, Deep Belief Network (DBN), and Deep Feedforward Network (DFN). Various models are trained through the disturbance signal data set, and the accuracy curve is drawn. Figure 8 is a graph of the accuracy curve.

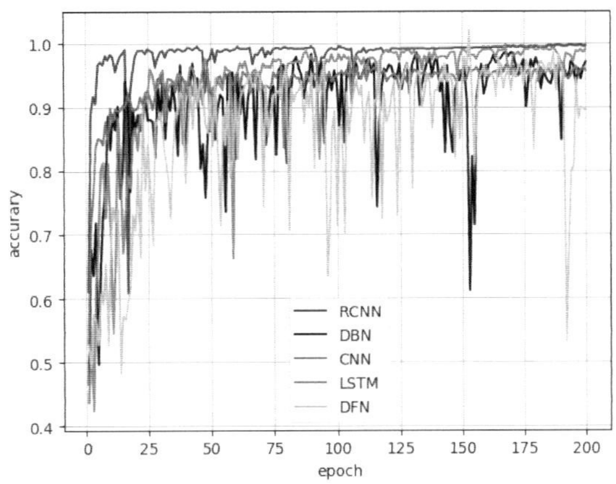

Figure 8. Accuracy curve.

Compared with other methods, the RCNN model achieves higher accuracy in a shorter time, and at the same time, with the increase in training times, the recognition accuracy of the model is maintained at a high level. The RCNN model proposed in this paper uses the LSTM layer to extract the temporal features of the disturbance signal, and the spatial features are extracted through the CNN layer. Compared with other methods, it can effectively improve the feature extraction ability of disturbance signals and achieve a high-precision recognition effect.

5. Conclusion

Different from traditional identification methods, this paper designs an RCNN model to achieve recognition of power quality disturbance signals by combining deep learning methods. This model combines the advantages of the CNN model and the LSTM model, using the LSTM model to encode the temporal correlation and the CNN model to encode the spatial correlation. Compared with other models, the feature extraction of disturbance signal is more sufficient, and the recognition accuracy is improved. Furthermore, the model adopts an end-to-end training method, which reduces the feature loss and improves the model classification effect. After many experiments and analysis, the model not only can classify the disturbance signal accurately, but also has a good recognition effect for the noise signal and has certain robustness.

Acknowledgments

This paper is funded by the State Grid Shanxi Electric Power Company Science and Technology Project (Research and Application of New Energy Voltage Transient Whole Process Analysis and Visualization

Technology Based on Online Data. No. 520530200011).

References

[1] Khokhar S, Zin A A, Mokhtar A S and Pesaran M 2015 A comprehensive overview on signal processing and artificial intelligence techniques applications in classification of power quality disturbances *Renewable and Sustainable Energy Rev.* **51** 1650-63.
[2] Li X, Wei W, Wang Y, Mu Z and Gu W 2009 Research on development technology of strong smart grid *Power System Protection and Control* **37** 1-7.
[3] Xu W and Yong J 2013 Analysis of Power Disturbance Data-New Application of Power Quality Monitoring Data *Proceedings of the Chinese Society for Electrical Engineering* **33** 93-101.
[4] Qiu X and Li F 2015 Analysis and identification of power quality disturbance based on ITD and K-means clustering *J. Electric Power System and Automation* **27** 54-59.
[5] Kochmann J, Manjunatha K, Gierden C, Wulfinghoff S, Svendsen B and Reese S 2019 A simple and flexible model order reduction method for FFT-based homogenization problems using a sparse sampling technique *Computer Methods in Applied Mechanics and Engineering* **347** 622-38.
[6] Sebastian P and DŚa P A 2015 Implementation of a power quality signal classification system using wavelet based energy distribution and neural network *Int. Conf. on Power and Advanced Control Engineering* p 157-61.
[7] Yi J 2010 Power Quality Disturbance Analysis Based on S-transform (Changsha: Hunan University) Doctor's Thesis.
[8] Zhang W, Huang N, Yang J, Yang Y and Wang X 2016 Power quality disturbance signal classification based on generalized S transform and DE-ELM *Electrical Measurement and Instrumentation* **53** 50-55.
[9] Yin B, He Y and Zhu Y 2015 A generalized S-transform and fuzzy SOM network power quality multi-disturbance detection and recognition method *Proc. CSEE* **35** 866-72.
[10] Zhang H, Zhao, Li L and Liu H 2018 Power system restoration overvoltage prediction based on multi-layer artificial neural network *Wisdom Electr. Power* **46** 67-73.
[11] Zhao L and Long Y 2016 Power quality compound disturbance classification based on improved SVM *New Technology of Electrical Engineering and Energy* **35** 63-68.
[12] Saini R, Mahela O P and Sharma D 2018 An algorithm based on Hilbert transform and rule based decision tree classification of power quality disturbances. *Int. Conf. IEEE 8th Power India* p 1-6.
[13] Lecun Y, Bengio Y and Hinton G 2015 Deep learning *Nature* **521** 436-44.
[14] Wang S and Chen H 2019 A novel deep learning method for the classification of power quality disturbances using deep convolutional neural network *Appl. Energy* **235** 1126-40.
[15] Chen W, He J and Pei X 2018 Classification for power quality disturbance based on phase-space reconstruction and convolution neural network *Power System Protection and Control* **46** 87-93.
[16] Cao M and Zhang Y 2020 Power quality disturbance classification based on convolution-long short-term memory network *Power System Protection and Control* **48** 86-92.
[17] Zhu R, Guo W and Gong X 2019 Power quality disturbance classification based on autoencoder and convolutional neural network *J. Electric Power System and Automation* **31** 70-75.
[18] Xu L, Li K, Xiao X, Zhao C, Yin J and Ni Y 2020 Identification of power quality complex disturbance based on deep feedforward network *Electrical Measurement & Instrumentation* **57** 62-69.
[19] Wu Z, Yang A and Zhu L 2020 Power quality disturbance identification based on recurrent neural network *Power system protection and control* **48** 88-94.
[20] IEEE Std 1159-2019 2019 IEEE Recommended Practice for Monitoring Electric Power Quality.
[21] Yang L, Wu Y, Wang J and Liu Y 2018 Summary of research on recurrent neural network *Comput. Appl.* **38** 1-6.

[22] Zhou F, Jin L and Dong J 2017 Summary of convolutional neural network research *Chinese J. Computers* **40** 1229-51.

Calculation Method of Ship's External Magnetic Field Based on Neural Network

Chao Zuo[1], Mingyu Ma[2], Maobin Li[2], Yang Pan[2], Hui Yan[3], Jianxun Wang[1], Pan Geng[1] and Jun Ouyang[2*]

[1] Wuhan Second Ship Design and Research Institute, Wuhan 430205, Hubei, China
[2] School of Optics and Electronic Information, Huazhong University of Science and Technology, Wuhan 430070, Hubei, China
[3] Big Data & Artificial Intelligence College, Xinyang University, Xinyang 464000, Henan, China
Email: *oyj@hust.edu.cn

Abstract. In view of the difficulties in the calculation of the external magnetic field in the closed-loop degaussing of the ship, this paper analyzes the relationship between the internal and external magnetic field of the ship, and builds a neural network model accordingly to realize the calculating model between internal and external magnetic field. Aiming at the problem of insufficient training set, a data augmentation method based on the linear nature of the induced magnetic field is proposed. By using this method to expand the training set, the training speed and accuracy of the model can be improved. Through experimental verification of the steel double-layer shell model, the proposed magnetic field estimation method can accurately predict the external magnetic anomaly when the induced magnetic field and the fixed magnetic field are changed. This method not only has the advantages of fast calculation speed and high real-time performance, but also solves the long-standing problem that the fixed magnetic field changes are difficult to predict.
Keywords. Magnetic field; closed-loop degaussing; fixed magnetic field.

1. Introduction

When a ship is sailing, due to its own fixed magnetism and the influence of the geomagnetic field, it will cause magnetic anomalies in the surrounding space [1-3]. In order to ensure the safety of ships, the current in the degaussing coil is usually adjusted to generate a magnetic field opposite to the magnetic anomaly in the external space to offset the magnetic anomaly [4]. This method needs to know the magnetic anomaly signal generated by the ship, which is often not obtained by direct measurement, so it is necessary to design a method to calculate the magnetic field in the outer space of the ship [5].

Vuillermet et al. [6] established a direct calculation model for the induced magnetic field of a double-hulled submarine from the perspective of magnetic potential, and reversed the fixed magnetic field value. This method has high accuracy, but it requires some important parameters of ferromagnetic materials, which are generally difficult to obtain in practice. Demilier et al. used the integral equation method to calculate the magnetic moment distribution of the ship, and then calculated the magnetic field in the outer space [7]. This method requires the hull to be divided into many small areas, which makes the calculation very complicated. Nguyen et al. used the fast multipole algorithm to improve the solution speed of the integral equation [8], but the calculation process still takes 3 minutes, which obviously cannot meet the needs of real-time demagnetization. In addition, this method is only suitable for situations where the fixed magnetism of the hull is not considered. Zhou et al. used the boundary element method to estimate the magnetic field at the position to be calculated from the magnetic field distribution of the measurement surface [9]. This method requires that there is no magnetic source and current element between the measurement surface and the calculation surface, so the measurement surface needs to be outside the hull, which is impossible for the ships in the water.

This paper proposes to arrange a certain number of sensors inside the hull, and establish a calculation model between the internal sensor measurement data and the external space magnetic field through neural network modeling. This method not only considers the practical feasibility, but also has good real-time performance.

2. Calculation Method of Ship's External Magnetic Field

Consider a ferromagnetic shell placed in the earth's magnetic field H0. At this time, the total magnetic field in space is:

$$H = H_0 + H_{ind} + H_{rem} \tag{1}$$

In equation (1), H_{ind} is the induced magnetic field generated by the shell under the influence of the geomagnetic field, and H_{rem} refers to the residual magnetic field accumulated during the construction and navigation of the ship. Both are magnetic fields generated by the magnetization of the shell. From the magnetic dipole model:

$$H_{ind} + H_{rem} = \frac{1}{4\pi}\int_V \frac{3(M\cdot r)r - r^2 M}{r^5} dV \tag{2}$$

$$M = M_{ind} + M_{rem} \tag{3}$$

where M_{ind} and M_{rem} are the induced magnetization and residual magnetization of the shell, r is the position vector from the magnetic source to the field point.

Assuming that the shell thickness e is small and uniform, then:

$$H_{ind} + H_{rem} = \frac{e}{4\pi}\int_S \frac{3(M\times r)r - r^2 M}{r^5} dS \tag{4}$$

A fluxgate sensor is placed at a certain point in space, and the measured value at this position should be:

$$B = \mu_0 H = B_0 + \frac{e\mu_0}{4\pi}\int_S \frac{3(M\cdot r)r - r^2 M}{r^5} dS \tag{5}$$

The shell is divided into N units. Assuming that each unit is small enough and the magnetization state is uniform, then equation (5) can be written as

$$B - B_0 = \frac{e\mu_0}{4\pi}\sum_{i=1}^{N}\int_{S_i} \frac{3(M_i\cdot r)r - r^2 M_i}{r^5} dS_i = \frac{e\mu_0}{4\pi}\sum_{i=1}^{N} k_i M_i \tag{6}$$

where k_i is a 3×3 matrix, which is only related to the division of the unit and the location of the measurement point. In other words, the magnetic flux density has a linear relationship with the magnetization of the shell.

We consider placing m fluxgate sensors inside the casing and n fluxgate sensors outside the casing, by equation (6):

$$B_{in} - B_0 = C_{in} M \tag{7}$$

$$B_{out} - B_0 = C_{out} M \tag{8}$$

In equation (7) and (8), B_{in} and B_{out} are the measured values of the internal and external sensors, the sizes are $3m$*1 and $3n$*1, B_{out} - B_0 is the magnetic anomaly generated by the shell, C_{in} and C_{out} are the corresponding conversion matrix, the size is $3m$*$3N$ and $3n$*$3N$, M is the magnetization matrix of the shell unit, the size is $3N$*1.

Considering that in actual situations, the number of sensors is far less than the number of units divided by the housing, that is, $m \ll N$, so it is impossible to directly solve B_{out} - B_0 through equation (7) and (8), so the neural network is used to establish the relationship between the measured values of the internal and external sensors.

B_{in} - B_0 is used as the input of the network, and B_{out} - B_0 is used as the output of the network. From equation (7) and (8), it can be seen that the internal and external magnetic fields are in a linear relationship with the magnetization state of the shell, which determines that the non-linear activation function is unnecessary. In order to improve the generalization performance of the network, Bayesian regularization [10] is used to constrain the model parameters.

3. Steel Model Experiment

A steel model with double shell was used to verify the magnetic field estimation method proposed in this article.

The model has a length of L = 5 m and a width of B = 2 m, as shown in figure 1. Forty three-axis fluxgate sensors are placed inside the model, and the sensor coordinate system is consistent with the model coordinate system. A row of single-axis fluxgate sensors is placed horizontally at 5 m above and below the model, as shown in figure 2, to measure the vertical component of the magnetic field.

When collecting external magnetic field data, the model is moved from -L to L in the longitudinal direction, and magnetic field data every 1/40 L is measured to obtain the magnetic field data of the $2L$*$2B$ plane 5 m above and below the model.

Under the condition of not demagnetizing the model, we measure the internal and external magnetic field data of the model in the geomagnetic background field of Sanya and Huludao regions along the east, west, north and south four directions respectively, a total of 8 sets of data, denoted as N1.

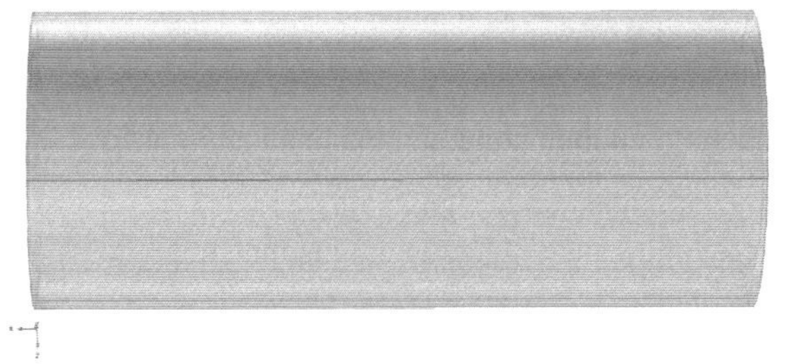

Figure 1. A model with double shell.

Figure 2. Schematic diagram of the installation position of the external sensors.

Then the model was demagnetized 5 times. At this time, the residual magnetization of the model changed. After each demagnetization was completed, the internal and external magnetic field data of the model along the east direction under the geomagnetic background field in Sanya province were measured. A total of 5 sets of data were recorded as N2.

Limited by the experimental conditions, it is extremely difficult to collect more data. However, training the neural network with only 13 sets of data will cause the model to overfit because the training set is too small. In order to solve this problem, the training set needs to be expanded.

Since the N1 group of data has no demagnetization processing, it can be considered that the fixed magnetism of the model has not changed, so only the induced magnetic field changes at this time. Using these 8 sets of data, the inductance of the induced magnetic field with the geomagnetic background field can be calculated. At the same time, the fixed magnetic field is separated. For the magnetic field data of the N2 group, the calculated inductance can also be used to separate the fixed magnetic field. The distribution of the separated 6 sets of fixed magnetic fields at 5 m above and below the model keel line is shown in figure 3.

After using these 13 sets of data to obtain the inductance and the fixed magnetic field in 6 different magnetic states, different induced magnetic fields can be obtained by randomly simulating the geomagnetic background field, and randomly combined with the known fixed magnetic field to obtain new training data, such as equation (9) shows:

$$B - B_0 = kB_0 + B_{pi} \tag{9}$$

where B_0 is the randomly simulated geomagnetic background field, k is the calculated inductance, B_{pi} is a random set of known fixed magnetic fields, and B is the generated internal/external magnetic field data.

In order to verify the effectiveness of the proposed magnetic field estimation method, each set of data in N2 is used as the test set, and the remaining four sets are used to generate the training set to train the neural network and predict the output of the test set. The relative error and Pearson correlation coefficient are used to evaluate the results. The relative error characterizes the peak error of the estimated result, and is defined as:

$$r = \frac{|Measured\ value - Predicted\ value\ |_{max}}{|Measured\ value|_{max}} \times 100\% \tag{10}$$

Pearson's correlation coefficient represents the similarity between the estimated result and the measured value.

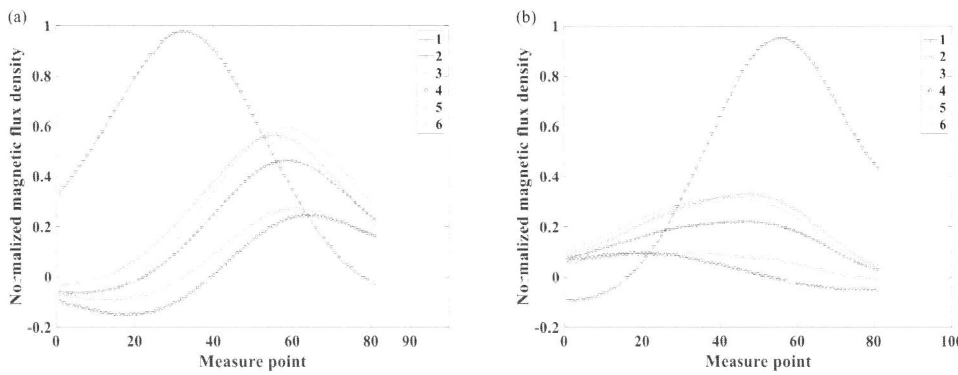

Figure 3. Distribution of fixed magnetic field. (a): 5 m above the keel line; (b): 5 m below the keel line.

The results are shown in table 1.

Table 1. The prediction results of 5 test sets.

Number	1	2	3	4	5
Peak value of normalized magnetic anomaly	0.83	1.00	0.57	0.92	0.68
Peak value of error	0.03	0.02	0.02	0.06	0.04
Relative error (%)	3.55	2.31	4.23	7.00	6.34
Pearson correlation coefficient (%)	99.91	99.94	99.83	99.64	99.70

In order to verify the generalization ability of the model, the model was demagnetized again, and the internal and external magnetic field data of the model along the south course under the geomagnetic background field in the Shanghai area were collected as the test set, and denoted as N3, and the training set was generated with the previous 6 fixed magnetic fields. The relative error of the prediction result is 7.96%, and the Pearson correlation coefficient is 0.9989. Figure 4 shows the comparison between the predicted value and the measured value on the port, keel, and starboard.

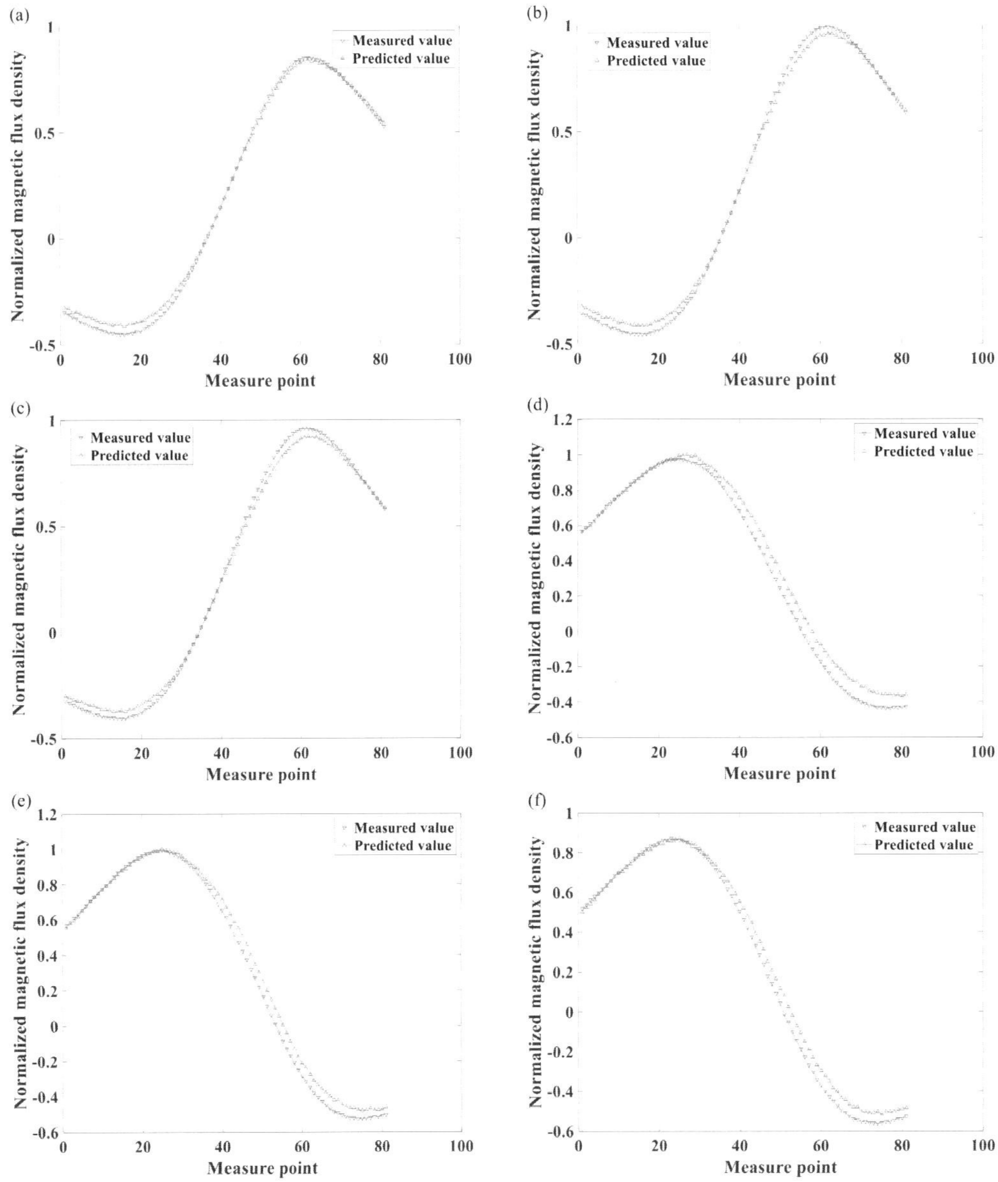

Figure 4. The prediction results. (a): 5 m above the port side of the model; (b): 5 m above the keel side of the model; (c): 5 m above the starboard side of the model; (d): 5 m below the port side of the model (e): 5 m below the keel side of the model; (f): 5 m below the starboard side of the model.

The fixed magnetic field is separated from the measured value, and it is compared with the previous 6 sets of fixed magnetic fields. The result is shown in figure 5.

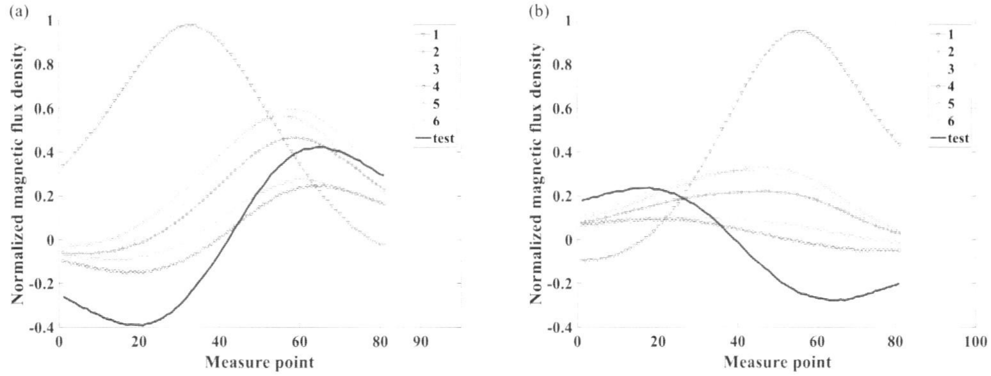

Figure 5. The fixed magnetic comparison between the test set and the training set. (a): 5 m above the keel; (b): 5 m below the keel.

It can be seen from figure 5 that the fixed magnetic field of the test set has a large change compared with the fixed magnetic field contained in the training set, but the magnetic anomaly outside the model can still be accurately predicted.

The experimental results show that the proposed external magnetic field estimation method has a good adaptability to the changes of the induced magnetic field and the fixed magnetic field. Even under the condition of very little prior data, this method can still use inductance to expand the training set and accurately predict external magnetic anomalies. In addition, once the neural network is trained, the prediction result can be obtained in a very short time, which can meet the real-time requirements under actual conditions.

To verify the effect of the proposed data augmentation method, the prediction results with and without data augmentation are compared. The results are shown in tables 2 and 3, where No. 1-No. 5 is N2, which is used to train model, No. 6 is N3, which is not used to train model.

Table 2. Result without data augmentation.

Test set	1	2	3	4	5	6
Peak of external fixed magnetic field	0.37	0.45	0.23	0.45	0.21	1.00
Peak of error	1.85×10^{-4}	2.26×10^{-4}	1.61×10^{-4}	2.71×10^{-4}	1.58×10^{-4}	0.45
Relative error (%)	0.05	0.05	0.04	0.06	0.07	44.69

Table 3. Result with data augmentation.

Test set	1	2	3	4	5	6
Peak of external fixed magnetic field	0.37	0.45	0.23	0.45	0.21	1.00
Peak of error	3.89×10^{-4}	3.39×10^{-4}	1.07×10^{-4}	5.30×10^{-4}	2.62×10^{-4}	0.08
Relative error (%)	0.11	0.07	0.05	0.12	0.12	7.92

Comparing the results in tables 2 and 3, it can be seen that when the data augmentation is not performed, since groups 1 to 5 participated in the training model, the prediction of the external magnetic field was very accurate, while for group 6, which did not participate in training the model, the error of the prediction result was very large, which indicated that the neural network model had obvious overfitting. After using data augmentation, although there is still a certain degree of overfitting, for group 6 that did not participate in training, the prediction accuracy was greatly improved, the degree of overfitting was suppressed, and the generalization performance of the model was significantly improved.

The method proposed in this paper is compared with machine learning algorithms such as ridge regression and support vector regression. The results are shown in table 4.

The results show that the proposed data augmentation method is equally effective for Ridge Regression and SVR, and has a greater boost to neural networks. After data augmentation, the neural network is significantly better than Ridge Regression and SVR.

Table 4. Comparison with ridge regression and SVR.

Method	Peak of external fixed magnetic field	Peak of error	Relative error (%)
Ridge Regression without data augmentation	1.00	0.36	36.17
SVR without data augmentation	1.00	0.36	36.14
Ridge Regression with data augmentation	1.00	0.11	11.05
SVR with data augmentation	1.00	0.11	10.98
Proposed method	1.00	0.08	7.92

4. Conclusion

Based on the magnetic dipole model, this paper analyzes the linear relationship between the internal and external magnetic fields of the ship and the magnetization of the hull. Based on this conclusion, a neural network is proposed to establish a calculation model between the internal and external magnetic fields, and the external magnetic anomaly is calculated according to the measured value of the internal fluxgate sensor. It is proposed to use the inductive coefficient to augment the training set data, which solves the problems of difficulty and high cost of training data acquisition. After the experimental verification of the steel double-shell model, the method has good adaptability to the changes of the induced magnetic field and the fixed magnetic field. The proposed data augmentation algorithm can effectively improve the performance of the network. Even if only a small amount of prior data is used to train the model, the accuracy rate can still reach more than 90%. In addition, the proposed data augmentation method is also applicable to methods such as ridge regression and SVR, but the improvement of neural network is greater. After data augmentation, the prediction effect of neural network is significantly better than that of ridge regression and SVR.

References

[1] Liu S, Chen Z, Pan M, Zhang Q, Liu Z, Wang S and Wan C 2019 Magnetic anomaly detection based on full connected neural network IEEE Access **7** 182198-182206

[2] Modi A and Kazi F 2020 Magnetic-signature prediction for efficient degaussing of naval vessels *IEEE Transactions on Magnetics* **56(9)** 1-6

[3] Wan C, Pan M, Zhang Q, Wu F, Pan L and Sun X 2018 Magnetic anomaly detection based on stochastic resonance *Sensors and Actuators A: Physical* **278** 11-17

[4] Wang Z F 2020 Overview on the ship degaussing system *Mar. Electr. Electron. Eng.* **40** (9) 4-7

[5] Isa M C, Nain H, Yusoff N H N, Manap A R A, Slamatt R and Anuar M H 2019 An overview of ship magnetic signature and silencing technologies *Defence S &T Technical Bulletin* **12** (2) 176-192

[6] Vuillermet Y, Olivier C and Jean-Louis C 2007 Closed Loop Degaussing Applied to Double Hull Submarine *Electromagnetic Silencing Symposium 2007 (EMMS 2007)*

[7] Demilier L, Cauffet G, Chadebec O, Coulomb J L and Rouve L L 2010 Validation of closed loop degaussing system for double hull submarines *Marine System and Technology*

[8] Nguyen T S,Guichon J M, Chadebec O, Labie P and Coulomb J L 2011 Ships magnetic anomaly computation with integral equation and fast multipole method *IEEE Transactions on Magnetics* **47** (5) 1414-17.

[9] Zhou J, Chen J and Shan Z 2017 Spatial signature analysis of submarine magnetic anomaly at low altitude *IEEE Transactions on Magnetics* **53** (12) 1-7.

[10] Burden F and Winkler D 2008 Bayesian regularization of neural networks *Artificial Neural Networks* 23-42.

Research on Intelligent Decision Method of Optimal Production Planning and AGV In-time Delivery in Mixed-Model Assembly Line

Li Jia[1], Lei Shi[1], Jianfeng Yao[2], Xiang Dai[2] and Gang Guo[1*]

[1] College of Mechanical and Vehicle Engineering, Chongqing University, Chongqing, China

[2] Chongqing Chang'an Automobile Company Limited, Chongqing, China

Email: *cquguogang@163.com

Abstract. An automated guided vehicle (AGV) system is an indispensable part of a mixed-model assembly line (MMAL) for flexible and efficient production, which is utilized for the material delivery. With the urgent needs of industry digitalization and market customization, along with the trend of corporate smart decision management, MMAL has been widely adopted in the manufacturing process of modularized products. To address the parameter matching and energy consumption optimization of the AGV system in MMAL, the overall architecture of an intelligent MMAL (IMMAL) is firstly proposed. And the model of this problem is constructed. Then, the research methodology is proposed, which comprehensively adopts the dispatching rule, genetic algorithm, design of experiment, BP neural network, and simulated annealing. Eventually, a real-life case is used to validate the practicality of the proposed solutions. The results show that the optimal solutions of MMAL production and AGV system delivery at the minimum energy consumption are obtained successfully. Furthermore, three factors: the quantity, capacity, and velocity of AGVs greatly impact the energy consumption of the AGV system, where the quantity of AGVs has the most negligible significance, while the capacity of AGVs has the most significance.

Keywords. Mixed-model assembly line; AGV system; Energy optimization.

1. Introduction

With the introduction of the Industry 4.0 concept, information technology has provided breakthrough technical means for the manufacturing industry [1]. Facing the increasing competition in the global manufacturing market, manufacturing enterprises urgently need to carry out comprehensive and profound transformation and upgrading, cater to the demand of market customization, and establish smart factories that fit the needs [2], making the whole manufacturing process automized and intelligent.

Mixed-model assembly line (MMAL) [3] refers to the assembly line that is capable of assembling multiple kinds of products continuously, as the Bill of material (BOM) for each product is similar, making it possible to assemble and produce multiple types of products according to customers' customization needs [4].

An automated guided vehicle (AGV) system [5] is an information system that controls AGVs for material handling and distribution. AGVs can replace manual forklifts and are widely used in various production environments such as processing workshops, assembly lines, and warehouses, significantly improving automation, informatization, and intelligence in production logistics [6-9].

This paper is dedicated to studying the model construction, parameter matching, and energy consumption optimization of AGV system in MMAL, to lay the theoretical and practical foundation for the development and application of AGV systems. The rest of the paper is organized as follows: Section 2 constructs the general architecture of the intelligent mixed-flow assembly line and its AGV system; Section 3 derives the mathematical model of the parameter matching and energy consumption optimization problem of the MMAL AGV system; Section 4 presents the research methods and steps to solve the problem, using a combination of allocation rules, genetic algorithms, experimental designs, BP neural networks, and simulated annealing; Section 5 validates the proposed theories, algorithms and models through discrete-event simulation and calculation cases; Section 6 summarizes the paper's contents.

2. Overall Architecture

The overall architecture design of the intelligent mixed-model assembly line (IMMAL) and its AGV system is the foundation for its optimization. The architecture of the AGV system integrates modern information technology, advanced manufacturing technology, and automated production technology to incorporate various isolated information and underlying physical subsystems scattered in the product manufacturing process in an organic and unified manner, satisfying the diverse, dynamic, and customized needs of the custom market. The proposed architecture of the intelligent mixed-flow assembly line and its AGV system, shown in figure 1, is built with the service-oriented architecture (SOA) [10] at the core of market customization activities, linking the underlying physical systems and the mid-to-high-level information systems with unified standard protocols, highlighting the critical role of the AGV system in this architecture. From top to bottom, the overall architecture of the IMMAL proposed in this paper is divided into a three-level structure of enterprise LAN, factory LAN, and fieldbus network.

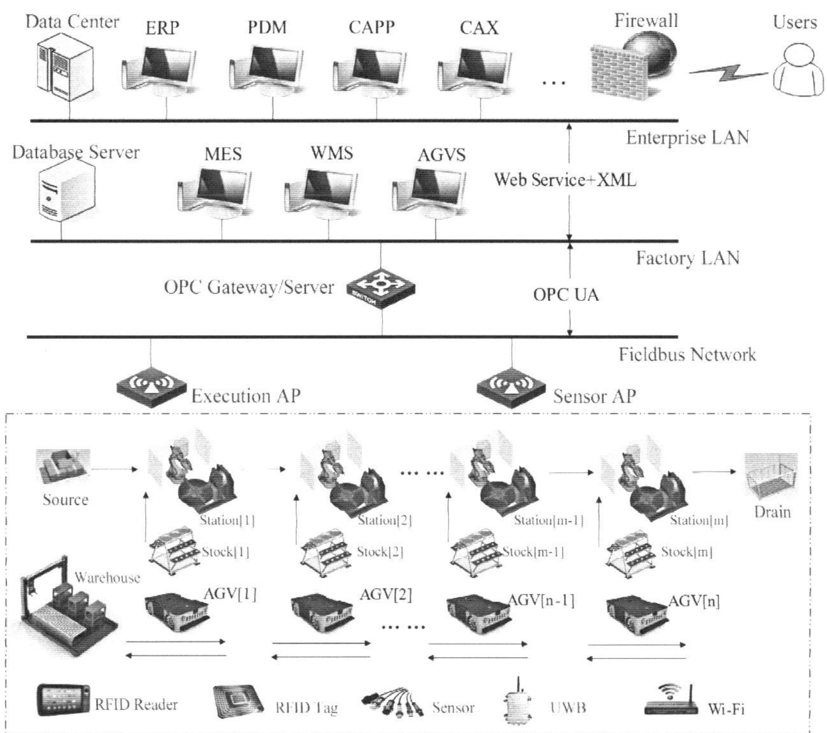

Figure 1. The architecture of the IMMAL and AGV system.

3. Mathematical Model

A complete mixed-integer programming (MIP) model consists of decision variables, optimization objectives, and constraints [11]. In this paper, a MIP model for the MMAL AGV system parameter matching and energy consumption optimization problem is established. The symbolic table of the MMAL AGV system optimization problem is shown in table 1.

Table 1. Symbolic table of the MMAL AGV system optimization problem.

Symbol	Meaning
I, l	Set and number of products in the minimum production cycle.
J, m	Set and number of processes in a MMAL.
A, n	Set and number of AGVs that distribute parts.
q	Capacity of AGVs that deliver the parts, assuming that all AGVs are identical.
v	Speed of AGVs that deliver the parts, assuming all AGVs are identical.
x_{ij}	0-1 variable, whether product i is assembled in the station of process j.
$y_{i'ij}$	0-1 variable, whether product i' is assembled in the station of process j immediately before product i.
z_{ajt}	0-1 variable, whether AGV a's destination is stock area j at the tth delivery.
S_{ij} °F_{ij}	Start and end time points of process j for product i.
$ST_{i'i}$	Preparation time for the product i to be assembled in a station immediately after product i'.
AT_{ij}	Assembly time for product i to be assembled in the station of process j.
WT_{ij}	Waiting time for product i to be assembled at process j's station due to missing parts.
q_{ij}	Number of parts consumed by product i's process j.
L	A sufficiently large positive number.
$\overline{p}_0$	Average base power of the AGV system during operation.
$\overline{p}$	Average system power of the AGV system during operation.
Q_j	Number of parts in the stock area j.
Q_j^{full}	Capacity of parts that can be stored in the stock area j.
D_j	Distance from the distribution start point to the stock area j.
DT_{ajt}	Time of the AGV a's tth delivery to the stock area j, loading and unloading time ignored.
C_{max}	Maximum completion time.
TWT	Total waiting time.
AEC	Energy consumption of the AGV system.
Pr_{ajt}	Priority value of the AGV a's tth delivery to the lineside staging area j.

3.1. Decision Variables

The MMAL AGV system optimization problem has three decision variables and three variable parameters. The three decision variables are x_{ij} for allocating the stations, $y_{i'ij}$ for sorting the production sequence, and z_{ajt} for scheduling the delivery.

$$x_{ij} = \begin{cases} 1, & \text{If } (i, j) \text{ is true;} \\ 0, & \text{otherwise.} \end{cases} \tag{1}$$

$$y_{i'ijk} = \begin{cases} 1, & \text{If } (i', j) \text{ preceding } (i', j) \text{ is true;} \\ 0, & \text{otherwise.} \end{cases} \tag{2}$$

$$z_{atj} = \begin{cases} 1, & \text{If}(a,t,j) \text{ is true;} \\ 0, & \text{otherwise.} \end{cases} \tag{3}$$

3.2. Goal of Optimization

The objective of the optimization is to minimize the AGV system energy consumption (AEC) value, which can be shown in equation (4) according to the practical experience in factories.

$$\begin{aligned} \min AEC &= \overline{p} \times C_{max} \\ &= (\overline{p}_0 + \alpha \cdot n \cdot q \cdot v) \times C_{max} \end{aligned} \tag{4}$$

where $\bar{p}_0$ is the average base power of the AGV system; n, v, q are the number, capacity, and speed of AGV carts in the AGV system; α is the scaling factor, in this paper $\alpha = 2.5$.

3.3. Constraints

The constraints for product assembly are shown in equations (5)-(9), given the condition $\forall \alpha \in A, \forall j \in J, \forall t = 1,2,3...$:

(1) All stations should undertake the same number of product assemblies.

$$\sum_{i \in I} x_{ij} = l \tag{5}$$

(2) All products are to undertake the same number of assembly processes

$$\sum_{j \in J} x_{ij} = m \tag{6}$$

(3) The process duration of each product is equal to the sum of its preparation, assembly and interval.

$$\begin{cases} F_{ij} = S_{ij} + y_{i'ij}(ST_{i'i} + AT_{ij} + WT_{ij}) \\ WT_{ij} = \text{Time required for } Q_j \geq q_{ij} \\ (\exists WT_{ij} \geq 0) \end{cases} \tag{7}$$

(4) No product can start assembly earlier than its end time in the previous process.

$$S_{ij} \geq F_{i(j-1)} \tag{8}$$

(5) Each station can only assemble one process of a product at any given time.

$$S_{ij} + L \times (1 - y_{i'ij}) \geq F_{i'j} \tag{9}$$

The constraints for AGV distribution are shown in equations (10)-(14):

(1) AGV's delivery target can only be the stock buffer.

$$\sum_{j \in J} z_{atj} = 1 \tag{10}$$

(2) Delivery of the stock buffer can only be arranged with single AGV.

$$\sum_{a \in A} z_{atj} = 1 \tag{11}$$

(3) Each AGV is fully loaded with a certain type of parts when in delivery.

$$Q_j^{full} - Q_j \geq z_{atj} \times q \tag{12}$$

(4) The inventory in stock buffer shall be within its capacity.

$$0 \leq Q_j \leq Q_j^{full} \tag{13}$$

(5) The delivery path of AGV is from the starting point to the stock buffer and then back.

$$DT_{\alpha tj} = z_{\alpha tj} \times \frac{2D_j}{v} \quad (\forall \alpha \in A, \forall j \in J, \forall t = 1,2,3...) \tag{14}$$

4. Research Methodology

4.1. Rule of Allocation
Considering the complexity of the studied problem, based on the heuristic algorithm [12], we proposed an allocation rule of the lowest stock rate (LSR), as shown in equation (15).

$$Pr_{\alpha tj} = \frac{Q_j}{Q_j^{full}} \tag{15}$$

The allocation rule of the LSR is a matching method for resource delivery. In the MMAL AGV system, the smaller the priority value of one stock buffer, the higher the matching degree between the AGV and this buffer area, and the AGV will prioritize the parts required by this buffer area for delivery. If the priority value of multiple stock buffers is the same minimum value, the AGV will choose the parts required by the nearest stock buffer for delivery.

4.2. Genetic Algorithm
Firstly, with both LSR allocation rules and AGV system parameter values set to the maximum, we use a genetic algorithm [13] to obtain the best product order for the MMAL, and the specific algorithm is shown in figure 2.

In this case, the genetic algorithm is encoded using symbolic coding, and a gene of the chromosome represents the type number of a product. The chromosomes are decoded to represent a feasible solution for the product order. The fitness function is the minimization AEC, we used the fitness proportionate selection with sequential crossover and random variation. The termination condition is that the number of generations reaches the set value.

Based on practical experience, the population size is set to 20, the number of generations is set to 50, the crossover probability P_c is 0.7, and the variation probability P_m is 0.1. To preserve the chromosome structure of the best individuals and improve the convergence speed, P_c and P_m are reduced by 0.001 in each generation.

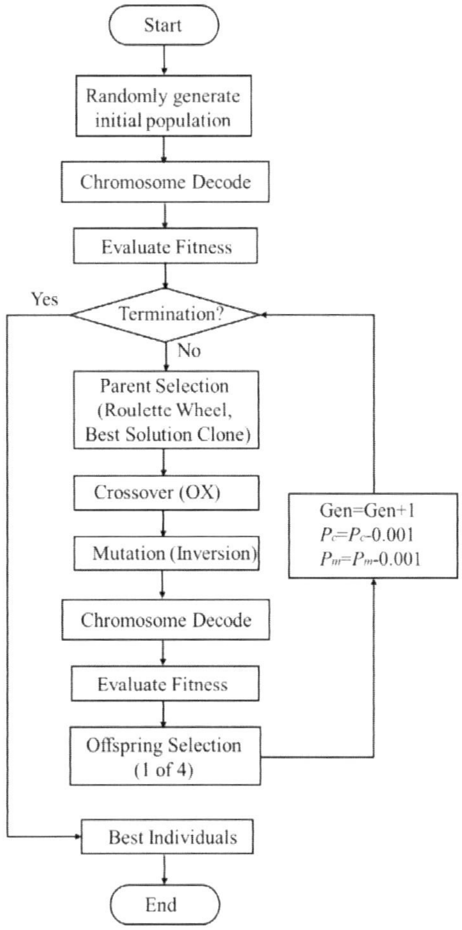

Figure 2. The genetic algorithm to obtain the best product order.

4.3. Design of Experiment and Analysis of Variance

Then, in the case of LSR allocation rules and the optimal product order, this paper uses design of experiment and ANOVA [14] to obtain the degree of influence of parameter changes on the AEC. The specific experiment points are shown in table 2.

Table 2. Multi-factor orthogonal experimental design of AGV system.

Input	Factor X	Factor Y	Factor Z
	Number n	Capacity q	Speed $v/m{\cdot}s$
Upper	1	1	0.01
Lower	5	10	0.50
Increment	1	1	0.01
Level number	5	10	50

According to the design of experiment in table 2, the number of carts n in the AGV system has five levels, the capacity q has ten levels, and the speed has 50 levels, for a total of 2,500 experiment points. The 2500 experiment points of the experiment design were substituted into the mathematical model for calculation to obtain 2500 samples under the three influencing factors so that the ANOVA of the three factors of AEC could be performed.

4.4. BP Neural Network
After obtaining 2500 samples under the three influencing factors, we use BP neural network [15] to fit the input and output values of AEC, the algorithm is shown in figure 3.

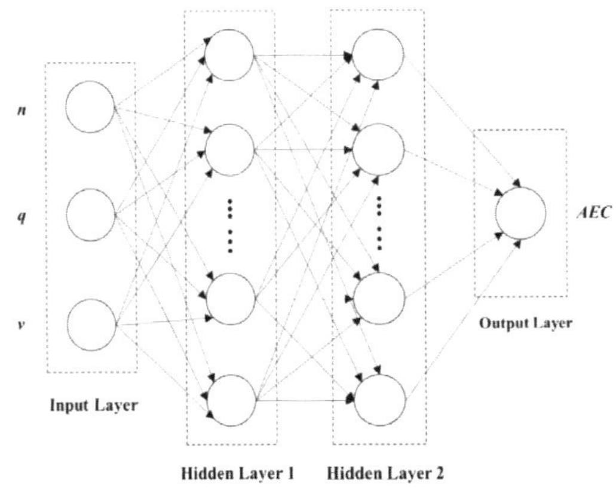

Figure 3. The BP neural network.

The neural network has two hidden layers, and the number of neurons in each hidden layer is set to 10. Among them, the input layer of this neural network has three neurons for the number n, capacity q, and speed v of the AGV system's carts. The output layer of this neural network has one single neuron for the output of AEC. According to the practical experience, the activation function of this neural network is set as a hyperbolic tangent function, the batch size is set as 8, the dropout is set as 10%, the training set accounts for 90%, the test set accounts for 10%, and the training method is gradient descent method with volume and adaptive learning rate (Traingdx), the minimum learning rate is set to 0.2, the maximum learning rate is set to 2, and the momentum coefficient is set to 2.

4.5. Simulated Annealing
Eventually, after training the BP neural network, this paper uses the simulated annealing algorithm [16] to match the input values and optimize the output values of the AGV system, as shown in figure 4.

The objective function $F(\boldsymbol{x})$ is the absolute value of the difference between the neural network output value $AEC(n,q,v)$ and the set value AEC_0, as shown in equation (16).

$$F(x) = F(n,q,v) = \left| AEC(n,q,v) - AEC_0 \right| \tag{16}$$

The starting point of the algorithm for simulated annealing is the minimum values of the AGV system parameters n_{min}, q_{min}, v_{min}; the end point is the matching values of the AGV system parameters n_{opt}, q_{opt}, v_{opt}. When setting AEC_0 to 0, the optimal value of AEC can be finally obtained. The new solution F' is generated from the old solution F by random perturbation, and its equations are shown in equations (17) and (18).

$$x' = x + \Delta x(r), {}^{\circ}F' = F(x'),\ random\ r \in \{1,2,3\} \tag{17}$$

$$\Delta x(r)\begin{cases} n' = random\ n \in \{1,2,.....5\}, if\ r = 1 \\ q' = random\ q \in \{1,2,.....10\}, if\ r = 2 \\ v' = random\ v \in [0.01, 0.50], if\ r = 3 \end{cases} \tag{18}$$

According to the Metropolis criterion, the formula for accepting new solutions by probability is shown in equation (19).

$$x = x', F = F', \; if\; e^{-\frac{F'-F}{T}} > p, \; random\; p \in [0,1) \tag{19}$$

From the practical experience, the initial temperature T_0 is set to 200, the termination temperature T_{min} is set to 0.1, the Markov chain length L is set to 10, and the cooling factor K is set to 0.95.

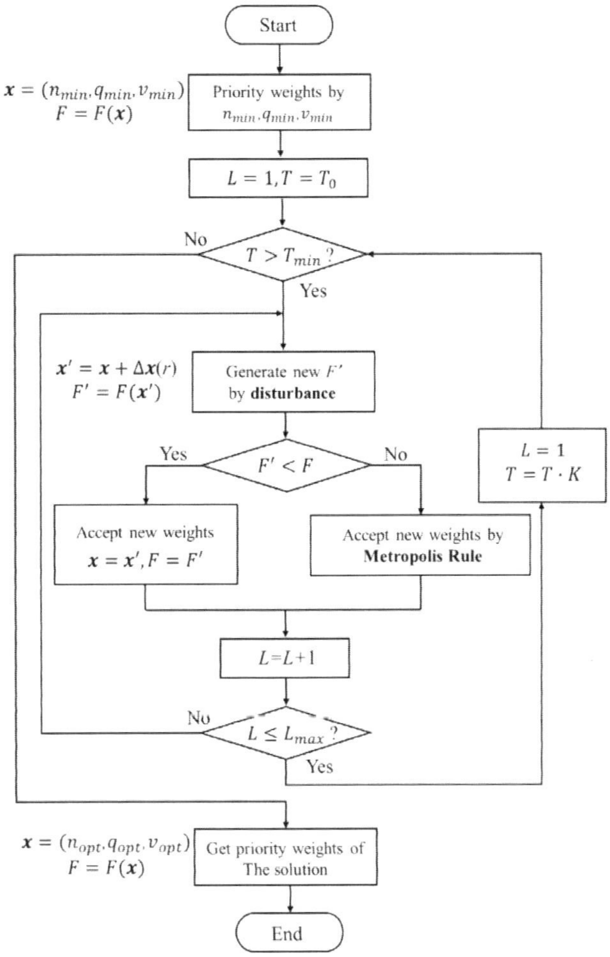

Figure 4. The simulated annealing algorithm.

5. Calculation Example

To test the practicality and validity of the proposed methods, these methods are applied to an example of a real life MMAL AGV system. In this example, the MMAL (l=30, m=6) considers five types of 30 products with random quantities, each with six processes executed in six stations, and each process requires consuming a certain number of parts. In this paper, we use the actual data of a domestic non-standard manufacturing company that come from the company's enterprise information system, and to increase the generality of the data, this paper only uses uniform distribution to generate the data set based on the upper and lower boundaries of these data.

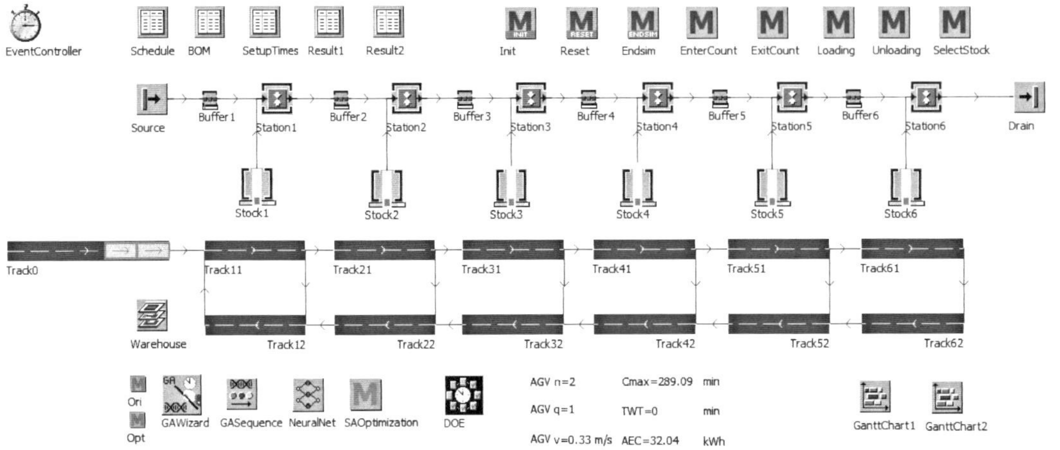

Figure 5. The overall architecture of the IMMAL and AGV system.

In this dataset, the relevant calculation parameters are set as follows: $ST_{i'i}{\sim}U(1,\ 2)$ min, $AT_{ij}{\sim}U(5,\ 10)$ min, $q_{ij}{\sim}U(2,\ 8)$ rounded, $\bar{p}_0$=5 kW, $n \in \{1, 2, \dots, 5\}$, $q \in \{1, 2, \dots, 10\}$, $v \in [0.01, 0.50]$ m/s, D_1=5 m, D_2=10 m, D_3=15 m, D_4=20 m, D_5=25 m, D_6=30 m, Q_j^{full} =20, initial Q_j=10. Using Siemens Tecnomatix Plant Simulation 13.1 software and SimTalk programming language, the modelling, and experiments of discrete-event simulation of AGV system [17] for MMAL were conducted, as shown in figure 5. The environment for the simulation is a Lenovo ThinkPad E570 computer with 2.50GHz/2.70GHz Intel Core i5-7200U CPU and 8.00GB of RAM.

Table 3. Assembly time of products on different stations.

Type	Quantity	Time of assembly AT_{ij}/min					
		Station 1	Station 2	Station 3	Station 4	Station 5	Station 6
A	5	9.46	5.19	6.52	5.44	8.44	6.32
B	1	5.15	7.55	9.46	9.64	8.03	8.17
C	8	6.23	7.32	5.85	6.29	6.45	9.96
D	10	6.01	6.87	8.66	8.23	8.45	7.09
E	6	6.51	8.15	9.88	6.9	7.67	7.18

The assembly time of 5 products on different stations is shown in table 3, and the number of parts required for different processes of 5 products is shown in table 4, which is the BOM table of the products. The preparation time for switching between 5 products is shown in table 5.

Table 4. Number of parts required for different processes of the product.

Type	Parts required q_{ij}/pcs					
	Process 1	Process 2	Process 3	Process 4	Process 5	Process 6
A	3	4	3	4	8	8
B	5	3	6	2	3	7
C	4	3	4	5	5	6
D	3	7	5	7	5	4
E	7	7	4	8	6	4

Table 5. Lead time for switching between products.

Type	Lead Time $ST_{i'i}$/min				
	A	B	C	D	E
-	1.69	1.14	1.06	1.04	1.5
B	0	1.09	1.24	1.7	1.04
C	1.09	0	2	1.93	1.37
D	1.24	2	0	1.78	1.25
E	1.7	1.93	1.78	0	1.7

Table 6. The best product order obtained by genetic algorithm.

Subject	The best product order				
Type	B	C	D	A	E
Quantity	1	8	10	5	6

Firstly, with the LSR allocation rule and AGV system parameter values set to the maximum, we use a genetic algorithm to obtain the best product order for the MMAL. The convergence curve of the genetic algorithm is shown in figure 6, and the optimal product order is shown in table 6. From figure 6 and table 6, it's clear that genetic algorithm successfully searches for the best product order for the MMAL, and the algorithm converges swiftly.

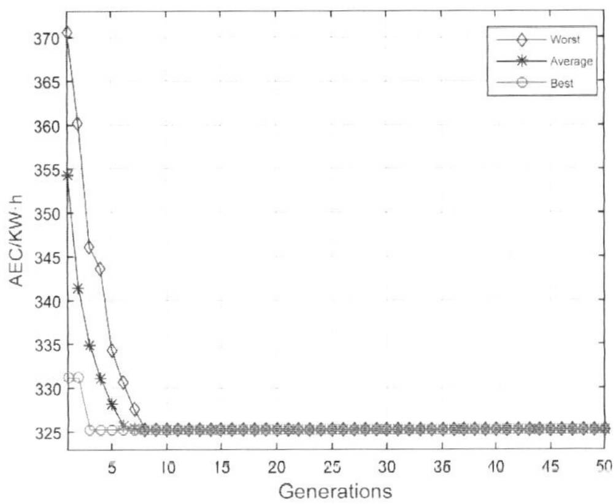

Figure 6. Convergence curve of genetic algorithm.

Then, in the case of LSR allocation rules and the optimal product order, this paper uses experimental design and ANOVA to obtain the degree of influence of AGV system parameter changes on the AEC. The result of the multiple-factor orthogonal experimental design of the AGV system is illustrated in figure 7. The result of ANOVA for the influence factors of the AGV system is shown in table 7.

Table 7. Result of ANOVA for the influence factors of the AGV system.

Factor of AGV	Number of levels	Number of samples	*F*-Value	*P*-value	F-boundary
Quantity *n*	5	2500	11.48381	<0.05	2.375502
Capacity *q*	10	2500	63.44376	<0.05	1.883649
Speed *v*	50	2500	43.60284	<0.05	1.359294

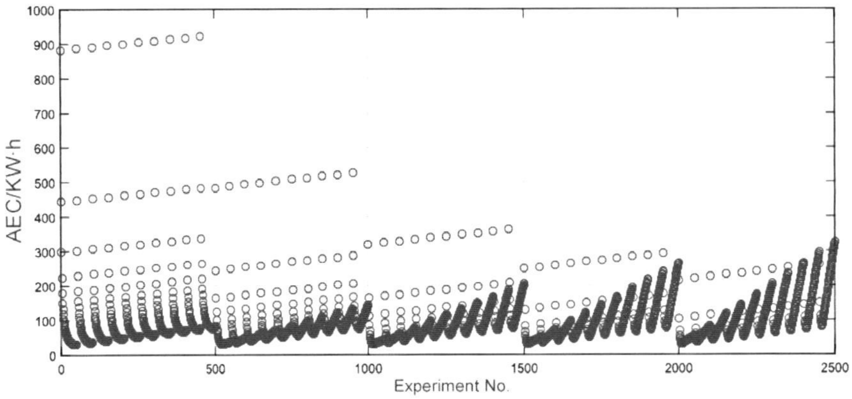

Figure 7. Result of the multiple-factor orthogonal experimental design of the AGV system.

From figure 7, we can see that at the beginning, when the number of AGVs is certain, as the capacity and speed of AGVs increase, the AEC starts to decrease gradually to reach the minimum value. However, as the capacity and speed of AGVs continue to increase, the AEC begins to gradually increase, which we call the performance saturation of the AGV system. Moreover, as the number of AGVs increases, the more prone the AGV system is to performance saturation.

From table 7, we can see that the test level P-values of the number of carts n, capacity q, and speed v in the AGV system are less than 0.05, and the significance level F-values are greater than the F-boundary value, indicating that all three factors have a significant impact on the AEC. Among them, the F-value of the number of AGVs n is the smallest, and the F-value of the AGV capacity q is the largest, indicating that the significance of the effect of n on the AEC is the smallest and the significance of the effect of q on the AEC is the largest.

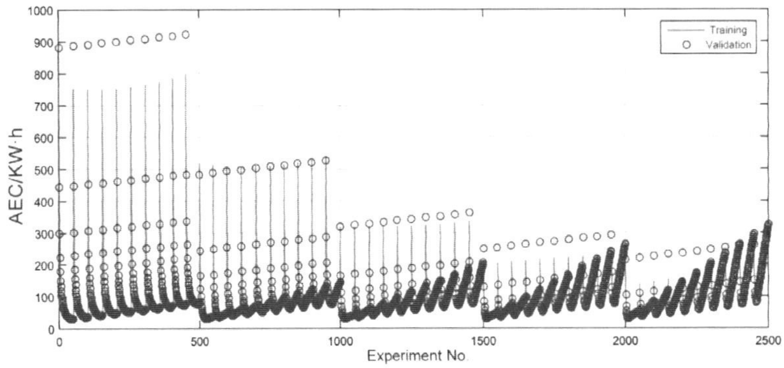

Figure 8. Fitting results of the AEC.

Then after obtaining 2500 samples under the three influencing factors, we use BP neural network to fit the input and output values of the AEC. The fitting results of the AEC are shown in figure 8. The average absolute error of the training result of this BP neural network about the test set is 4.698%, and the average absolute error of the training result about the validation set is 5.047%, which indicates that this BP neural network has achieved a good training result.

Finally, after training the BP neural network fitted with the input and output values of the AGV system, this paper uses the simulated annealing algorithm to match the input values and optimize the output values of the AGV system. The matching and prediction results of the AEC parameters are shown in table 8.

Table 8. Matching and prediction results of the parameters of the AEC.

Target AEC	Prediction n	Prediction q	Prediction v	Predicted AEC	Calculated AEC	Error
500	1	10	0.02	499.21	483.44	3.26%
400	5	10	0.5	321.2	325.22	1.24%
300	3	2	0.01	300.05	324.38	7.50%
200	4	8	0.01	199.74	222.9	10.39%
100	2	8	0.42	100.05	105.03	4.74%
min	2	1	0.33	30.88	32.04	3.62%

From table 8, the predicted parameters of the AGVs corresponding to each target AEC can be easily obtained using the simulated annealing algorithm. Also, by setting the target AEC as small as possible (or setting it to 0), the optimized value and its optimized parameters for the AGVs can be obtained. The error values between the predicted and calculated values are relatively small in the whole prediction results, which indicates the high accuracy of the whole prediction results.

The optimized parameters of the AGV system carts obtained from table 8 are the number n=2, capacity q=1, and speed v=0.33. Substituting them into the discrete-event simulation model, the Gantt chart of the optimal production solution for the MMAL can be obtained, as shown in figure 9. and the Gantt chart of the optimal distribution scheme for the AGV system is as shown in figure 10.

From figures 9 and 10, it can be seen that the research method and steps proposed in this paper successfully achieved the goal. Ultimately the optimal production scheme of the MMAL and the optimal distribution scheme of the AGV system with minimal AEC is obtained.

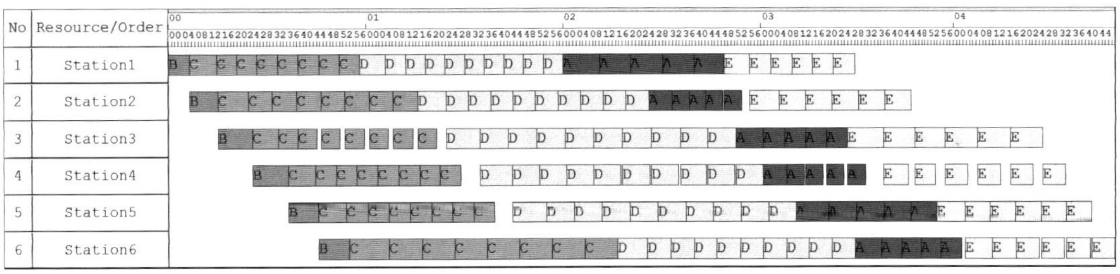

Figure 9. Gantt chart of the optimal production solution for the MMAL.

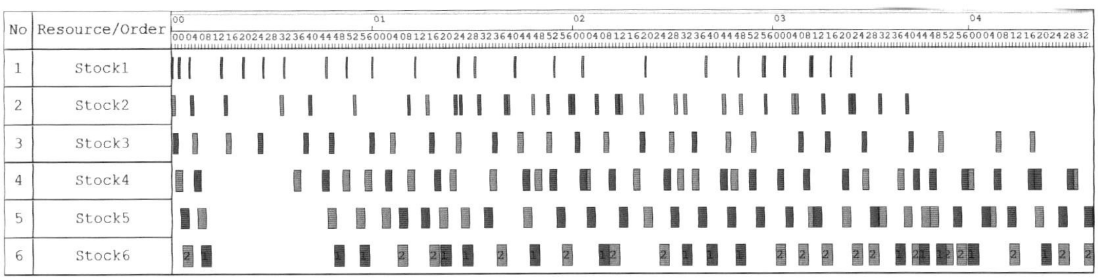

Figure 10. Gantt chart of the optimal distribution scheme for the AGV system.

6. Conclusion

- This paper constructs the general architecture of the intelligent mixed-model assembly line and its AGV system. It derives their mixed-integer programming model, which is the basis for studying the problem of parameter matching and energy consumption optimization of the mixed-model assembly line AGV system. Then, using Siemens Tecnomatix Plant Simulation

13.1 software and SimTalk programming language, we successfully conducted modelling and experiments for discrete event simulation of AGV systems on MMALs.

- In the computational example of discrete-event simulation, the proposed research methods and steps integrate distribution rules, genetic algorithm, experimental design, BP neural network, and simulated annealing algorithm to achieve the parameter matching and energy consumption optimization of the mixed-model assembly line AGV system. The optimal production solution and delivery scheme of the MMAL and the AGV system under the minimum AGV system energy consumption is obtained.
- In the process of experimental design and ANOVA, we found the performance saturation phenomenon of the AGV system, and with the increase in the number of AGV carts, the more likely the AGV system is to have performance saturation. At the same time, we found that the number of carts, capacity, and speed of the AGV system significantly impact the energy consumption of the AGV system. Among them, the number of carts has the most negligible impact, and the capacity has the most significant impact.

Acknowledgments
This work is supported by the National Key R&D Program, China (2019YFB1706102).

References

[1] Xu L D, Xu E L and Li L 2018 Industry 4.0: state of the art and future trends *Int. J. Prod. Res* **56** (8) 2941-62.
[2] Zhang Y, Feng Y and Rong G 2016 Reference model and key technology of smart factory *Computer Integrated Manufacturing Systems* **22** (01) 1-12.
[3] Schmid NA, Limère V and Raa B 2021 Mixed model assembly line feeding with discrete location assignments and variable station space *Omega* **102** 102286.
[4] Yuan M, Li D B and Yu M J 2008 Research on mixed-model assembly line balance for mass customization *Computer Integrated Manufacturing Systems* **14** (1) 79.
[5] De Ryck M, Versteyhe M and Debrouwere F 2020 Automated guided vehicle systems, state-of-the-art control algorithms and techniques *Journal of Manufacturing Systems* **54** 152-73.
[6] Udhayakumar P and Kumanan S 2010 Task scheduling of AGV in FMS using non-traditional optimization techniques *International Journal of Simulation Modelling* **9** (1) 28-39.
[7] Rao Y Q, Wang M C, Wang K P and Wu T M 2013 Scheduling a single vehicle in the just-in-time part supply for a mixed-model assembly line *Computers & Operations Research* **40** (11) 2599-610.
[8] Rahman H F, Janardhanan M N and Nielsen P 2020 An integrated approach for line balancing and AGV scheduling towards smart assembly systems *Assembly Automation* **40** (2) 219-234.
[9] Nunes V A and Barbosa G F 2020 Simulation-based analysis of AGV workload used on aircraft manufacturing system: a theoretical approach *Acta Scientiarum-technology* **42** (1).
[10] Erl T 1900 Service-oriented architecture *Upper Saddle River: Pearson Education Incorporated.*
[11] Tonelli F, Paolucci M, Anghinolfi D and Taticchi P 2013 Production planning of mixed-model assembly lines: A heuristic mixed integer programming based approach *Production Planning & Control* **24** (1) 110-27.
[12] Kalra M and Singh S 2015 A review of metaheuristic scheduling techniques in cloud computing *Egyptian informatics journal* **16** (3) 275-95.
[13] Affenzeller M, Wagner S, Winkler S and Beham A 2009 *Genetic Algorithms and Genetic Programming: Modern Concepts and Practical Applications* Chapman and Hall/CRC.
[14] Wu C, Yan Y, Cao Q, Fei F, Yang D, Lu X, Xu B, Zeng H and Song A 2020 sEMG measurement position and feature optimization strategy for gesture recognition based on ANOVA and neural networks *IEEE Access* **8** 56290-9.

[15] Yalcin U, Karaoglan A D and Korkut I 2013 Optimization of cutting parameters in face milling with neural networks and Taguchi based on cutting force, surface roughness and temperatures *International Journal of Production Research* **251** (11) 3404-14.
[16] Stefanoiu D, Borne P, Popescu D, Filip F G and El Kamel A 2014 *Optimization in Engineering Sciences: Metaheuristic, Stochastic Methods and Decision Support* John Wiley & Sons.
[17] Padhi S S, Wagner S M, Niranjan T T and Aggarwal V 2013 A simulation-based methodology to analyse production line disruptions *International Journal of Production Research* **51** (6) 1885-97.

A Novel Privacy-Preserving Computing System Based on VAE Federated Meta-Learning

Wei Zheng[1], Shaojie Yang[2*] and Xueyang Zhang[2]

[1] Information Department, Beijing University of Technology, Beijing 100124, China
[2] China Academy of Information and Communications Technology, Beijing 100191, China
Email: zhengwei@caict.ac.cn; yangshaojie@caict.ac.cn; zhangxueyang@caict.ac.cn

Abstract. In recent years, the release of data value under privacy-preserving has always been the research focus. As an essential data protection technology, encrypted computing provides a good solution. Based on this, a federated learning model is first constructed to meet the needs of participants using local data for joint modeling. Secondly, the meta-learning method accelerates the federated learning model's training effect and improves the accuracy of model detection. Finally, the variational auto-encoder is used to optimize the federated learning model, which speeds up the training of the federated learning model and improves the system's overall security. Experimental results show that the proposed system can provide secure computing conditions for each data provider and better performance.
Keywords. Federated learning; meta-learning; privacy-preserving; neural network; variational auto-encoding.

1. Introduction

With the maturity of Internet technology, big data-driven artificial intelligence has completed large-scale and complex learning tasks in many fields, such as government affairs, finance, medical care, and autonomous driving. Traditional machine learning models need to collect local data and train them intensively. However, many data are closely related to individual users and may directly contain identifying information such as names, IDs, mobile phone numbers, and disease information. This information has a severe risk of leakage during traditional machine learning models' data collection or training [1, 2]. Federated learning (FL) was first proposed in 2016 [3]. As a particular distributed machine learning model, federated learning can realize the joint training task of multiple devices without the data being in the local domain. Local data privacy can be protected to a certain extent. Since then, FL has developed rapidly and has become a research hotspot at the intersection of artificial intelligence and privacy protection [4, 5].

Meta-learning is a promising approach with low resource requirements. In particular, model-agnostic meta-learning frameworks can quickly adapt to various neural network models to solve different learning tasks [6]. Reference [7] serves the bank's anti-fraud detection work by building a federated meta-learning model (FML). The experiment proves that the federated meta-learning model performs better than the traditional federated learning model. However, the security of the federated meta-learning model still faces many challenges. For example, some information is often hidden in the gradient value and can be used to speculate the original data. The private training data can also be reconstructed [8, 9]. A variational auto-encoder (VAE) is a directed model that uses learned approximate inference and can be trained purely using gradient-based methods [10]. Since VAE has blurring samples and back-propagating gradient values [11, 12], it is theoretically possible to clarify that the federated meta-learning model optimized by VAE can further improve the security and accuracy of the algorithm and the training speed.

This article is organized as follows. Section 2 presents the architectural design of a federated meta-learning system optimized by a variational auto-encoder. Section 3 introduces the algorithm realization of the system. Section 4 compares and analyzes the experimental results of References [13, 14]. Section 5 concludes this paper.

2. System Framework

Assuming that each data provider is clear about the data labels and features to carry out related work, the system architecture is shown in figure 1. First, the coordination center issues a general global model for each device for subsequent local training. Second, the local device trains the downloaded model, optimizes the model using a variational auto-encoder and uploads the optimized gradient information to the coordination center using the SHA-256 algorithm [15]. Then, the coordination center obtains the gradient information uploaded by each local node through the meta-learning method for optimization training and sends the optimal global model to each local model again for updating. Finally, the above process is repeated until the global models delivered in the system are entirely converged.

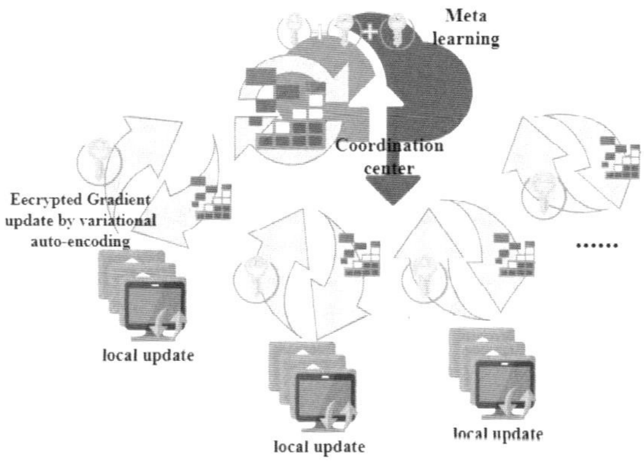

Figure 1. VAE federated meta-learning design.

3. Function Implementation

3.1. Implementation of Federated Learning

This system adopts the training method of horizontal federated learning (HFL), and the underlying algorithm is implemented by CNN [16, 17]. HFL can be applied to scenarios where datasets in different locations share overlapping feature spaces but different sample spaces. Specifically, the preconditions for the realization of horizontal federated learning with two participants are:

$$X_i = X_j, Y_i = Y_j, I_i \neq I_j \tag{1}$$

In equation (1), (X_i, X_j) and (Y_i, Y_j) indicate that the data feature labels in the data of both parties are the same as the label space, and (I_i, I_j) indicates that the user identifiers of both parties are different. In addition, the sum of data sets of both parties $\forall D_i, D_j, i \neq j$ should be satisfied.

3.2. Variational Auto-Encoding & Federated Meta-Learning

After the federated learning model is constructed, the federated learning model is optimized using the VAE algorithm in combination with Reference [18], and the stochastic gradient descent (SGD) is used as the meta-learning optimizer [19] in order to calculate the optimal gradient value, the specific implementation method is as follows:

$$\theta(X_i) = \frac{V_{random}}{2max(i)} \sum_{i=1}^{max(i)} \left(Y_i^{tru} - Y_i^{pre}\right)^2 + \varepsilon \quad (2)$$

$$\theta(X_j) = \frac{V_{random}}{2max(j)} \sum_{j=1}^{max(j)} \left(Y_j^{tru} - Y_j^{pre}\right)^2 + \varepsilon \quad (3)$$

Equations (2) and (3) is the calculation method of the gradient value of each participant's local federated learning model. V is the smooth random parameter between 0.01 and 0.05, $\left(Y_i^{pre}, Y_j^{pre}\right)$ obtained by the training of (X_i, X_j) and identified as the real label space value$\left(Y_i^{tru}, Y_j^{tru}\right)$in the data set ($D_i$, D_j). The optimal gradient value$\theta_{best}$is calculated by taking the weighted average based on $\left(\theta(X_i), \theta(X_j)\right)$.

4. Experimental Results and Analysis

4.1. Datasets & Experimental Environment

This section uses the ORL face dataset in Reference [20] as training data and uses CNN as the underlying neural network model for testing. In addition, to further ensure users' privacy, we have fuzzed the dataset and trained it. The entire experiment uses two host environments, and the coordination center is deployed in computer 1 using docker technology. The specific experimental configuration is as follows:

Computer 1
- CPU: Inter core i7-6700HQ CPU @ 2.60 GHz.
- GPU: NVIDIA GeForce MX130
- RAM: 16 GB.
- OS: Windows 10.

Computer 2
- CPU: Inter core i7-6700HQ CPU @ 2.60 GHz.
- GPU: NVIDIA GeForce GTX 960M 8G.
- RAM: 8 GB.
- OS: Windows 10.

Comparison methods. By comparing Accuracy and Convergence, we made a comparative analysis with References [13, 14] and reached the experimental conclusion. In order to ensure the objectivity of the experimental results, the limited computing power of References [13, 14] was taken into account, we set the number of local epochs to be 1 for all methods.

4.2. Results & Analysis

The face recognition database ORL was used for prediction experiments in this experiment. ORL data set contains 400 images of 40 different users, each user has 10 images. Each face image has 10304 pixels, graphically displayed in 112 columns and 92 rows, and all images are grayscale. In our experiment, we set the system to an 8:2 ratio for training and prediction and compared the results of the local deployment model Computer 2.

Figure 2. Images for privacy preserving.

As shown in figure 2, federated learning can ensure that local data does not go out of the domain to achieve distributed training with other hosts. However, there is a risk that the data will be reversed in the training process, and there is still a compliance risk in the face image without desensitization. In order to further protect face privacy, we convert face images in the data set into binary grayscale images to confuse the images to protect personal privacy further. On this basis, experiments are carried out, and the training results are shown in figure 3.

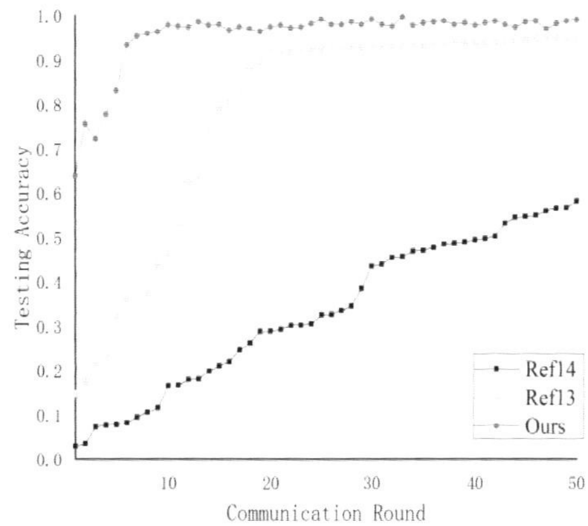

Figure 3. Accuracy based on ORL dataset.

The proposed VAE federated meta-learning was trained on the ORL dataset for 50 epochs. As shown in figure 3, the results show that the VAE Federated meta-learning method has good performance in the face recognition system, and the accuracy rate can reach 99.96%. Although Reference [14] was optimized in each epoch and its accuracy gradually increased, the experimental results showed that Reference [14] needed much training to get better results, and the model could not converge within 50 epochs, resulting in low accuracy. Compared with our method, the model convergence of Reference [13] is relatively slow, requiring 20 epochs to obtain relatively good results, and there is a gap in accuracy compared with our method. To further prove that our method is superior to Reference [13], we compared it with the loss function, and the results are shown in figure 4.

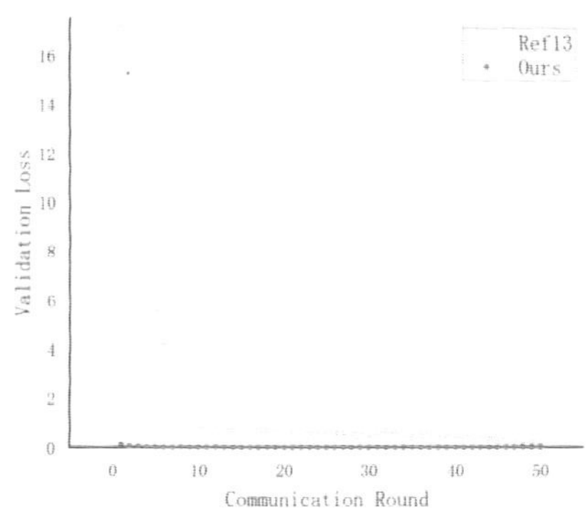

Figure 4. Loss function compared with Reference [13].

As shown in figure 4, the minimum loss function in Reference [13] is 0.12, while the lowest value obtained by training with our method is 0.0032. In addition, Reference [13] needs to go through 10 epochs to reach the lowest loss value, and our loss value shows good performance at the initial stage, which also shows that our method has high accuracy, achieves relatively advanced precision performance, and can realize the protection of personal privacy on the premise of high performance.

The above experiments show that the proposed method has significantly improved the accuracy and convergence, and the security in the data use process is also a significant advantage. This method can achieve multi-party cooperation and obtain relatively accurate results while protecting data privacy.

5. Conclusion

A federated meta-learning system based on the VAE algorithm is proposed in this paper. Using the ORL data set to test, the accuracy of the trained system can reach 99.96%. It is proven that the system can achieve high-performance result output while protecting face information security. The method proposed in this paper can be applied to current face recognition scenarios based on privacy protection. In the future, we will devote ourselves to introducing more privacy protection mechanisms into deep learning algorithms and conduct in-depth studies on algorithm security and data security.

References

[1] Kissner L and Song D 2005 Privacy-preserving set operations *Annual International Cryptology Conf.* pp 241-257.

[2] Kreso I, Kapo A and Turulja L 2021 Data mining privacy preserving: Research agenda *Wiley Interdisciplinary Reviews: Data Mining and Knowledge Discovery* **11** (1) e1392.

[3] McMahan B, Moore E, Ramage D, et al. 2017 Communication-efficient learning of deep networks from decentralized data *Artificial Intelligence and Statistics* 1273-1282.

[4] Yin X, Zhu Y and Hu J 2021 A comprehensive survey of privacy-preserving federated learning: A taxonomy, review, and future directions *ACM Computing Surveys (CSUR)* **54** (6) 1-36.

[5] Ng K L, Chen Z, Liu Z, et al. 2021 A multi-player game for studying federated learning incentive schemes *Proc. of the Twenty-Ninth Inter. Conf. on International Joint Conf. on Artificial Intelligence* pp 5279-5281.

[6] Finn C, Abbeel P and Levine S 2017 Model-agnostic meta-learning for fast adaptation of deep networks *Inter. Conf. on Machine Learning* pp 1126-1135.

[7] Zheng W, Yan L, Gou C, et al. 2021 Federated meta-learning for fraudulent credit card detection *Proc. of the Twenty-Ninth Inter. Conf. on Inter. Joint Conf. on Artificial Intelligence* pp 4654-4660.

[8] Zhu L, Liu Z and Han S 2019 Deep leakage from gradients *Advances in Neural Information Processing Systems* **32**.

[9] Wainakh A, Ventola F, Müßig T, et al. 2021 User label leakage from gradients in federated learning *arXiv preprint arXiv:2105.09369*

[10] Kingma D P and Welling M 2013 Auto-encoding variational bayes *arXiv preprint arXiv:1312.6114*

[11] Takiddin A, Ismail M, Zafar U, et al. 2021 Variational auto-encoder-based detection of electricity stealth cyber-attacks in AMI networks *28th European Signal Processing Conf. (EUSIPCO)* pp 1590-1594.

[12] Sun X and Buettner F 2021 Hierarchical variational auto-encoding for unsupervised domain generalization *arXiv preprint arXiv:2101.09436*

[13] Said Y, Barr M and Ahmed H E 2020 Design of a face recognition system based on convolutional neural network (CNN) *Engineering, Technology & Applied Science Research* **10** (3) 5608-5612.

[14] Miakshyn O, Anufriiev P and Bashkov Y 2021 Face recognition technology improving using convolutional neural networks *IEEE 3rd Inter. Conf. on Advanced Trends in Information Theory (ATIT)* pp116-120.
[15] Gilbert H and Handschuh H 2003 Security analysis of SHA-256 and sisters *International Workshop on Selected Areas in Cryptography* pp 175-193.
[16] Yang Q, Liu Y, Cheng Y, et al. 2019 Federated learning *Synthesis Lectures on Artificial Intelligence and Machine Learning* **13** (3) 1-207.
[17] Aloysius N and Geetha M 2017 A review on deep convolutional neural networks *2017 Inter. Conf. on Communication and Signal Processing (ICCSP)* pp 0588-0592.
[18] Kingma D P and Welling M 2014 Stochastic gradient VB and the variational auto-encoder *arXiv:1312.6114*
[19] Bottou L 2012 Stochastic gradient descent tricks *Neural networks: Tricks of the Trade* (Berlin, Heidelberg: Springer) pp 421-436.
[20] AT&T Database of Faces: ORL face database, available at: http://cam-orl.co.uk/facedatabas.

Toward Efficient Geolocation in SAGIN: A Single-Satellite Geolocation Method Combined with Bandlimited Signal Extrapolation

Yusi Zhang[1], Xuejun Sha[1], Yong Li[2] and Xu Lin[1,3]

[1] School of Electronics and Information Engineering, Harbin Institute of Technology, Harbin 150001, Heilongjiang, China
[2] Science and Technology on Communication Networks Laboratory, Shijiazhuang 050051, Hebei, China
[3] ZTE Corporation, Shenzhen 518057, Guangdong, China
Email: yszhang@hit.edu.cn; shaxuejun@hit.edu.cn; young_li_54@126.com; linxu99@126.com

Abstract. The satellite network, as an essential part of space-air-ground integrated network (SAGIN) with high coverage and mobility, not only expects a promising development but faces increasing interference and tenser satellite spectrum resources. The geolocation function of single satellite is widely used in the low-orbit satellite network but limited to the long measuring duration, which cannot adapt the flexiblity in SAGIN scenario. In this paper, an efficient single satellite geolocation method is proposed combined with bandlimited signal extrapolation in order to save more operation time and ensure the positioning accuracy. Due to the bandlimited characteristic of the radiated signal, information received from the satellite can be further extrapolated so that more doppler frequency values are obtained. The estimated accuracy can be improved with the extra information input to grid search method. Besides, short-term single-satellite geolocation can be realized to meet the higher efficiency requirement of SAGIN. Simulation results demonstrate the superiority of the proposed method in terms of the positioning accuracy compared to the geolocation method without extrapolation.
Keywords. Single-satellite geolocation method; signal extrapolation; radiation source location.

1. Introduction

Space-air-ground integrated network (SAGIN), as an emerging enhanced architecture for future 6G network, makes full use of the advantages of different dimensional networks as well as their respective strengths to realize the exchange and sharing of information and data between different dimensional networks with improved coverage, reliability, mobility and usability [1, 2]. As an extension of the terrestrial communication system before, satellite communication is bound to move along with SAGIN. At present, the application of satellites not only relates to communication, but penetrates into various fields such as geolocation, navigation, military reconnaissance and weather forecasting [3]. The wider-range application brings an increase in the number of satellites in orbit, which in turn leads to tenser satellite spectrum resources and prominent interference problems, thus seriously affects the normal and stable operation of satellite networks [4]. Besides, recognising random unauthorized connectivity provisioning or radiations is one of the key links to guarantee information security [5]. Based on the above considerations, satellite geolocation methods towards radiation source are necessary to be studied.

The satellite passive geolocation system enables to position terrestrial radiation sources, with an advantage of all-weather, all-day, long operating distance and wide coverage. Interferometer measurement is often used to locate the terrestrial radiation source by time difference of arrival in the system. Though the geolocation algorithm is simple, phases of the channels are strictly required to be consistent [6]. The spatial spectrum estimation method can achieve high positioning sensitivity and accuracy, but it has high requirements on the spaceborne antenna array [7]. Due to the high speed of low-orbit satellites (LEO), single-satellite frequency measurement geolocation makes use of the Doppler effect generated by the relative motion between the satellite platform and the radiation source, which is beneficial to enhance positioning accuracy and reliability as well as reduce the systematic cost [8]. However, the single-satellite geolocation method generally requires multiple measurements to ensure accuracy, leading a contradiction between the long operation time and the flexibility of SAGINs. Besides, the measurement accuracy of the Doppler variation rate is expected to reach the order of Hertz, which produces high cost on the instantaneous frequency measurement receiver. Considering the deficiency, research on short-term single-satellite geolocation is quite worthwhile.

In terms of the bandwidth of most terrestrial radiated signals, the bandlimited characteristic enables the usage of extrapolation methods, which can recover the entire signal from a part of it in the time domain. The Gerchberg-Papoulis algorithm (GPA) is a classic iterative extrapolation method for frequency bandlimited signal [9, 10]. For more practical application, GPA is extended to discrete signals and the feasibility of a discrete GPA is analysed in case of uniform samples [10]. Since its simple core computation of discrete Fourier transform (DFT) for discrete signals Fourier transform (FT) for analytic signals, accelerated algorithms and extended applications are studied according to GPA [11, 12]. While the interference signals may present more apparent bandlimited characteristic in some other domain, such as the LFM signals in the fractional domain. In those cases, extrapolation methods have also been modified and analysed [12-15].

In this paper, a novel single-satellite geolocation method combined with bandlimited signal extrapolation is proposed to realize short-term positioning process. The main contributions are summarized twofold:

(1) The extrapolation method is integrated into the geolocation process considering the bandlimited characteristic of emitted signal, so as to realize short-term positioning.

(2) The accuracy of geolocation is improved due to the information received from the extrapolated signal, compared with the grid search method within the same measurement slots.

2. Problem Statement

2.1. Solution and Processing of Single-Satellite Geolocation Method

When the satellite flies over the vicinity of the radiation source, the distance from the radiation source will gradually decrease and then increase. Figure 1 indicates the frequency variation of the received signal on the satellite near the radiation source in the earth centered earth fixed (ECEF) coordinates. For a radiation source transmitting the signal with the frequency f_0, the signal reaches the satellite with varying frequency at different moments during the flight. It can be seen that the frequency of the received signal at the satellite decreases to f_0 when the satellite approaches to the right above location of the radiation source, and then keep decreasing with the satellite flying away.

In addition, due to the radial acceleration between the high-speed satellite and the terrestrial radiation source, doppler shift exists in the signal received by the satellite receiver [16]. Generally, the emitter can be considered stationary with respect to the satellite's velocity. When the satellite is closest to the radiation source (directly above the radiation source), the Doppler shift at that moment reaches the minimum. Assuming a stationary radiation source on the earth's ellipsoidal surface, the earth centred earth fixed (ECEF) coordinates are (x, y, z) and those of satellite at that moment t_n are (x_n, y_n, z_n). The velocity vector of the satellite at t_n is $\vec{v}_n = (v_{n,x}, v_{n,y}, v_{n,z})$. In the ECEF model in figure 1, the orientation of the radiation source (R) and satellite (S) at t_n can be expressed with vector $\vec{r} = \overrightarrow{OR}$

and $\vec{s}_n = \overrightarrow{OS}$, respectively. The doppler shift as well as ephemeris information such as the position and speed of the satellite at several moments enable to estimate the location of the radiation source.

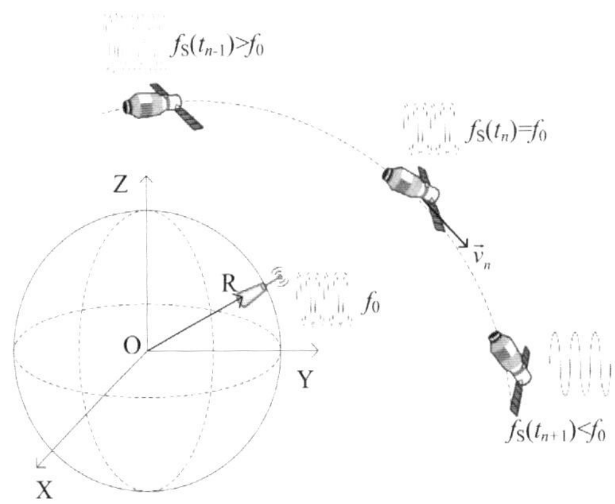

Figure 1. The frequency variation received on the satellite near the radiation source in ECEF.

The relative position between R and S at t_n can be denoted as $\vec{r}_n = \overrightarrow{SR} = \overrightarrow{OR} - \overrightarrow{OS}$, with the unit vector $\vec{u}_n = \vec{r}_n / r_n$ where r_n is the modulus of $\vec{r}_n$. At the moment t_n, the angle between the velocity of S ($\vec{v}_n$) and the unit vector from the satellite to the surface radiation source $\vec{u}_n$ is denoted by α_n as

$$\cos\alpha_n = \frac{(\vec{v}_n \cdot \vec{u}_n)}{|\vec{v}_n|} \tag{1}$$

The modulus of the relative velocity between R and S is expressed as $|\vec{v}_n|\cos\alpha_n = \vec{v}_n \cdot \vec{u}_n$. The relative velocity influences the delay of the moment when satellite receives the signal compared to the moment when the radiation source emits the signal, calculated as

$$\tau_n \approx \frac{r_n - (\vec{v}_n \cdot \vec{u}_n)(t_{n+1} - t_n)}{c} \tag{2}$$

where τ_n denotes the delay between the emitted moment and the received moment and c is the velocity of light.

It can be explained that if R emits a signal $S(t)$ at the moment t_n, the signal received at S at the moment t_{n+1} will be $S_n(t) = S(t - \tau_n)$. For simplicity, assume the R emits a signal $S(t) = A(t)\cdot\cos[\omega t + \varphi(t)]$ with the narrow bandwidth σ at $t = 0$, noted that the $A(t)$ and $\varphi(t)$ both vary slowly. The received signal at the nth measurement slot ($\tau_n \sim \tau_{n+1}$) will be

$$S_n(t) \approx A_n \cos\left\{\omega\left[1 + \frac{(\vec{v}_n \cdot \vec{u}_n)}{c}\right]t + \varphi_n - \omega\frac{r_n}{c}\right\} \tag{3}$$

where A_n denotes the initial value of $A(t)$ at the beginning of the slot (τ_n), φ_n denotes the initial value of $\varphi(t)$ at τ_n and ω is the frequency of the emitted signal from R. The doppler phase shift of the nth frequency measurement slot ω_n^d can be expressed as

$$\omega_n^d = \omega\left[1+\frac{(\vec{v}_n \cdot \vec{u}_n)}{c}\right]+\varepsilon_n \tag{4}$$

where ε_n is the error at the nth measurement slot.

An iso-frequency conic line can be obtained according to the doppler phase shift in equation (4), with the position of the satellite as the vertex, the velocity vector as the axis and the cone passing through the radiation source, shown in figure 2. Several iso-frequency conic lines which all cover the radiation source could be received from several frequency measurements, hence the overlapping area gradually shrinks and the accuracy of geolocation gradually lifts. Ideally, those iso-frequency cones would have two intersection points, which are the position of the radiation source (R) and the symmetry point of R with respect to the satellite's trajectory.

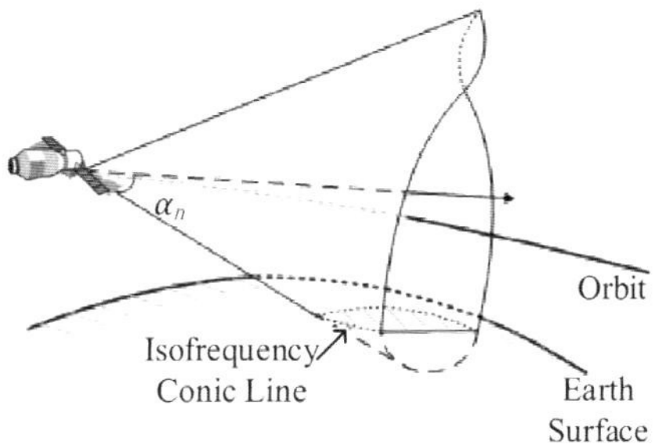

Figure 2. The diagram of the iso-frequency conic line received at the nth measurement slot.

To estimate the location of R, namely (x, y, z), equation (4) should be rewritten as

$$\omega_n = \omega\left(1+\dot{x}_n\frac{x-x_n}{cr_n}+\dot{y}_n\frac{y-y_n}{cr_n}+\dot{z}_n\frac{z-z_n}{cr_n}\right)+\varepsilon_n, \qquad n=1,\cdots N \tag{5}$$

where the location of the satellite (x_n, y_n, z_n) can be obtained from the satellite ephemeris, the velocity at nth measurement slot $\vec{v}_n = (\dot{x}_n, \dot{y}_n, \dot{z}_n)$. Since only the radiation source on the ground is considered, the ECEF coordinates (x, y, z) meet the earth surface reference model as

$$\frac{x^2}{a^2}+\frac{y^2}{a^2}+\frac{z^2}{b^2}=1 \tag{6}$$

where a denotes the length of Earth's semimajor axis (roughly equals to 6378 kilometres) and b denotes the length of Earth's semi-minor axis (roughly equals to 6357 kilometres).

Due to the difficulty of solving N nonlinear equations as equations (5) and (6) which is always ill-posed, the grid search method is often taken into use. To reduce the number of unknowns, the equation (6) is substituted into equation (5) as

$$\begin{aligned}\omega_n &= \omega\left(1+\dot{x}_n\frac{x-x_n}{cr_n}+\dot{y}_n\frac{y-y_n}{cr_n}+\dot{z}_n\frac{\pm\frac{b}{a}\sqrt{a^2-x^2-y^2}-z_n}{cr_n}\right)+\varepsilon_n \\ &\triangleq \omega g_n(x,y)+\varepsilon_n, \qquad n=1,\cdots N\end{aligned} \tag{7}$$

and can be expressed in matrix form as

$$\mathbf{\Omega}=\mathbf{G}\omega+\mathbf{E} \tag{8}$$

where $\mathbf{\Omega}=[\omega_1,\cdots,\omega_N]^T$, $\mathbf{G}=[g_1(x,y),\cdots,g_N(x,y)]^T$ and $\mathbf{E}=[\varepsilon_1,\cdots,\varepsilon_N]^T$. The objective function can be expressed as

$$J(\hat{\omega})=\mathbf{\Omega}-\mathbf{G}\hat{\omega} \tag{9}$$

The solution of equation (8) is equivalent to the MMSE solution of the objective function as

$$\arg\min_{\hat{\omega}}\left\|\mathbf{\Omega}-\mathbf{G}\hat{\omega}\right\|_2 \tag{10}$$

The minimum squares solution for the estimated value $\hat{\omega}$ is

$$\hat{\omega}=(\mathbf{G}^T\mathbf{G})^{-1}\mathbf{G}^T\mathbf{\Omega} \tag{11}$$

It is obvious that more measurement values increase the accuracy of geolocation but spend longer measuring time. To estimate the location of a radiation source more efficiently, an improved geolocation method combined with bandlimited signal extrapolation is proposed.

2.2. GPA for Frequency Bandlimited Signals

Assuming a frequency bandlimited signal $f(t)$ satisfying

$$F(\omega)=0,\quad \forall\omega\notin[-\sigma,\sigma] \tag{12}$$

where σ is real and $F(\omega)$ is the Fourier transform result of $f(t)$. The initial known signal is

$$g(t)=\mathcal{U}_T(t)f(t) \tag{13}$$

where $\mathcal{U}_T(t)$ denotes a time-domain filter only remaining the part within [-*T*, *T*] and *T* is positive.

To extrapolate $f(t)$ from $g(t)$, GPA iterations are conducted as follows. The initial signal of the first iteration is $f_0(t)=g(t)$ and the *i*th iteration consists of 4 steps:

- $f_{i-1}(t)$ is input as the initial signal of the *i*th iteration;
- The signal with the lowpass bandwidth σ in the frequency domain is filtered;
- The part of the filtered signal within [-*T*, *T*] is replaced with $g(t)$;
- The replaced signals $f_i(t)$ is output as the result of *i*th iteration.

3. Methodology

3.1. Geolocation Method Combined with Bandlimited Signal Extrapolation

In this paper, the geolocation for a radiation source on the ground is considered. In SAGINs, the satellite may not be used for geolocation specifically, hence the received signal from a radiation source could be partial. Due to the bandlimited characteristic of the emitted signal, a geolocation method combined with extrapolation are described in table 1.

Table 1. The description of proposed geolocation method.

	Description
Step 1	Satellite ephemeris data are read and they are marked with $\tau_1,\ldots,\tau_N$. The coordinates (x_n,y_n,z_n) are converted to (x_n,y_n) according to equation (6), $n=1,2,\ldots,N$;
Step 2	The measured frequency ω_n is calculated according to equation (7), $n=1,2,\ldots,N$;

	Description
Step 3	The difference of $\{\omega_n\}_{n=1}^{N}$ and received $\{\Delta\omega_n\}_{n=1}^{N-1}$ is calculated;
Step 4	The variation trend of $\{\Delta\omega_n\}_{n=1}^{N-1}$ is judged: (1) If $\{\Delta\omega_n\}_{n=1}^{N-1} > 0$, the satellite has passed radiation source (R) and is moving away from the R, named condition (1); (2) If $\{\Delta\omega_n\}_{n=1}^{N-1} < 0$, the satellite has not passed R and is moving towards the R, named condition (2); (3) If $\{\Delta\omega_n\}_{n=1}^{N-1}$ begins from lower than 0 and keep increasing to more than 0, the satellite can be regarded at right above R during the slot when $\Delta\omega_n$ is closest to 0. In this paper, we only consider the condition (1) and (2).
Step 5	Condition (1): The received signal is zero-padded at the beginning with the same sampling rate and duration (from $\tau_{-(N-1)}$ to τ_0); Condition (2): The received signal is zero-padded at the end with the same sampling rate and duration (from $\tau_{(N+1)}$ to τ_{2N+1});
Step 6	The signal at the zero-tapped area is extrapolated according to Subsection 3.2. The frequency values at corresponding slots are obtained as well.
Step 7	Satellite ephemeris data at the extrapolated slots are read and the coordinates are converted as conducted in Step 1.
Step 8	The current sub-satellite point is taken as the centre and meshed with specific degree of latitude and longitude.
Step 9	The satellite position, speed, and measurement frequency are substituted into equation (9) to calculate the error of each grid point.
Step 10	The sub-satellite point corresponded to the grid point with minimum error is calculated. It is taken as the new centre and meshed with finer degree.
Steps 9 and 10 are repeated until the reach the stopping condition of meshing.	
Step 11	The location corresponding to the minimum module value of the error is output.

In Step 11, two sets of location coordinates will be obtained as aforementioned, which are the estimated position of the radiation source and the symmetrical point about the satellite's trajectory. By adding direction measurement information, mostly DoA, the symmetrical point can be eliminated [17].

3.2. Extrapolation Process

In the geolocation scenario, the processed signals are always with high central frequency. In order to facilitate the sampling and filtering in the extrapolation process, the signal should pass through a digital downconverter.

Assume the received signal has been converted to baseband. Since the slowly-varying characteristic as described in equation (3), the known signal can be spliced as

$$g(t)=\sum_{n=1}^{N} g(t;\tau_n,\bar{\omega}_n) \tag{14}$$

where $\bar{\omega}_n$ is the down-converted result of ω_n. From the received signal, the type of the signal can be roughly estimated through AI technology [18].

For instance, the radiation signal is a linear frequency modulation (LFM) signal with the form of $f_{LFM}(t) = A_0 e^{j(2\pi f_0 t + \pi k t^2)}$. Though the phase of $f(t)$ varies at a speed of $2\pi kt$ and the variation is continuous, the phase of the spliced signal $g_{LFM}(t)$ according to equation (14) does not vary smoothly. During the nth slot, the phase $\bar{\omega}_n$ is considered invariable. The entire spliced $g_{LFM}(t)$ can be expressed as

$$g_{LFM}(t) = A_0 \sum_{n=1}^{N} \mathcal{U}_{\Delta\tau}\left(t - \left(n + \frac{1}{2}\right)\Delta\tau\right) e^{j(\bar{\omega}_0 + n2\pi k \Delta\tau)t} \tag{15}$$

where A_0 is the amplitude, k is the frequency modulation rate and $\Delta\tau$ is the uniform measurement interval. Considering the practical extrapolated signal is discrete, $g(t)$ is discretized to g with a sampling frequency higher than 2 times of its maximum estimated bandwidth.

$$g = \left[g_0, g_1, \ldots, g_{M-1}\right]^T \tag{16}$$

For condition (1), the initial signal after zero-padding can be expressed as

$$g_{zp,1} = \left[\underbrace{0, \ldots, 0}_{M}, g_0, g_1, \ldots, g_{M-1}\right]^T \tag{17}$$

For condition (2), the initial signal after zero-padding can be expressed as

$$g_{zp,2} = \left[g_0, g_1, \ldots, g_{M-1}, \underbrace{0, \ldots, 0}_{M}\right]^T \tag{18}$$

According to the zero-padding location, the index matrix of the initial signal is $\Psi = diag(\varphi)$, where φ is a vector consisting of 0 and 1. When $g_{zp}(m)$ is zero-tapped with 0, $\varphi(m) = 0$; otherwise, $\varphi(m) = 1$. It can be estimated that the original signal is bandlimited in some domain, and the transfer matrix from time domain is $\mathcal{F}$. A lowpass filter $\mathbf{H}_L$ is generated with bandwidth L, which is not less than the estimated maximum bandwidth. The extrapolated result of ith iteration can be expressed as

$$\begin{aligned} \hat{g}_i &= (I - \Psi)\mathcal{F}^{-1}\mathbf{H}_L \mathcal{F}\hat{g}_{i-1} + g_{zp}, \quad i > 1 \\ \hat{g}_0 &\triangleq g_{zp} \end{aligned} \tag{19}$$

where I denotes the unit matrix with $2M$ orders. The stopping condition of the iteration can be set as a maximum iteration [12].

4. Experimental Analysis

Simulations are conducted to verify the feasibility of the proposed method. The central frequency of emitted signal is 10 GHz. The height of a circular LEO orbit is 500 km. As shown in figure 3, a radiation source is set at (37.9386° N,115.378° E). Table 2 demonstrates a part of the satellite ephemeris.

Assume the radiation source stationary (R) sends the LFM signal. At a certain moment near the sub-satellite point of the satellite, the grid is meshed according to the latitude and longitude density of 1° firstly. The stopping condition of the refined grid is that the grid density reaches 0.0001° of latitude and longitude. Note that the latitude and longitude coordinates of each grid point should be converted into Cartesian coordinates, as (x, y, z) described in equation (5). The reciprocal of the module value of the error E has been curved for the observation convenience. According to equation (10), the minimum error should be taken, namely the peak value.

It can be seen in figure 4a that when meshing for the first time, there are two sets of latitude and longitude coordinates corresponding to the highest peak. It is attributed to the axial symmetry of the iso-frequency cone. The iso-frequency cone will intersect at two points, which are the radiating source and the symmetric point of radiating source about the satellite's sub-satellite point. According to the last peak position, the grid is refined on both sides of the satellite sub-satellite trajectory and continue to calculate the measurement error, as shown in figure 4b.

Table 2. The location and velocity in the satellite ephemeris.

Time (UTCG)	x (km)	y (km)	z (km)	v_x (km/s)	v_y (km/s)	v_z (km/s)
08: 02: 42.814	-725.647017	6302.318730	2657.635206	-5.122619	-2.490740	4.507852
08: 03: 42.000	-1027.799448	6142.674363	2918.902161	-5.082678	-2.896272	4.305345
08: 04: 42.000	-1331.143310	5956.676720	3170.941469	-5.023544	-3.295152	4.081154
08: 05: 42.000	-1630.385453	5747.140321	3408.963370	-4.945783	-3.680139	3.838922
08: 06: 42.000	-1924.414098	5514.945949	3631.915669	-4.849606	-4.049702	3.579719
08: 07: 42.000	-2212.132183	5261.064290	3838.812785	-4.735295	-4.02378	3.304693
08: 08: 42.000	-2492.461419	4986.551818	4028.740113	-4.603195	-4.736774	3.015057

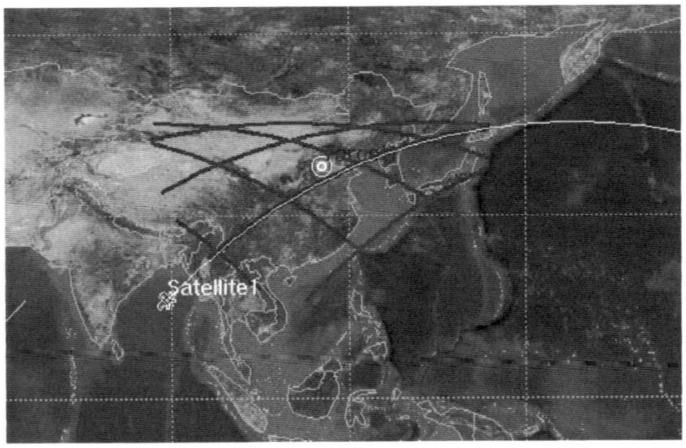

Figure 3. The diagram of the satellite and radiation source.

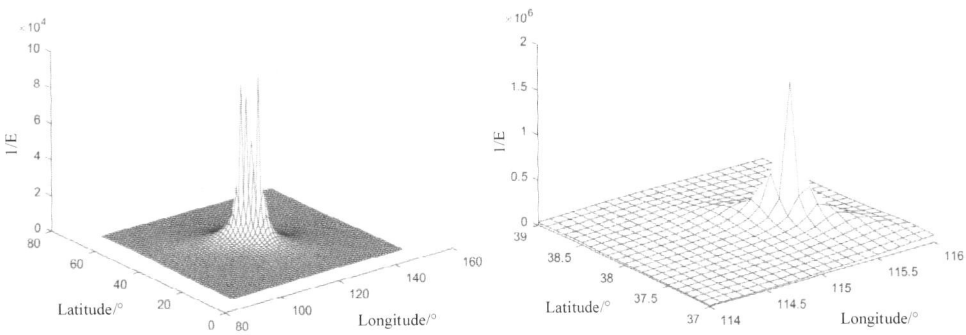

(a) Values at meshing for the first time (b) Values at meshing for the second time

Figure 4. The frequency measurement error at each grid point (proposed method).

The geometric coordinates received from the proposed method are $\hat{R}_1 = (37.9072°N, 115.3834°E)$ and $\hat{R}_2 = (34.4158°N, 117.9483°E)$. The DoA information could help discard the result $\hat{R}_2$.

As a benchmark, the error values at the first/second mesh of a direct grid search method [17] are indicated in figure 5. It can be seen from figure 5a that beyond the symmetrical high values, a peak exists solely which should be neglected due to its asymmetry. Similar process for the choice of peak values should also be conducted in the next meshing level.

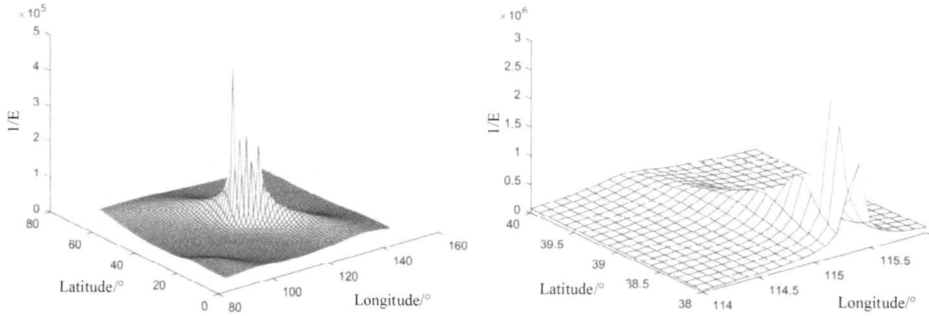

(a) Values at meshing for the first time (b) Values at meshing for the second time

Figure 5. The frequency measurement error at each grid point (traditional method).

With the same stopping meshing condition, the geometric coordinates are respectively $\hat{R}_{bm,1} = (38.0890°N, 115.3296°E)$ and $\hat{R}_{bm,2} = (34.3240°N, 118.0110°E)$, of which $\hat{R}_{bm,2}$ can be discarded as well. Table 3 compares the ECEF coordinates received from proposed method and other grid search method without extrapolation, of which the geolocation errors are 9.3354 km and 17.1785 km, respectively. The error of the proposed method reduces to nearly half, demonstrating the superiority in terms of the accuracy of geolocation.

Table 3. The comparison of the ECEF coordinates of R and estimated results.

	x (km)	y (km)	z (km)
Radiation source	-2158.6	4550.6	3900.1
Results of the proposed method	-2162.9	4558.5	3902.6
Results of the traditional method	-2150.4	4543.1	3913.2

5. Conclusions

In this paper, a single-satellite passive geolocation method combine with extrapolation is proposed to meet the efficiency requirement of SAGIN. Considering the bandlimited characteristic of the emitted signal, the idea of bandlimited extrapolation is adopted in the proposed method to obtain the signal information at more time slots. The initial signal to be extrapolated is generated by splicing the signal received by satellites in different measuring slots, thus more measurement frequencies can be obtained through longer time-domain signal. The Doppler frequency values of the known/extrapolated signal and the satellite ephemeris information are then substituted into the grid search method to get the equivalent solution to the equations set of the radiation source's ECEF coordinates. When the measuring duration is limited, the proposed method possesses lower estimated error compared to the geolocation method without extrapolation. Moreover, the efficiency advantage of the proposed method is also obvious since the estimated error with short-term geolocation is still tolerable. Simulation results show that the error of the proposed method can reduce to nearly half of the contrast method.

Acknowledgments
This research was funded by National Natural Science Foundation of China under Grant 62171151 and by Harbin Institute of Technology Original Frontier Exploration Fund CLASS A under Grant HIT.OCEF.2021012.

References

[1] Fang X, Du Z, Yin X, Liu L and Sha X 2022 Toward physical layer security and efficiency for SAGIN: A WFRFT-based parallel complex-valued spectrum spreading approach *IEEE Trans. Intell. Transp. Syst.* **23** (3) 2819-2829.
[2] Li W, Liu Y, Li X and Shen Y 2022 Three-dimensional cooperative localization via space-air-ground integrated networks *China Commun.* **19** (1) 253-263.
[3] Feng Y, Fang X, Li C and Sha X 2020 Frequency-domain dual component computation diversity design for air-ground integrated mobile communication *2020 Inter. Conf. on Space-Air-Ground Computing (SAGC)* pp 38-42.
[4] Zhou J, Yang H and Anderson K 2019 SNPP ATMS On-orbit geolocation error evaluation and correction algorithm IEEE Geosci. Remote Sens. 57 (6) 3802-3812.
[5] Fang X, Yin X, Sha X, Qiu J and Zhang H 2021 Catalyzing Random Access at Physical Layer for Internet of Things: An Intelligence Enabled User Signature Code Acquisition Approach *China Commun.* **18** (10) 181-192.
[6] Tian D, Wang S and Yang L 2018 Research on the influence of stations distribution to HF TDOA geolocation accuracy *2018 12th Inter. Symp. on Antennas, Propagation and EM Theory (ISAPE)* pp 1-4.
[7] Adeogun R 2014 A robust MUSIC based scheme for interference location in satellite systems with multibeam antennas *Int. J. Compu. Appl.* **82** (12) 1-6.
[8] Li S, Zhang Q, Deng B, Wu B and Gao Y 2022 A fast and accurate LEO satellite-based direct position determination assisted by TDOA measurements *China Commun.* **19** (1) 92-103.
[9] Papoulis A 1975 A new algorithm in spectral analysis and band-limited extrapolation *IEEE Trans. Circuits Syst.* **22** 735-742.
[10] Gerchberg R 1974 Super-resolution through error energy reduction *Opt. Acta Int. J. Opt.* **21** 709-720.
[11] Sanz J and Huang T 1983 Some aspects of band-limited signal extrapolation: Models, discrete approximations, and noise *IEEE Trans. Acoust. Speech Signal Process* **31** (6) 1492-1501.
[12] Zhang Y, Fang X, Sha X and Li X 2022 Piecewise iterative extrapolation method for bandlimited signal *Electronics* **11** (8) 1175-1188.
[13] Zhang Y, Li Y, Fang X, Sha X, Feng Y and Wang W 2022 Towards spectral efficiency enhancement for IoT aided smart transportation: A compressive OFDM transmission and regularized recovery approach *EURASIP J. Adv. Signal Process* **29** 1-16.
[14] Xu S, Tao S, Chai Y, Yang X and He Y 2018 The extrapolation of bandlimited signals in the offset linear canonical transform domain *Optik* **180** 626-634.
[15] Shi J, Sha X, Zhang Q and Zhang N 2011 Extrapolation of bandlimited signals in linear canonical transform domain *IEEE Trans. Signal Process* **60** (3) 1502-1508.
[16] Nuhoglu M, Kemal A, Bayri A and Cirpany H 2020 A new iterative method for passive doppler geolocation based on semi-definite programming *2020 28th European Signal Processing Conference (EUSIPCO)* pp 1812-1816.
[17] Ellis P, Rheeden D and Dowla F 2020 Use of doppler and doppler rate for RF geolocation using a single LEO satellite *IEEE Access* **8** 12907-12920.
[18] Carrino F, Janka A, Abou O and Mugellini E 2019 LoRaLoc: Machine learning-based fingerprinting for outdoor geolocation using LoRa *2019 6th Swiss Conf. on Data Science (SDS)* pp 82-86.

A Deep Learning-Based Cascade Automatic Classification System for Malignant Thyroid Nodule Recognition in Ultrasound Image

Haoyuan Zhou[1], Rui Wang[1*], Meng Zhou[1], Peng Fu[2] and Yang Bai[3]

[1] School of Instrumentation and Optoelectronic Engineering, Beihang University, Beijing, China
[2] Department of Ultrasound, Peking University Third Hospital, Beijing, China
[3] Department of general surgery, Peking University Third Hospital, Beijing, China
Email: *wangr@buaa.edu.cn

Abstract. As the most common malignancy in the endocrine system, thyroid cancer is often determined based on ultrasound images by professional radiologists based on ultrasound images. In this study, a cascade automatic classification system (CACS) is proposed to accurately achieve automatic segmentation and classification of thyroid nodules, reducing over-reliance on experienced doctors in the diagnosis. Firstly, the Res-UNet is used to segment the region of interest (ROI) of thyroid nodules in ultrasound images. Then, to overcome the limitations of the small sample size of medical training data as well as category imbalance, the modified Conditional Variational Auto-Encoder (CVAE), learning the probability distribution of the ultrasound images, is trained to produce new medical ultrasound images for ROI data augmentation. Finally, we perform the classification task by a three-branch ensemble network combined with clinical diagnostic features of thyroid nodule such as the aspect ratio, internal and external echoes, and edges. The proposed CACS can identify benign and malignant thyroid nodules with an accuracy of 0.892 and an AUC of 0.887, according to testing on real patient data.
Keywords. Thyroid nodules; ultrasound image; cascade classification; data augmentation.

1. Introduction

Thyroid nodules commonly appear in a variety of thyroid diseases, such as nodular goiter, benign and malignant thyroid tumors. As a common endocrine cancer, the incidence of thyroid cancer has increased globally over the last 30 years [1]. Moreover, researches have demonstrated that high-resolution ultrasonography can detect thyroid nodules in 19%-68% of populations chosen at random, while the incidence of malignant nodules, i.e., thyroid cancer, is 7%-15% [2]. Therefore, accurate diagnosis of malignant thyroid nodules has always been the focus of clinical attention.

Currently, ultrasonography has become one of the most common methods for clinical diagnosis of thyroid nodules owing to its safe, convenient, non-invasive and inexpensive. However, ultrasound images are affected by speckle noise, and the image quality is relatively low. Thus, accurate identification of nodule location by radiologists is a time-consuming process. Meanwhile, the identification of benign and malignant nodules is heavily dependent on the clinical experience of radiologist. Consequently, the subjectivity of the doctor has a great impact on the diagnosis.

However, computer-aided diagnosis (CAD) is a computer technique that will considerably increase diagnostic accuracy and minimize over-reliance on professional sonographers to assist radiologists in the diagnosis of thyroid cancers. Therefore, based on computer-aided technology, a cascade automatic classification system (CACS) for thyroid nodule in ultrasound image is crucial.

2. Related Works

2.1. Segmentation Method

The segmentation methods commonly employed in thyroid nodules are mainly divided into the following categories: active contour [3], feature extraction and classification [4], and deep learning [5], etc. Although the traditional segmentation methods have made some progress in CAD, the feature selection and parameter setting have more subjective factors. Therefore, in order to meet clinical needs, it is vital to achieve automatic segmentation of thyroid nodules to obtain higher segmentation accuracy.

Recently, convolutional neural networks (CNNs) have shown desirable results in the segmentation task of medical ultrasound images. Chen et al proposed a UNet bases marker-guided segmentation model of thyroid nodules, with a segmentation accuracy of 98% [5].

2.2. Classification Method

Ttraditional machine learning methods [6] and deep learning methods [7] make up the majority of computer-aided ultrasound image classification approaches. The traditional machine learning methods use prior knowledge to extract the texture and other features, and then put them into the machine learning classifier for classification. Although traditional pattern recognition methods have high accuracy, it is difficult to match the large amount of clinical data due to the diversity of thyroid nodules in texture and morphology. Thus, it is crucial to develop automatic classification models.

The deep learning-based methods are automatic methods that do not require manual feature extraction. Additionally, deep learning models have been shown to be efficient in medical image classification task. Sun et al combined deep features extracted by a CNN with handmade several features [7]. By using support vector machine to classify thyroid nodules, the accuracy reached 92.5%.

Training deep CNNs requires a large number of images, however, the amount of medical image data is inadequate. A good alternative is to employ the pre-trained CNNs. In addition, increasing the number of images by data augmentation is also a solution. Traditional data augmentation methods are often employed to enhance the data of thyroid nodules. However, it is difficult to determine which augmentation strategy can best capture the data features of medical images, and these methods cannot consistently improve the performance of CNNs. Thus, in our study, we employed Conditional Variational Auto Encoder (CVAE) to achieve data augmentation [8].

The remainder of this paper is organized as follows. The proposed CACS is described in detail in Section 3. Since then, Section 4 summarizes the data, materials and specifically outlines the experimental results. Finally, Section 5 provides the conclusion.

3. Methods

3.1. ROI Extraction

The UNet is a fully convolutional network and can be trained end-to-end [9]. UNet is effective in segmentation for medical images, due to its lightweight and cascading deep features and shallow features.

Res-UNet is to add residual connection on the basis of UNet [10]. The structure we utilized for thyroid nodule segmentation is shown in figure 1. The residual connection is introduced between every two convolutional layers, which can speed up network convergence and avoid degradation.

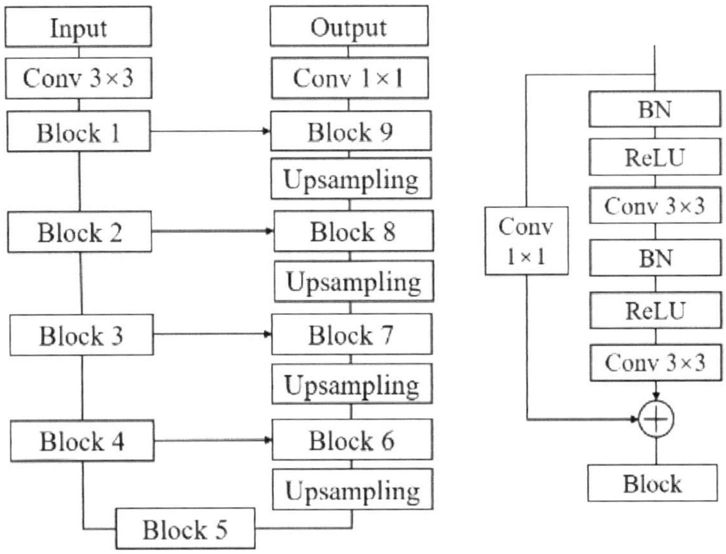

Figure 1. The architecture of Res-Unet.

3.2. Data Augmentation

3.2.1. Variational Auto Encoder. A directed probability graph generation model called the Variational Auto Encoder (VAE) consists of an encoder $q_\varphi(z|x)$ and a decoder $p_\theta(x|z)$ [11]. The encoder with parameter φ maps the input x to the latent variable z, and the decoder with parameter θ maps the latent variable to reconstructed data $\hat{x}$.The Auto-Encoding Variational Bayes (AEVB) method can be utilized to optimize the parameters φ and θ [11]. Equation (1) describes the variational lower bound of the VAE model.

$$L(\theta,\varphi;x) = -KL(q(z|x)\|p(z)) + E_{q(z|x)}(\log p(x|z)) \tag{1}$$

where the Kullback-Leibler divergence of $q(z|x)$ and $p(z)$ is donated by $KL(q(z|x)||p(z))$. To reduce the discrepancy between the encoder distribution and previous distribution, KL divergence is employed in the first term in equation (1). In addition, the decoder must learn to reconstruct the input data by using the second item to indicate the reconstruction loss. Consequently, increasing the variational lower bound $L(\theta,\varphi;x)$ is the aim of training VAE.

3.2.2. Independent Conditional VAE. Conditional Variational Auto Encoder (CVAE), a variation of VAE, models by conditioning the encoder and decoder to label y [12]. However, it is preferred to make the encoding and labeling of images independent of each other for data augmentation. As seen in figure 2, it enables independent control over the features of the image [8]. The encoder converts the input x into the latent variable z, and the decoder reconstructs the input $\hat{x}$ from the labels y and the latent variable. Moreover, the encoder parameterized by ω compresses y into a vector $\hat{y}$ with the same size as z to balance the effects of y and z on $\hat{x}$. The variational lower bound of ICVAE is given by the equation (2).

$$L(\varphi,\theta,\omega,x,y) = E_{q_\varphi(z|x)}[\log p_{\theta,\omega}(x|y,z)] - KL(q_\varphi(z|x)\|p(z)) + \log p(y) - \log(y\,|\,x) \tag{2}$$

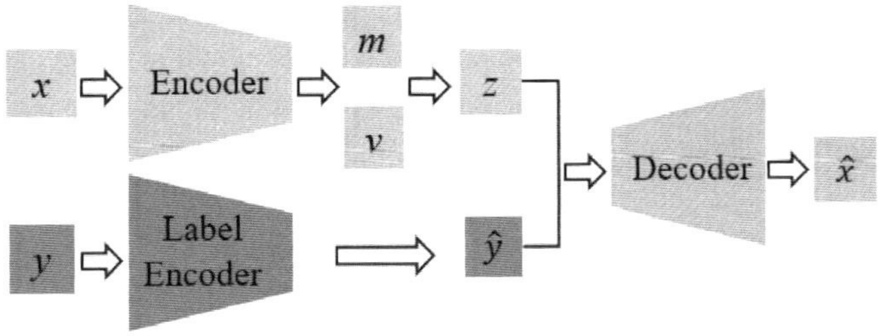

Figure 2. The structure of CVAE.

3.3. The Pre-trained SE-ResNet-50

In the medical image classification task, transfer learning is commonly used. Menegola et al contrasted fine-tuning training from scratch to a pre-trained network in a series of tests [13]. The results showed that the fine-tuned CNN should always be the first choice, regardless of the size of the data set.

SE-Net obtains the weights of feature channels by Squeeze and Excitation operations and introduces the channel attention mechanism into the classification network [14]. The SE module can be embedded in an existing classification network. Figure 3 shows it embedded in ResNet50.

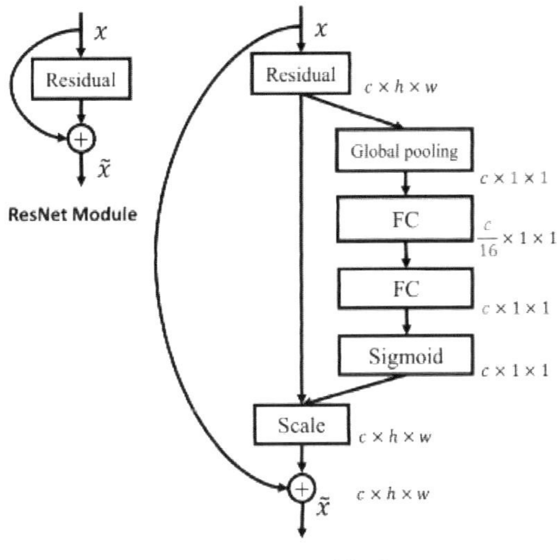

Figure 3. SE block.

3.4. The Three-Branch Ensemble Network Based on Ultrasonic Features

The shape and boundary of thyroid nodules are the vital features to identify nodules, especially the aspect ratio and burr-like edges are the two main features to determine malignant thyroid nodules. Therefore, we employed a three-branch ensemble network to classify thyroid nodules, combining with the ultrasound characteristics of thyroid nodules. The structure is shown in figure 4. The network structure of each branch is the same, which is SE-ResNet50.

In addition to the information within the nodules, the characteristics of the tissues around the nodules also have an impact on the classification of nodules [15]. In order to consider this contextual information, 1.5 times the size of the original nodule ROI is cropped as the input of the first branch (called the context branch). The input of the second branch is the ROI area of the nodule, that is obtained according to the segmentation result of Res-UNet. Moreover, taking into account the difference between the borders and edges of thyroid nodules, the input image of the third branch only contains the edges of the nodules to enhance the learning of edges. We use the nodule contour line obtained by Res-UNet to automatically segment, expand the contour, and obtain the band-shaped area. The band-shaped area covers the boundary of the nodule and contains the ultrasonic features inside and outside the nodule. For network training, firstly, we train each branch independently, then assign weights to the features of each branch according to the classification accuracy. Finally, in order to achieve cascade automatic classification system, the three branch networks are integrated and the classification result is obtained through a full connection layer.

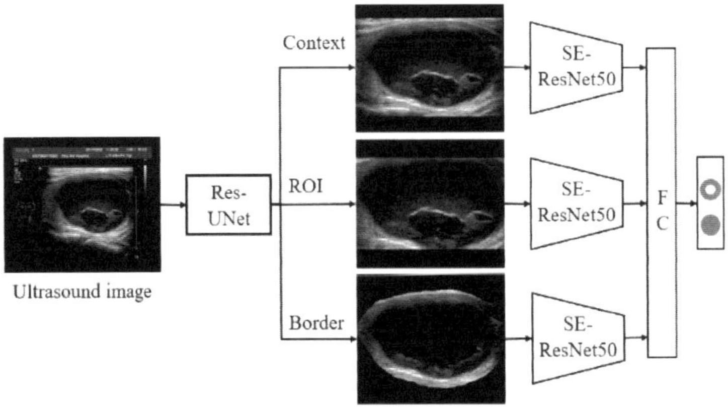

Figure 4. The overview of Multi-branch network.

4. Experiments

4.1. Data and Materials

The experiments in this study are performed on the database TND-PUH3. A total of 3771 thyroid nodule ultrasound images are collected from 2360 patients, among which 1316 are benign and 2455 were malignant. Meanwhile, all images are labeled by the radiologists of Peking University Third Hospital. Figure 5 shows an example of an ultrasound image and its nodule contour labelling.

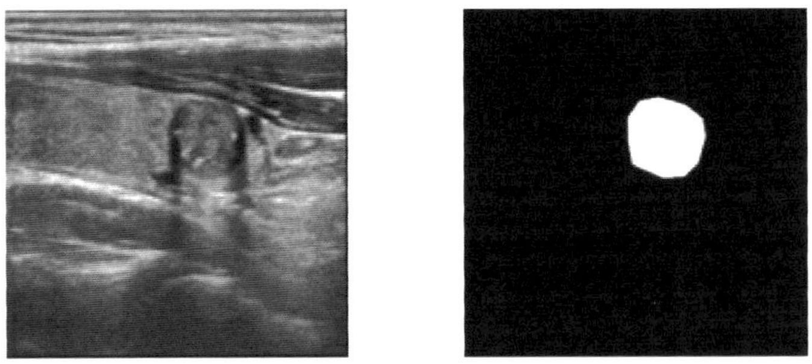

(a) Ultrasound image (b) Mask image

Figure 5. An example of image data.

4.2. Evaluation Criteria

In this paper, Dice coefficient and Sensitivity are employed as the evaluation index of segmentation.

$$\text{Dice} = \frac{2TP}{FN+2TP+FP} \quad (3)$$

$$\text{Sensitivity} = \frac{TP}{TP+FN} \quad (4)$$

The classification results are measured using accuracy, sensitivity, specificity, F1-score, AUC and ROC curve.

$$\text{Accuracy} = \frac{\text{TP+TN}}{\text{TP+FP+FN+TN}} \tag{5}$$

$$\text{Specificity} = \frac{\text{TN}}{\text{TN+FP}} \tag{6}$$

$$F_1 - \text{score} = \frac{\text{2TP}}{\text{2TP+FP+FN}} \tag{7}$$

4.3. The Segmentation of Thyroid Nodules

The Dice coefficient and Sensitivity of the two segmentation networks on the TND-PUH3 test sets are shown in table 1. The Dice coefficient of Res-UNet segmentation is 1.3 percentage higher than UNet, reaching 78.3%, indicating the effectiveness of the residual structure.

Table 1. The quantitative segmentation results of the UNet and Res-UNet.

	Dice	Sensitivity
UNet	0.770	0.825
Res-UNet	0.783	0.833

4.4. The Classification of Thyroid Nodules

The classification results on the TND-PUH3 test set are shown in table 2 and figure 6. SE-ResNet50 achives 0.863 accuracy and is used as the basic network of the multi-branch network. When the training set is not augmented, the sensitivity of the network is 20% higher than the specificity, which proves that there is a serious category imbalance. After data augmentation, the specificity of the network increases, and the AUC after traditional data augmentation (random cropping, flipping, rotating, and distortion) increases by 2.3%. The AUC of CVAE-SE-ResNet50 is 0.880, which is 5.1% higher than 0.829 without augmentation which shows that the image-rich database generated by CVAE can improve the performance of the network. In order to alleviate the problem of category imbalance, we use CVAE to generate samples of the categories with a small number of samples, that is, benign samples. It is emphasized that ultrasound images generated by CVAE are only used for training the SE-ResNet50. In process of verification and testing, we only use raw ultrasound images.

Table 2. The quantitative classification results of SE-ResNet50 on the TND-PUH3 test set.

	Accuracy	Sensitivity	Specificity	AUC	F1-score
SE-ResNet50	0.863	0.94	0.72	0.829	0.900
AUG-SE-ResNet50	0.865	0.90	0.81	0.852	0.897
CVAE-SE-ResNet50	**0.881**	**0.91**	**0.83**	**0.880**	**0.926**

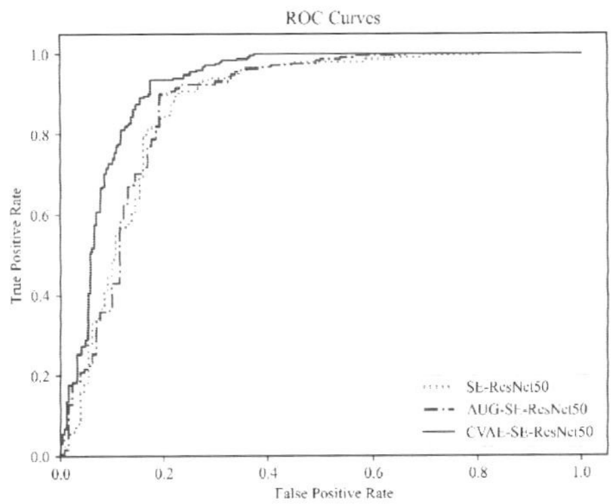

Figure 6. The ROC curves.

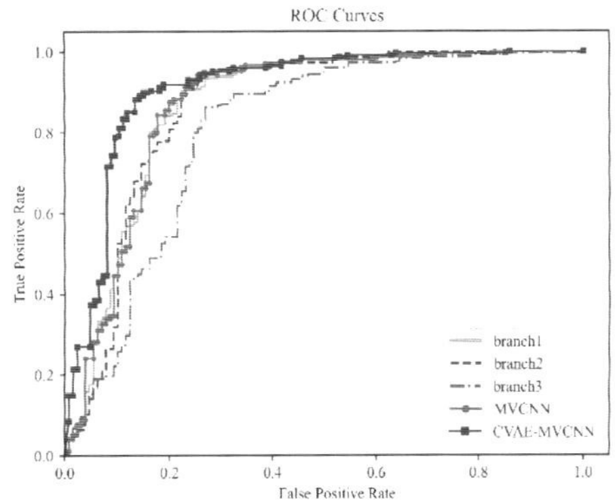

Figure 7. The ROC curves.

Table 3 and figure 7 show the classification results of the multi-branch network on the TND-PUH3. It can be seen that branch 1 has the highest classification accuracy. It shows that the ultrasound features outside the nodules will be beneficial to the classification of benign and malignant. After integration, the accuracy rate of MVCNN is increased by 1.3%, and the AUC is increased by 1.6%. In addition, the AUC obtained by CVAE-MVCNN is 4.2% higher than that of MVCNN when the data augmented samples are incorporated into the training process.

Table 3. The quantitative classification results of multi-branch CNN on the TND-PUH3 test set.

	Accuracy	Sensitivity	Specificity	AUC	F1-score
Branch-1	0.863	0.94	0.72	0.829	0.900
Branch-2	0.855	0.90	0.78	0.836	0.890
Branch-3	0.815	0.86	0.73	0.795	0.859
MVCNN	0.876	0.94	0.75	0.845	0.910
CVAE-MVCNN	**0.892**	**0.92**	**0.84**	**0.887**	**0.928**

5. Conclusion

This paper mainly studies the intelligent diagnosis of thyroid nodules to assist clinical practice. The proposed cascade automatic classification system (CACS) mainly consists of two parts: ROI acquisition and classification. Automatic segmentation of ROI with Res-UNet can help doctors to find the specific location of nodules. In addition, since the number of malignant nodules in the database is much greater than that of benign nodules, we employed a generative network that can learn the distribution of the input for ROI data augmentation and balance the data set. Moreover, a multi-branch classification model is constructed by combining the obvious ultrasonic characteristics that distinguish benign from malignant nodules: aspect ratio, echo, and boundary. Compared with the single-branch network, the classification accuracy of multi-branch model is improved. The diagnosis accuracy and consistency of thyroid nodules are improved by our CACS.

Acknowledgment

This work was supported in part by a grant from National Natural Science Foundation of China (61673039).

References

[1] Zhou M, Wang R, Fu P, et al. 2020 Automatic malignant thyroid nodule recognition in ultrasound images based on deep learning *E3S Web of Conferences* **185** (2) 03021.
[2] Ka rkada M, Costa A F, Imran S A, et al. 2018 Incomplete thyroid ultrasound reports for patients with thyroid nodules: Implications regarding risk assessment and management *American Journal of Roentgenology* **211** (6) 1348-1353.
[3] Sa velonas M A, Iakovidis D K, Legakis I, et al. 2009 Active contours guided by echogenicity and texture for delineation of thyroid nodules in ultrasound images *IEEE Transactions on Information Technology in Biomedicine* **13** (4) 519-527.
[4] Poudel P, Illanes A, et al. 2019 Thyroid ultrasound texture classification using autoregressive features in conjunction with machine learning Approaches *IEEE Access* **7** 79354-79365.
[5] Chu C, Zheng J and Zhou Y 2020 Ultrasonic thyroid nodule detection method based on U-Net network *Computer Methods and Programs in Biomedicine* **199** (1) 105906.
[6] Acharya U R, Chowriappa P, Fujita H, et al. 2016 Thyroid lesion classification in 242 patient population using Gabor transform features from high resolution ultrasound images *Knowledge-Based Systems* **107** (1) 235-245.
[7] Sun C, Zhang Y, Chang Q, et al. 2020 Evaluation of a deep learning-based computer-aided diagnosis system for distinguishing benign from malignant thyroid nodules in ultrasound images *Medical Physics* **47** (9) 3952-3960.

[8] Pesteie M, Abolmaeşumi P and Rohling R N 2019 Adaptive augmentation of medical data using independently conditional variational auto-encoders *IEEE Transactions on Medical Imaging* **38** (99) 2807-2820.
[9] Ronneberger O, Fischer P and Brox T 2015 U-Net: Convolutional networks for biomedical image segmentation *Medical Image Computing and Computer-Assisted Intervention (MICCAI)* 234-241.
[10] Drozdzal M, Vorontsov E, Chartrand G, et al. 2016 The importance of skip connections in biomedical image segmentation *ArXiv preprint arxiv 1608.04117.*
[11] Kingma D P and Welling M 2014 Auto-encoding variational bayes *ArXiv preprint arxiv1312.6114.*
[12] Sohn K, Yan X, Lee H, et al. 2015 Learning structured output representation using deep conditional generative models *Inter. Conf. on Neural Information Processing Systems* pp 3483-3491.
[13] Tajbakhsh N, Shin J Y, Gurudu S R, et al. 2016 Convolutional neural networks for medical image analysis: Full training or fine tuning? *IEEE Transactions on Medical Imaging* **35** (5) 1299-1312.
[14] Hu J, Shen L, Albanie S, et al. 2020 Squeeze-and-excitation networks *IEEE Transactions on Pattern Analysis and Machine Intelligence* **42** (8) 2011-2023.
[15] Liu T, Guo Q, Lian C, et al. 2019 Automated detection and classification of thyroid nodules in ultrasound images using clinical-knowledge-guided convolutional neural networks *Medical Image Analysis* **58** 101555.

3D CNN Hardware Circuit for Motion Recognition Based on FPGA

Shidong Lv[1,2], Tao Long[1], Zhixian Hou[1], Liang Yan[1,3] and Zhenzhong Li[4]

[1] School of Integrated Circuits and Electronics, Beijing Institute of Technology, Beijing, China
[2] The Second Institute of China Aerospace Science and Industry Corporation, Beijing 100039, China
[3] Amileyuan Intelligent Technology (Beijing) Co., Ltd., Beijing, China
[4] CATARC Automotive Test Center (Ningbo) Co., Ltd., Ningbo, Zhejiang, China
Email: 921180762@qq.com; longtao1905@163.com; 3220210820@bit.edu.cn; yanliangbj@163.com; lizhenzhong@catarc.ac.cn

Abstract. In recent years, three-dimensional convolutional neural network (3D CNN) has been widely used in the fields of action recognition and video analysis. The general purpose processors are difficult to achieve efficient and intensive computing, and the deployment of 3D CNN based on FPGA has the advantages of low power consumption, high energy efficiency, and customizability, and has gradually become a hot choice for deploying convolutional neural networks in many embedded scenarios. This paper designs a small 3D convolutional neural network based on the classic 3D convolutional neural network C3D, and uses the general matrix multiplication (GEMM) to map the 3D convolution calculation to the 2D matrix multiplication calculation. The matrix is divided into blocks and transmitted to the FPGA through the AXI bus, and the multiplication operation of the block matrix is realized through a two-dimensional multiply-accumulate array. A System on Chip (SoC) architecture is built on the PYNQ platform, using ARM Cortex-A9 as the process control core, and the calculation of the entire matrix is completed under reasonable block and scheduling on the ARM. The IP core of matrix calculation is designed using High Level Synthesis (HLS), and the corresponding parallel optimization scheme is given. Experiments have verified that the prototype design of the hardware acceleration circuit achieves low power consumption, high energy efficiency, and high precision motion recognition while using less hardware resources.
Keywords. 3D CNN; FPGA; matrix multiplication; HLS.

1. Introduction

Before 3D CNN was proposed, 2D CNN (two-dimensional convolution neural network) achieved good results in 2D image processing, but it was not satisfactory for video analysis. In 2010, Ji et al. innovatively proposed a three-dimensional convolution model based on two-dimensional plane convolution, and used 3D CNN to realize action recognition. The model can simultaneously extract information in both temporal and spatial dimensions through 3D convolution, so it can extract action information contained between multiple consecutive frames, and the model achieves 90.2% accuracy on the KTH dataset [1]. In 2015, Du et al. proposed C3D convolutional network, which is a milestone of 3D convolutional network, and explored the optimal convolution kernel size suitable for 3D convolution. When the size of all convolution kernels of the model is determined to be $3\times3\times3$, the model can obtain the highest inference accuracy [2]. Since then, a lot of action recognition work will be carried out based on C3D model, and the research work of this paper is based on C3D model. FPGA has the advantages of parallel computing compared with CPU and low power consumption compared with GPU, so it is suitable as an embedded hardware acceleration platform for convolutional neural network. Previously, some accelerated operations were performed on the FPGA platform based on the C3D network, but these research efforts were based on high-performance and high-price FPGA platforms [3-7], and the characteristics of this paper are that the 3D CNN is deployed based on low-cost FPGA platforms. At the same time, this paper uses the convenience of HLS tool to quickly realize the hardware design and synthesis of the algorithm.

Compared with 2D CNN, 3D CNN has a geometric multiple growth in the amount of parameters and calculation. The amount of weight parameters and intermediate data far exceed the on-chip storage capacity of general FPGA, and even the on-chip storage cannot complete the caching of single-layer input features and weights. The excessive number of parameters makes it difficult for 3D convolutional neural networks to design hardware acceleration on FPGA. In view of this design difficulty, this paper uses General Matrix Multiplication (GEMM) to realize the convolution calculation of 3D CNN. Specifically, by mapping the three-dimensional convolutional expansion into two-dimensional matrix multiplication, and then through matrix cutting, the multiplication of the weight and feature matrix is realized in chunks, and the problem of insufficient storage in the FPGA can be well solved through this scheme. Previously, Caffeine proposed by Zhang et al. used GEMM to realize the hardware acceleration of 2D CNN on FPGA [8].

2. Overview of Three-Dimensional Convolutional Neural Network

There is no essential difference between 3D CNN and 2D CNN. Compared with 2D convolution, the most prominent feature is that the convolution kernel, input feature and output feature of 3D convolution all add a time dimension. The convolution kernel of 3D CNN is three-dimensional, and the input is a video segment, that is, a continuous sequence of multiple pictures, and the output feature after three-dimensional convolution calculation is also a video segment. In C3D, for single-channel video segments, the size of the convolutional kernel is 3×3×3. When performing a 3D convolution operation, the 3D convolution kernel slides first in the row and column directions of the input feature, and then slides on the temporal dimension of the feature. From figures 1 and 2, we can clearly see the difference in convolution mode between 2D CNN and 3D CNN, in which the black dotted line represents the sliding direction of convolution kernel. Because the spatial information and time dimension information of video data are extracted at the same time, the 3D CNN can learn the information between frames in video segments more comprehensively, and the accuracy of video classification and action recognition is better than that of 2D CNN. Assume that the size of the input feature X is CHin×Tin×Hin×Win, where Tin represents the number of frames of the video segment, and the size of the convolution kernel Wt are CHout×CHin×K_t×K_y×K_x, the size of the output feature is CHout×Tout×Hout×Wout. For convolutional layers, there is Tout=Tin, Hout=Hin, Wout=Win. Thus the elements in the hth row and wth column of the mth channel in frame f^{th} of the output video segment after performing three-dimensional convolution calculation can be expressed by equation (1), where s function is a nonlinear activation function.

$$Out[m][f][h][w] = s\left(\sum_{i=0}^{CHin-1}\sum_{d=0}^{K_t-1}\sum_{j=0}^{K_y-1}\sum_{i=0}^{K_x-1} X[c][f+d][y+j][x+i] * Wt[m][c][d][j][i] + b\right) \quad (1)$$

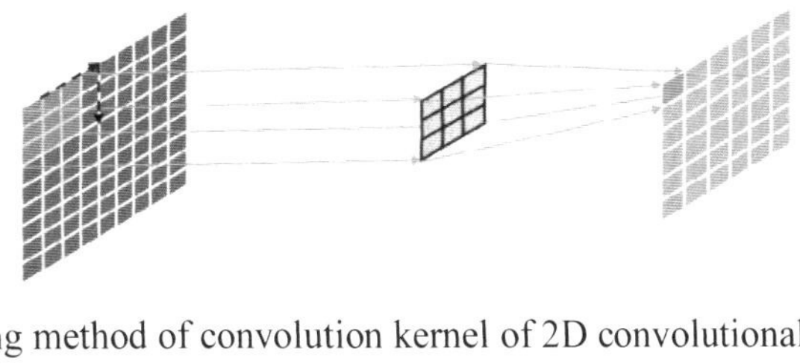

Figure 1. Sliding method of convolution kernel of 2D convolutional neural network.

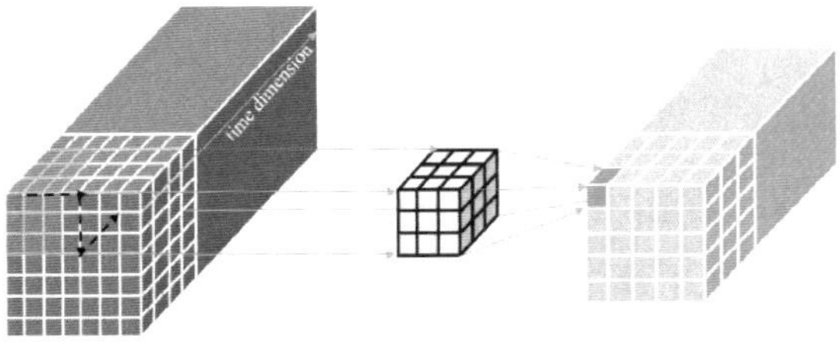

Figure 2. Sliding method of convolution kernel of 3D convolutional neural network.

3. Design and Train of Mini-C3D

The test set in this paper uses the Weizmann dataset [9], which includes 9 categories, 10 videos in each category, and a total of 90 video segments. The C3D model is proposed for the UCF101 dataset with 101 categories, so this paper implements channel compression on the basis of C3D, which we call mini-C3D, which has been able to achieve the classification of this dataset well. Using 70 videos in the Weizmann dataset as the training set and 20 videos as the test set, the accuracy of Mini-C3D on the test set after training on the training set is as high as 95%. The network structure of Mini-C3D is shown in figure 3.

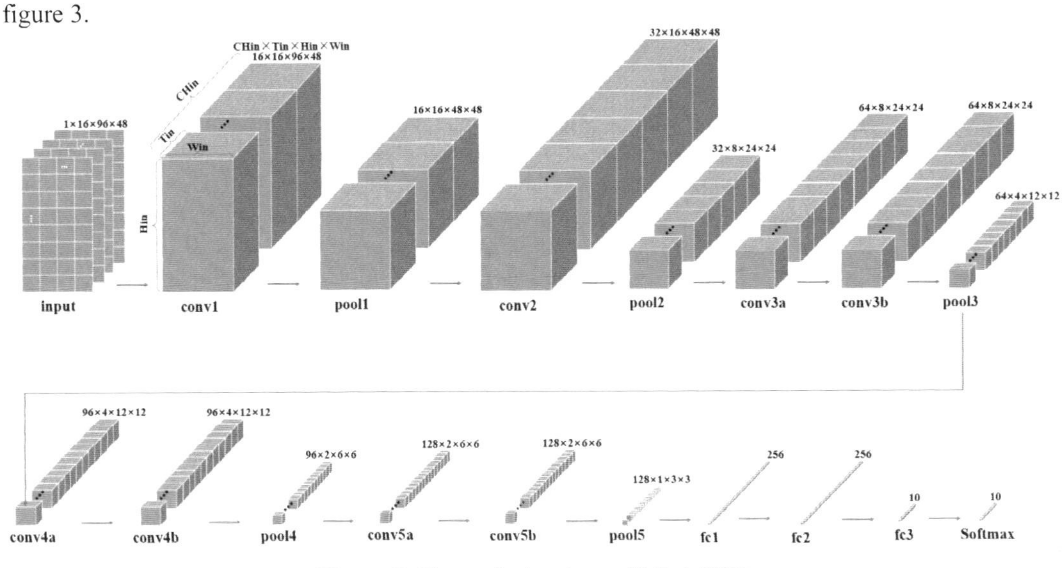

Figure 3. Network structure of Mini-C3D.

In model training, the video data is preprocessed first. The method of extracting one frame every four frames from video segments makes the actions contained in each input feature of the model have a larger amplitude range. At the same time, the three-channel image is converted into a single-channel image to initially reduce the size of the input features, at this time, the picture resolution is 120 pixels wide and 96 pixels high. In order to make the feature size as small as possible and reduce the computational complexity of the network, the previous work removed the background of the character, and then cut out part of the image containing the action outline of the human action. Use the Video Frame Difference Method to extract the contours of the moving human body in two adjacent frames. The specific algorithm is as follows: On the original image with a width of 120 pixels, select consecutive 48 columns and 96 rows of pixels, sum them, and select a new image composed of 48×96 pixels corresponding to the maximum value as the input of the neural network. The three pictures shown in figure 4 are the first and second frames of two adjacent frame pictures, and the pictures after the action contour is cut. Through the preprocessing of Weizmann data, the size of the data set is reduced as much as possible, the training process of Mini-C3D is accelerated, and the accuracy of model inference is guaranteed, and the forward inference on the FPGA platform is also reduced amount of calculation.

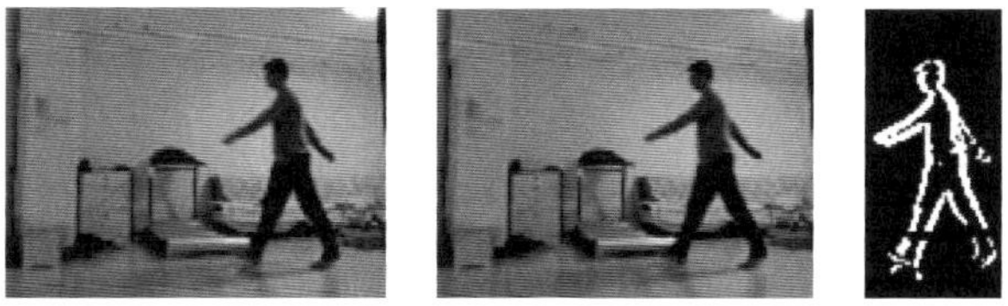

Figure 4. The first frame, second frame, and action contour of two adjacent frame pictures.

4. Design of Hardware Accelerator

4.1. Overall Architecture Design of Hardware Accelerator

The hardware platform of this article adopts the PYNQ-Z2 board of Xilinx Company, which has a ZYNQ XC7Z020-1CLG400C chip composed of a processor system (PS) and programmable logic (PL). The core of the PS-side is a dual-core ARM Cortex-A9 hard-core processor, and the core of the PL-side is an Artix-7 series FPGA which has 630 KB of fast block RAM and 220 DSPs. In the Mini-C3D network designed after C3D compression, the parameter is still 6.6MB, which is still much larger than the on-chip storage space of Zynq-7020. In this paper, GEMM is used to solve this problem by mapping three-dimensional convolution into two-dimensional matrix multiplication. Since the mapped matrix will have a $3\times3\times3$ times expansion in the data space, it is necessary to divide the mapped large matrix into blocks, and then load the divided feature matrix and weight matrix into the FPGA cache through DMA. The matrix calculation result is obtained after calculation by the matrix multiplication parallel computing IP core in the FPGA, and then the calculation result is sent back to the ARM through the AXI bus. After the single-layer convolution calculation is completed, ARM recover the original convolution feature data from the matrix result. After that, the feature is pooled and activated, and then the calculation result is divided into matrix blocks again to prepare for the next layer of convolution calculation. Figure 5 shows the overall design process of deploying Mini-C3D based on Zynq. Data preprocessing, Matrix partitioning, Feature recovery and Process control are allocated to the PS-side for execution, and the parallel calculation of the block matrix is allocated to the PL-side for execution, giving full play to advantages of the Zynq architecture.

The overall SoC design architecture is shown in figure 6. The PS-side uses processing_system7_0 as the control IP core, and the PL-side designs matrix_mult_0 as matrix multiplication parallel computing IP core. When the PS-side calls the PL side module for matrix hardware acceleration calculation, the block matrix data stored in the DDR is cached on-chip from the HP0 port of the Zynq IP core through the master input interface M_AXI_MM2S and the master output interface M_AXI_S2MM of the axi_dma_0 core to the matrix_mult_0 IP. After the matrix_mult_0 IP calculation, the calculation result is transmitted to the slave interface S_AXI_S2MM of the axi_dma_0 IP core through the output interface out_axi of the matrix_mult_0 IP core, and then the master interface M_AXI_S2MM of the axi_dma_0 IP core transmits the result data to the DDR through the HP0 port of the Zynq IP core. So far, the closed loop of the data path is completed. To use DMA, we need to open the interrupt port IRQ_F2P of Zynq. When the DMA completes the data transfer from the DDR, an interrupt will be triggered, and the TxDone signal on the PS-side is set to 1, indicating the completion of the input data transfer on the PL-side. When the DMA completes the data collection from the PL-side end and transfers the data to the DDR, an interrupt will be triggered, and the RxDone signal of the PS-side end is set to 1, indicating that the PL-side end output data transmission is completed. In addition, the Zynq core implements register configuration through the GP0 port, and directly transfers the configuration data to the s_axi_AXILites port of matrix_mult_0.

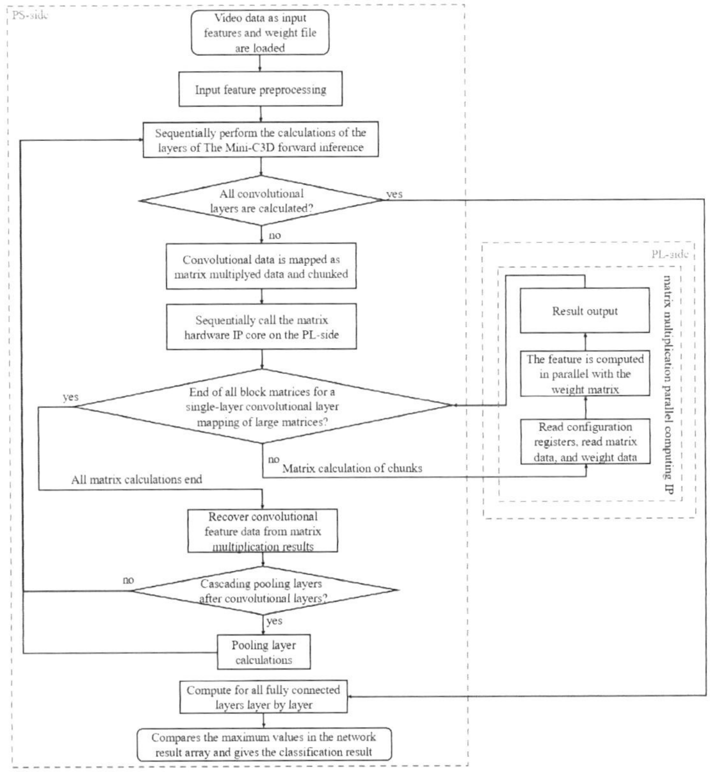

Figure 5. Overall design process of deploying Mini-C3D based on Zynq architecture.

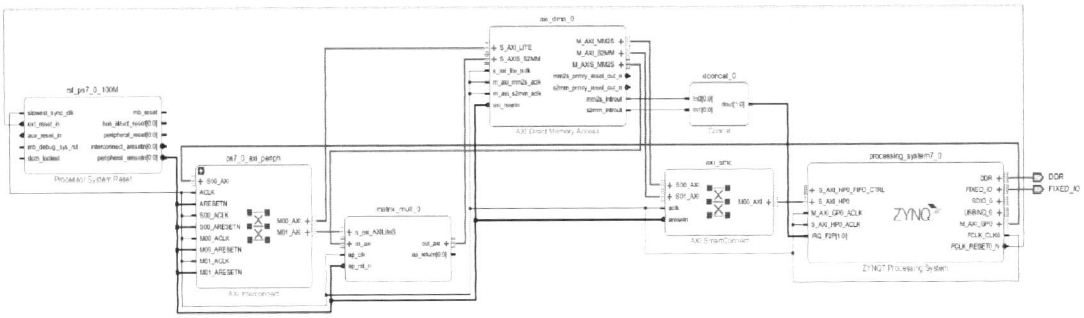

Figure 6. Overall design of SoC.

4.2. Software Design of 3D CNN Accelerator

The first step is to store the weight parameters of the neural network and the video segments to be classified in the SD card. After the system is started, the weight parameters and video data are loaded from the SD card to the array through the library function provided by Xilinx. After that, the video input features are preprocessed according to the preprocessing methods of the image sequence during training, that is, Video Frame Difference Method, Binarization, and Portrait Contour Cropping.

The second step is the mapping between convolutional data to matrix multiplication data, which was originally proposed in 2006 [10], and the mapping process is called Im2row. The specific operation of this method is to expand the feature pixels under the convolution kernel window into rows on the matrix into which the feature is mapped. In this paper, this idea is extended to the 3D CNN, and the feature X in the 3D CNN is converted into a two-dimensional matrix M_X, and the weight array W is also converted into a two-dimensional matrix M_W. In figure 7, taking the conv2 of Mini-C3D as an example. The input feature of conv2 has 16 input channels, the time dimension is 16, and the plane size is 48×48. The convolution kernel of conv2 has 32 output channels, 16 input channels, and the dimensions of the plane and time dimensions are 3×3×3. When the convolution kernel performs sliding convolution on the input feature, the 27 data on the corresponding feature of the convolution kernel window are flattened into a row of M_X. The number of times the convolution kernel W slides on the feature X is the row of M_X. After the feature mapping of a single input channel is completed, the mapping of the next input channel is carried out, and the conversion of the entire X to M_X is finally completed, so the number of lines of M_X is 16×16×48×48 which is 589824 and the dimension of the matrix M_X after the final feature X is mapped is 589824×27. When the convolution kernel is mapped, a single 3×3×3 convolution kernel is pulled into a column, and the number of output channels of the convolution kernel is mapped to the number of columns of the matrix, so the number of columns of M_W is 32 columns. After the convolution kernel of a single input channel is mapped, the next input channel is mapped, and finally the conversion of the entire W to M_W is completed. The final weight matrix M_W has a dimension of (16×27)×32 which is 432×32. Because the number of mapped feature matrix elements is expanded to 27 times, it is difficult to transfer the weights and features of a single layer to the on-chip cache of the FPGA at one time, and the matrix X and W need to be calculated in blocks.

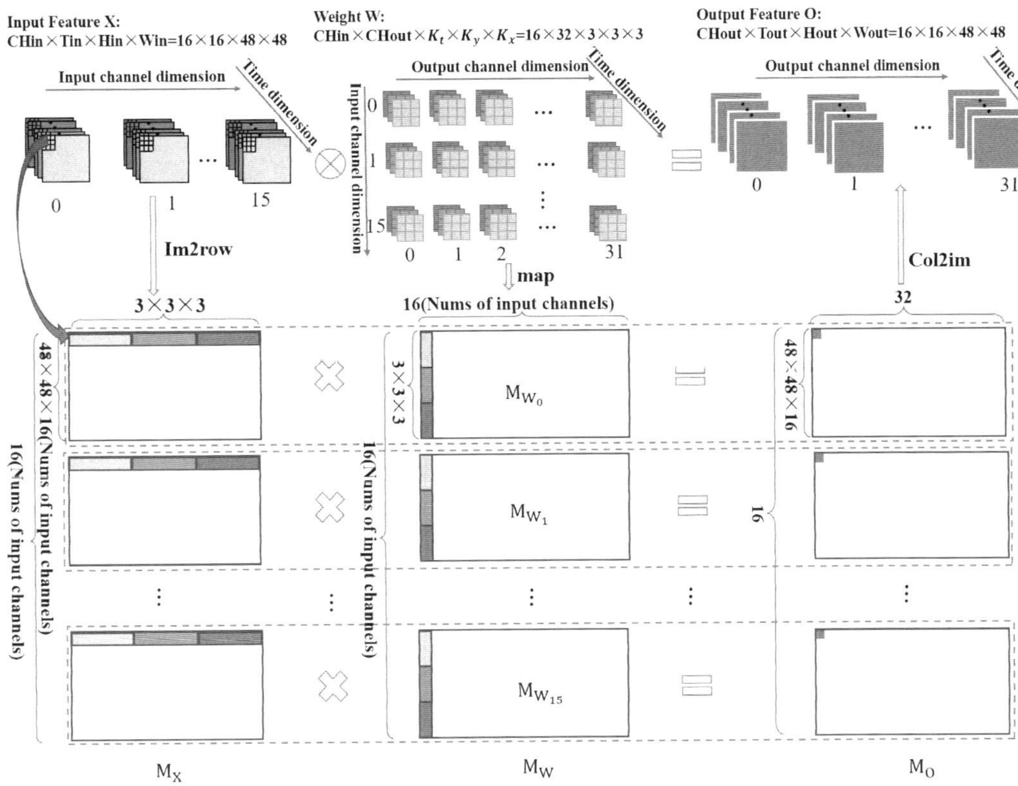

Figure 7. The process of converting 3D convolution calculation to GEMM method calculation.

The third step is to divide the mapped matrix into blocks, and transfer the block feature matrix and weight matrix to the PL-side through DMA. First, the greatest common factor of the height value of the matrix is calculated into which the input feature data of the 1, 2, 3a, 3b, 4a, and 4b convolutional layers are mapped, and find it as 576, and 576 is selected as the block small matrix, the width of the small matrix is 27. For the 5a and 5b convolution layers, the number of pixels on an input channel of the convolution feature has been reduced to 72, so 72 is selected as the height of the block small matrix. Matrix mapping and block are combined to complete, and padding zero is realized by judging whether the pixels during mapping are on the edge of the feature. Taking conv2 of Mini-C3D as an example, by adding a counter to the C code, every 576×27 data is stored in the last dimension of the feature matrix, The input feature before mapping is X[16][16][48][48], and the divided matrix after mapping is M[16][64][576×27]. After the matrix is divided into blocks, the dimension of each small matrix is 576×27. Figure 8 shows the process of dividing the two-dimensional matrix M_X mapped by conv2 into 1024 blocks. By calling the library function XAxiDma_SimpleTransfer() provided by Xilinx, the block matrix and weight matrix data can be transferred from the DDR to the PL-side on-chip cache.

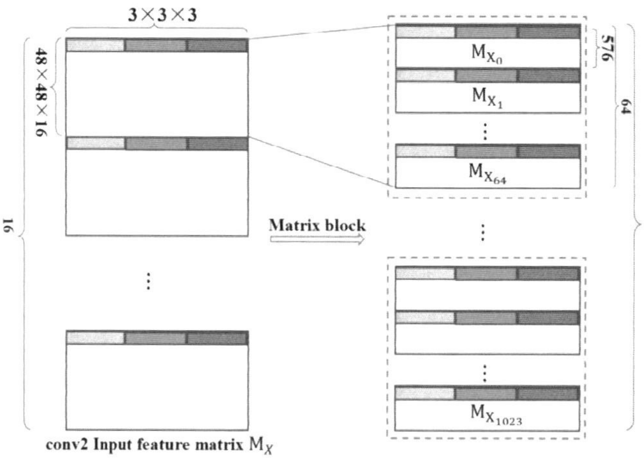

Figure 8. The conv2 matrix is divided into 1024 small matrices.

Figure 9 shows the block matrix calculation process on the FPGA. When the calculation on the FPGA is completed, the calculation result is sent back to the PS-side through DMA, and the Xil_DCacheInvalidateRange() function is used to read the calculated data by refreshing the output cache. The PS-side completes the single-layer convolution calculation by cyclically calling the hardware multiplication IP core of the PL-side, and then completes the calculation of all convolution layers in sequence according to the convolution layer order.

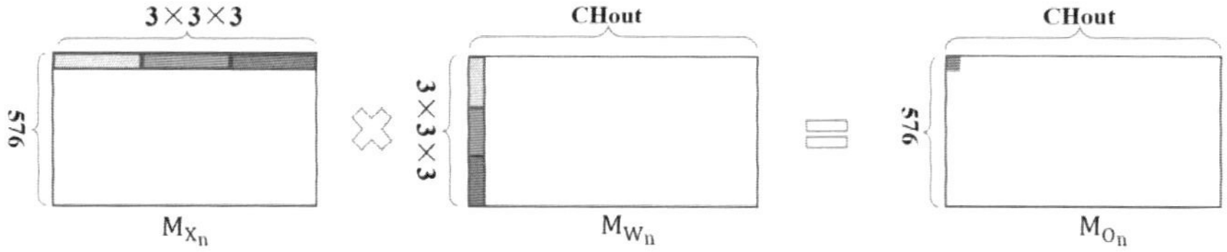

Figure 9. FPGA performs block matrix multiplication.

The reorganization of the result of the multiplication of the small matrix is the inverse process of the matrix block. Each block of the result matrix only needs to be arranged in the order of the input channels to complete the reorganization. That is, the output matrix M_O[16][64][32][576] is pre-defined, when the ARM accepts the matrix calculation result from the FPGA, the index address of M_O is assigned to the corresponding parameter of the XAxiDma_SimpleTransfer() function with 576×32 as the accumulation unit, after all the blocks of M_X are calculated, M_O is also reorganized immediately. Figure 10 is a schematic diagram of the reorganization of the calculation result of the conv2 block matrix.

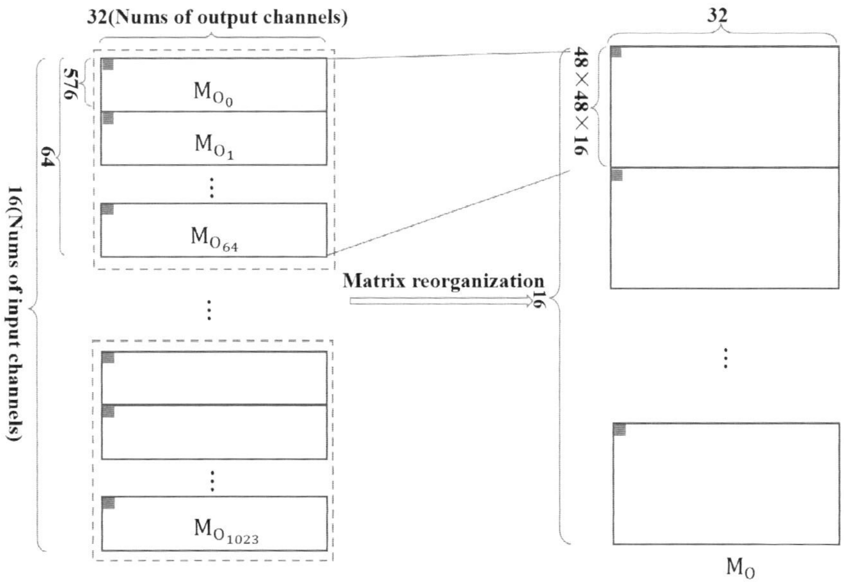

Figure 10. Schematic diagram of the reorganization of the calculation result of the conv2 block matrix.

The reorganized output matrix M_O contains the intermediate data of the input channel. It is necessary to accumulate the intermediate data of different input channels of each pixel to obtain the final output matrix M_{Out}. At the same time Col2im can be parsed out to get the final original convolution output result O. The columns of the output matrix M_{Out} represent each channel of the output feature, so the data of each column of the matrix represents all the data of one channel of the original convolution output feature. The mapping of the data in this column in the order of the width, height and the time dimension of the output features is reversed. Specifically, the Ow elements in the first row of the first frame of the first channel of the output feature O correspond to the first Ow elements of the first column of M_{Out}, and the Ow elements in the second row Ow elements of the first frame of the first channel of the output feature O correspond to the $Wout^{th}$ to $(2\times Wout-1)^{th}$ elements of the first column of M_{Out}, and so on to realize the reverse mapping in this order, and finally the original convolution output feature O is parsed out. As shown in figure 11, the channel accumulation and matrix feature de-mapping process are performed for the conv2 result matrix M_O.

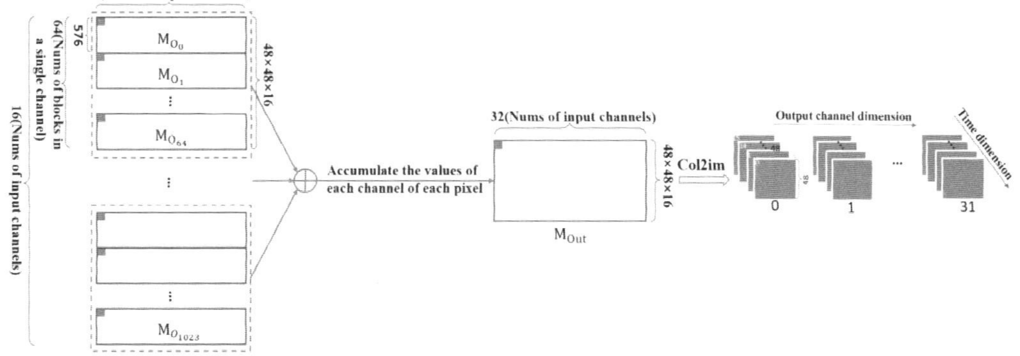

Figure 11. Conv2 result matrix M_O performs channel accumulation and matrix feature de-mapping processes.

4.3. Hardware Design of 3D CNN Accelerator

The design of the hardware part includes five parts: hardware interface design, on-chip cache design, weight and feature matrix loop boundary control design, weight and feature matrix tiling module design, and matrix parallel acceleration calculation and output module. The design of the hardware acceleration core uses High Level Synthesis tools and implements algorithms in C/C++.

(1) The hardware interface design

The interface of the matrix multiplication parallel computing IP core consists of an AXI4-Stream type data input interface, a 4 bit wide AXI4-lite type register configuration input interface and an AXI4-Stream type data output interface composition. The data input interface is used to receive the small block matrix and the weight matrix data of the corresponding channel transmitted from DDR3. The output interface writes the matrix calculation result back to the DDR after the calculation. Table 1 gives the definition of the small block matrix multiplication hardware interface.

Table 1. Small block matrix multiplication hardware interface definition and design pseudo code.

Definition and design of hardware interface for small block matrix multiplication
ap_uint<1> matrix_mult(
//Configure the height and weight of the small matrix and the number of output channels
ap_uint<4> config,
//Input port
ap_float in_axi[128*27+576*27];
//Output port
ap_float out_axi[576*96])
//Define the in_axi and out_axi port as the axis type interface
#pragma HLS INTERFACE axis register both port=in_axi
#pragma HLS INTERFACE axis register both port=out_axi

(2) On-chip cache design

By defining an array of set size, it is used to open up buffer space for feature matrices and weight matrices. The buffer space is reused by different convolutional layers, so the maximum buffer space required by each layer should be calculated to open up the buffer space. The array defined in HLS will be mapped to single-port BRAM by default after synthesis. Single-port BRAM cannot read data of multiple different addresses in a single cycle. Therefore, the PARTITION optimization of the defined array needs to be performed again in the subsequent implementation of matrix parallel design.

(3) Weight matrix and feature matrix loop boundary control design

When reading the data of the small block matrix and the weight matrix of the corresponding channel from the input interface, all the data will be loaded into the chip for caching at one time in the form of cyclic tiling. Because the output channels CHout of different convolution layers are different, it is necessary to set different cyclic boundaries for the number of channels of different conv layers when realizing cyclic tiling. In addition, conv5 is the last layer of convolution, and the pixel points of each channel of its input feature have been reduced to 72. Therefore, the height M_H of the block matrix of this layer convolution is set to 72. The height of the small block matrix of conv5 is different from that of other layers. The cycle boundary of the input characteristic matrix to realize on-chip cache also needs to be set separately. The settings of these two boundary parameters are controlled by the configuration register. The configuration register has 6 states, which are the stop working state and the working state of the first to fifth layers of convolution. Table 2 shows the configuration of matrix multiplication parallel computing IP core. In the calculation of each layer of convolution, the matrix multiplication parallel computing IP core needs to be called for the corresponding number of times, and the number of calls is defined as Blocks.

Table 2. Matrix multiplication parallel computing IP core configuration parameter table.

Config	M_H	CHout	Blocks
1	576	16	128
2	576	16	64
3	576	64	8
4	576	96	1
5	72	128	1

(4) Weight matrix and feature matrix tiling

After obtaining the height M_h of the small block matrix and the channel CHout, the next step is to cache the small block feature matrix data and the weight matrix data of the corresponding channel through circular tiling. Pipeline optimization through #pragma HLS PIPELINE II=1 enables one assignment operation to be completed per cycle.

(5) Matrix multiplication and result output

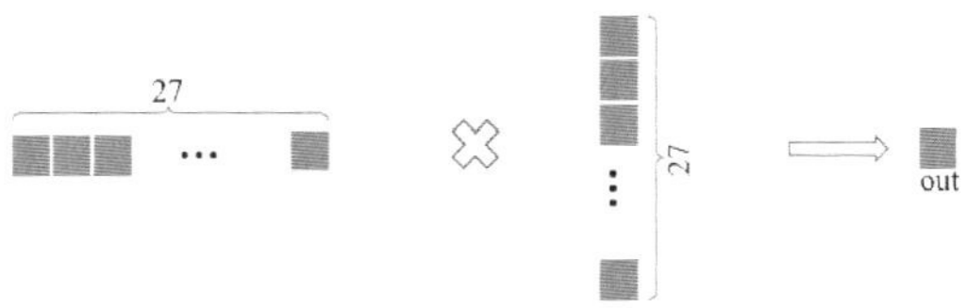

Figure 12. Completes 27 multiply-accumulates in one cycle.

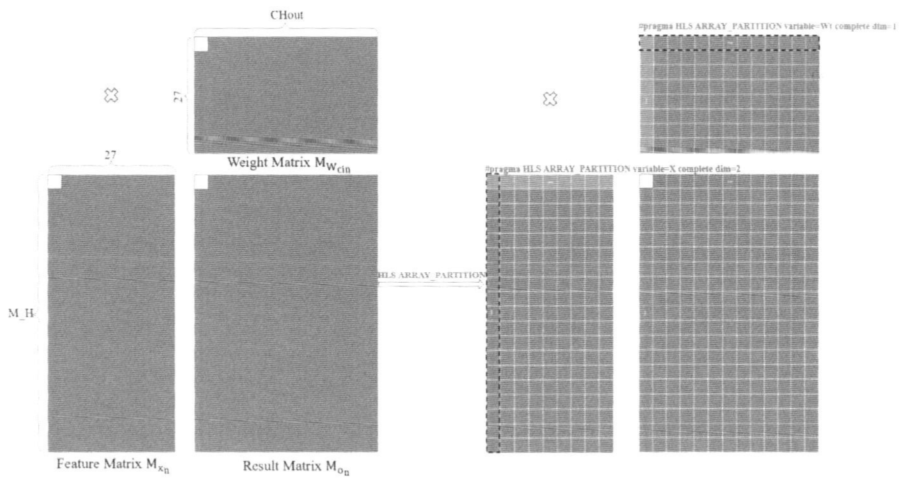

Figure 13. Schematic diagram of multiply matrix multiplication.

This module implements matrix multiplication and data output of the calculation results. Our parallel computing goal is to complete a 27 multiply-accumulate calculation in one cycle and output the result through the out_axi port, as shown in figure 12. The innermost matrix multiplication loop realizes loop unrolling through the HLS constraint statement #pragma HLS UNROLL. Since the HLS tool will automatically unroll the loop after adding PIPELINE optimization, UROLL optimization can be omitted. Add #pragma HLS PIPELINE II=1 in the second layer loop to implement pipeline optimization of multiply module. In order to achieve the goal that the Interval parameter of the pipeline startup interval is 1 cycle, it is necessary to complete the PARTITION of the array. The HLS constraint statement for array optimization is marked in figure 13. The second dimension of the feature matrix M_{X_n} and the first dimension of the weight matrix M_{W_c} are "complete" PARTITION. The optimized circuit can read 27 elements in a row of features and 27 elements in a column of weights in one cycle, and complete the parallel multiply-accumulate calculation of 27 multiplications to obtain an element of the output matrix M_{O_n}.

5. Performance Evaluation

This chapter evaluates the system performance from four aspects: Inference time, Resource utilization, Computing power and Power consumption.

(1) Inference time

As shown in table 3, when this model uses ARM Cortex-A9 alone to implement the mini-C3D forward inference model, it takes 241.13s to complete under the system clock of 650MHz. In this paper, based on the FPGA-based GEMM block deployment scheme, when the clock frequency of the FPGA is 125MHz, the inference time of the network is reduced to 13.40s, which is 17.99 times faster than that of the ARM. And the convolution calculation time in the whole system is only 0.97s, and the calculation time of the convolution part is 9.46s on the PC-side AMD 3550H CPU using 3.7GHz (turbo frequency), which shows that the effect of hardware acceleration in this paper is obvious.

Table 3. Inference time of Mini-C3D under different platforms.

Inference platform	Time cost/s
650MHz ARM CPU	241.13
125MHz Zynq 7020	13.40

(2) Resource utilization

In terms of resource utilization, table 4 shows the FPGA resource usage of the hardware platform designed in this paper. It can be seen that after realizing 27 multiply-accumulate and parallel expansion at the same time, the usage of DSP resources accounts for 62.27%, and the on-chip multiply calculation unit is fully utilized. In addition, the usage of other resources is less than 50%, indicating that the hardware acceleration platform still has sufficient space to further expand the scale of parallel matrix computing.

Table 4. FPGA resource usage.

Resource	Utilization	Available	Utilization (%)
LUT	13695	53200	25.74
LUTRAM	975	17400	5.60
FF	20720	106400	19.47
BRAM	42.50	140	30.36
DSP	137	220	62.27

(3) Computing power

Table 5 shows the operands and computing power of each conv layer on the PL-side. The number of convolution operations of each layer on the PL-side can be calculated by the following formula. One assignment calculation of cyclic tiling counts as one operation, and one multiply-accumulate calculation counts as two operations. The calculation formula of the computing power on the PL-side is (M_H×27+27×CHout + M_H×27×CHout×2)×Blocks, where Blocks is the number of times the convolution kernel is called for each layer. GOPS is the number of operations that the accelerator can process per second, which is used to measure the computing power of the PL-side.

Table 5. Operands and computing power of each conv layer on the PL-side.

Conv Layer	Time of calculating/s	PL-side operations	Computing power/GOPS
conv1	0.033	65,746,944	1.992
conv2	0.366	1,036,025,856	2.831
conv3a	0.141	514,031,616	3.645
conv3b	0.282	1,028,063,232	3.659
conv4a	0.047	192,264,192	4.091
conv4b	0.071	288,396,288	4.062
conv5a	0.014	48,294,144	3.450
conv5b	0.019	64,392,192	3.390
all	0.973	3,237,214,464	3.327

Table 6. Operands of each layer on the PS-side and the PL-side.

Layer	PL operations	PS operations	All operations
conv1	65,746,944	3,170,304	68,917,248
pool1	0	1,179,648	1,179,648
conv2	1,036,025,856	34,799,616	1,070,825,472
pool2	0	9,437,184	9,437,184
conv3a	514,031,616	13,418,496	527,450,112
conv3b	1,028,063,232	26,836,992	1,054,900,224
pool3	0	589,824	589,824
conv4a	192,264,192	4,534,272	196,798,464
conv4b	288,396,288	6,801,408	295,197,696
pool4	0	55,296	55,296
conv5a	48,294,144	1,071,360	49,365,504
conv5b	64,392,192	1,428,480	65,820,672
pool5	0	9,216	9,216
flatten	0	1,152	1,152
fc1	0	590,080	590,336
fc2	0	131,328	131,584
fc3	0	5,120	5,120
softmax	0	50	50
all	3,237,214,464	104,059,826	3,341,274,290

For the PS-side, the calculation operations mainly include im2row, col2im, pooling and fully connected layer operations. The im2row operand calculation formula is CHin×Tin×Hin×Win×27, and the col2im operand calculation formula is CHout×CHin×Tout×Hout×Wout. The calculation of the number of pooling operations is divided into two cases: the pooling of pool1 is only pooled in the height direction, and the ReLU operation is added at the same time, the calculation formula is CHout×Tout×Hout×Wout×2; The pooling of pool2, pool3, pool4, and pool5 layers is performed in both plane and time dimensions. The size of the pooling kernel is 2×2×2, and the ReLU operation is added at the same time, so the calculation formula of the operand is CHout×Tout×Hout×Wout×8. For conv3a, conv4a, and conv5a, there is no pooling layer behind, and the ReLU operation is performed synchronously when col2im is executed, and CHout×Tout×Hout×Wout needs to be added separately. The calculation formula of the calculation operand of the fully connected layer is Outlen×Inlen×Inlen×3, where Inlen is the one-dimensional vector length of the fully connected input feature, and Outlen is the one-dimensional vector length of the fully connected output feature. Since each fully connected output node both multiply and accumulate and perform ReLU operations, so multiply by 3. Table 6 shows the operands of each layer on the PS-side and the PL-side.

For the entire Zynq accelerator, the computing power is: 3,341,274,290/13.40=0.249GOPS. If only the ARM core on the PS-side is used to implement the forward inference of the Mini-C3D, the inference time is 241.13s. It can be seen from the calculation that the computing power of the forward inference using the ARM alone is 3,341,274,290/241.13=0.014GOPS. It can be seen that the coordination of ARM and FPGA greatly improves the computing power of the entire hardware system.

(4) Power consumption

The power consumption analysis tool that comes with Vivado is used to view the power consumption of Zynq, as shown in figure 14, the overall power consumption is only 1.602W, and the power consumed by the FPGA part of the hardware acceleration IP core is only 0.341W. The dynamic power consumption on the ARM side is 1.261W. Compared with the CPU power consumption and GPU power consumption on the PC-side, the power consumption of this system is very low, and the low power consumption target of deployment is achieved. In this paper, the Zynq deployment is used to achieve a power consumption ratio of: 0.249/1.602=0.155GOPS/W, and the power consumption ratio of pure ARM deployment is: 0.014/1.261=0.011GOPS/W. It can be seen that the Zynq deployment achieves a higher power consumption ratio.

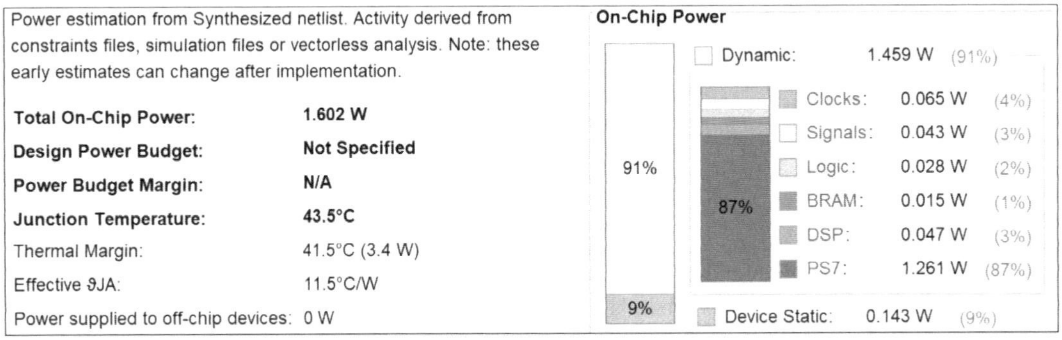

Figure 14. Zynq power consumption.

6. Summarize

In this paper, the channel compression of the C3D model is carried out, and the Mini-C3D network is designed for the Weizmann data set with a guaranteed accuracy rate of 95%, and the Mini-C3D model is deployed on the Zynq platform. Compared with the previous work, the feature of this paper is to realize the deployment of 3D CNN on a low-cost and low-resource FPGA platform. A Zynq platform-based rapid deployment solution is provided for computing large-scale 3D convolutional networks. Compared with pure ARM reasoning, it achieves a speed improvement of 17.99 times, and the system has lower power consumption and higher energy consumption, which has certain engineering reference significance for the low-power deployment of large-scale convolutional neural networks.

References

[1] Ji S, Xu W, Yang M and Yu K 2012 3D convolutional neural networks for human action recognition *IEEE Transactions on Pattern Analysis and Machine Intelligence* **35** 221-31.

[2] Tran D, Bourdev L, Fergus R, Torresani L and Paluri M 2015 Learning spatiotemporal features with 3D convolutional networks *IEEE International Conf. on Computer Vision* pp 4489-97.

[3] Liu Z, Chow P, Xu J, et al. 2019 A uniform architecture design for accelerating 2D and 3D CNNs on FPGAs *Electronics* **8** (1) 65.

[4] Fan H, Ng H C, Liu S, et al. 2018 Reconfigurable acceleration of 3D-CNNs for human action recognition with block floating-point representation *2018 28th Inter. Conf. on Field Programmable Logic and Applications* pp 287-2877.

[5] Fan H, Luo C, Zeng C, et al. 2019 F-E3D: FPGA-based Acceleration of an Efficient 3D Convolutional Neural Network for Human Action Recognition *2019 IEEE 30th Inter. Conf. on Application-specific Systems, Architectures and Processors* **2160** pp 1-8.
[6] Wang J, Deng H, Ye H, et al. 2021 FWUA: A flexible winograd-based uniform accelerator for 1D/2D/3D CNNs *2021 IEEE Inter. Conf. on Integrated Circuits, Technologies and Applications* pp 72-3.
[7] Deng H, Wang J, Ye H, Xiao S, Meng X and Yu Z 2021 3D-VNPU: A flexible accelerator for 2D/3D CNNs on FPGA *2021 IEEE 29th Annual Inter. Symp. on Field-Programmable Custom Computing Machines* pp 181-5.
[8] Zhang C, Sun G, Fang Z, Zhou P, Pan P and Cong J 2019 Caffeine: Toward uniformed representation and acceleration for deep convolutional neural networks *IEEE Transactions on Computer-Aided Design of Integrated Circuits and Systems* **38** (11) 2072-85.
[9] Blank M, Gorelick L, Shechtman E, Irani M and Basri R 2005 Actions as space-time shapes *Tenth IEEE Inter. Conf. on Computer Vision (ICCV'05)* **1** pp 1395-402.
[10] Chellapilla K, Puri S, Simard P 2006 High performance convolutional neural networks for document processing *Tenth Inter. Work. on Frontiers in Handwriting Recognition*.

Spatial Relationship Extraction of Geographic Entities Based on BERT Model

Jiannan Yang, Hong Jia and Hanbing Liu

The 15th Research Institute of China Electronics Technology Group Corporation, Beijing, China
E-mail: yangjiannan149162@outlook.com

Abstract. Geographic entity relationship extraction from text is an important way to acquire geographic knowledge. Entity relations in Chinese text are difficult to discover because of implicit representations of relations between entities in Chinese text. Therefore, using existing pattern matching and machine learning methods to extract entity relations from Chinese text often has problems such as insufficient artificial features, poor generality, inability to resolve word polysemy, and difficulty in making full use of contextual information. However, deep learning methods can better solve the above problems. This paper takes the spatial relation between geographic entities as the main research object to build a geographic entity spatial relation corpus, and propose a geographic entity relationship extraction method that combines BERT and attention mechanism. The method is trained and tested on the corpus. The results show that the F1 value of the BERT-BiGRU-Attention model proposed in this paper reaches 85.2% on the test set, which is a great improvement in relation extraction ability compared with other baseline models. The geographic entity relationship extraction method based on the BERT model can effectively learn the context information in Chinese text sentences, improve the accuracy of relationship extraction, and provide support for geographic knowledge graph construction and geographic entity information retrieval.
Keywords. BERT; relation extraction; attention mechanism; BiGRU.

1. Introduction

Chinese text containing geographic entity relations is an important source for obtaining geographic entity relations [1]. The main sources of traditional geographic information include map data, remote sensing images, experimental observation data, etc. These data are highly specialized, fixed in data types, and single in attribute relation information. Chinese text expressed in natural language contains rich entity relations, which makes the extraction of geographic entity relation with Chinese text as the object an effective supplement to traditional geographic information [2].

There are both explicit and implicit representations of entity relations in Chinese test. Using traditional methods to extract geographic entity relations from Chinese texts is highly dependent on artificial features, which requires massive manpower, and has the problem of poor generality. With the development of deep learning technology, relation extraction based on deep learning have been widely used in general fields and some specific fields such as medical field. These methods can automatically mine and learn high-dimensional features from annotated text, and extract the relations between entities. However, in order to train models with superior performance, a large-scale annotated corpus is indispensable, and the quality of data in the corpus also directly affects the training results.

This paper determines the classification of spatial relations between geographic entities, constructs a corpus of spatial relations of geographic entities corresponding to the classification system, and proposes a relationship extraction model that integrates Bidirectional Encoder Representations from Transformers (BERT) and attention mechanism. This method uses BERT to vectorize the input text to generate sentence embeddings that fuse token embeddings and position embeddings, which can better represent the structural and semantic features of sentences. At the same time the introduction of attention mechanism enables the relation extraction model to better capture the contextual information in sentences.

2. Related Work

Relation extraction methods in natural language processing (NLP) can be roughly divided into four types: unsupervised learning methods, semi-supervised learning methods, supervised learning methods and open domain extraction methods [3]. Among them, the first three methods are mostly used in relation extraction in restricted domain.

Unsupervised learning method does not rely on pre-defined relation classification system, and has strong generality, which is extremely advantageous when dealing with large-scale network text data. Xie et al. proposed an open relation extraction method based on unsupervised inheritance clustering, which realized relational fact extraction in open domain corpus [4].

Semi-supervised learning requires a small amount of annotated data and therefore it is suitable for relation extraction tasks with small scale corpus. Semi-supervised relation extraction methods mainly include methods based on distant supervision and bootstrapping principle. Wang Bin et al. used the existing relation triples in the knowledge base to extract entity relations from un-annotated Chinese texts by using a relation extraction model based on distant supervision that fused multiple features [5]. Xiang et al. adopted a bootstrapping-based semi-supervised relation extraction method to extract entity relations from professional and regular water conservancy texts [6].

The relation extraction method that based on supervised learning is the most commonly used, and is currently the fastest growing. Both pattern matching methods and deep learning methods belong to supervised learning methods. For text with particular content structure, pattern matching methods can extract target information quickly and efficiently. Tan et al. proposed location information extraction model based on finite state machine (FSM) for relatively regular location description patterns in microblog texts [7]. In recent years, with the rapid development of NLP and deep learning, more and more deep learning methods have been applied to relation extraction tasks. In 2013, Liu et al. first proposed the application of Convolutional Neural Network (CNN) in relation extraction tasks [8]. Zhou et al. adopted the Bi-directional Long Short-Term Memory (BiLSTM) model with the introduction of attention mechanism, which performed close to the models with artificial feature [9]. At the same time, with the application of Google's Transformer and the proposal of the BERT pre-training model [10], BERT has also been gradually applied to relation extraction tasks. Chen et al. proposed a relationship extraction method based on ALBERT-BiGRU-Attention, which realized the task of extracting the relationship between garden plants in Chinese text, and proved through experiments that the F1 value of this method reached 0.9161, which is better than other baseline models [11].

3. BERT-BiGRU-Attention Model

3.1. Summary

Considering that there is no public relation corpus in the geographic field at present, and the low accuracy rate of general NLP tools for geographic entity recognition, this paper proposes a spatial relation extraction model of geographic entity in network text, which is based on BERT and attention mechanism. The structure of the model is shown in figure 1, which can be divided into 4 layers. The first layer uses the BERT model to represent the input text as a vector; the second layer uses a Bi-directional Gated Recurrent Unit (BiGRU) to construct a neural network, and extracts high-dimensional features from the sentence feature vector input from the first layer. The third layer of attention layer captures the context information in the sentence. Finally, the softmax classifier is used to calculate the conditional probability of each relation type, and then output the classification result.

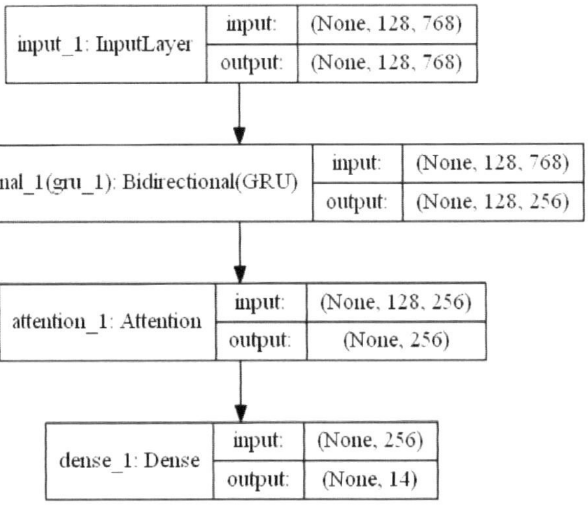

Figure 1. BERT-BiGRU-Attention model structure.

3.2. Vector Representation

In the relation extraction task using deep learning method, the text in natural language format must be vectorized before it can be input into the neural network for operation. Therefore, the word vector processing of text sentences is an essential work.

The traditional word vector generation models that are widely used mainly include one-hot, word2vec [12], Elmo [13] and so on. Compared with these methods, the BERT model uses a deep bidirectional Transformer component, which has stronger feature extraction capabilities and contextual information capture capabilities, and can also effectively solve the problem of polysemy. The structure of BERT is shown in figure 2.

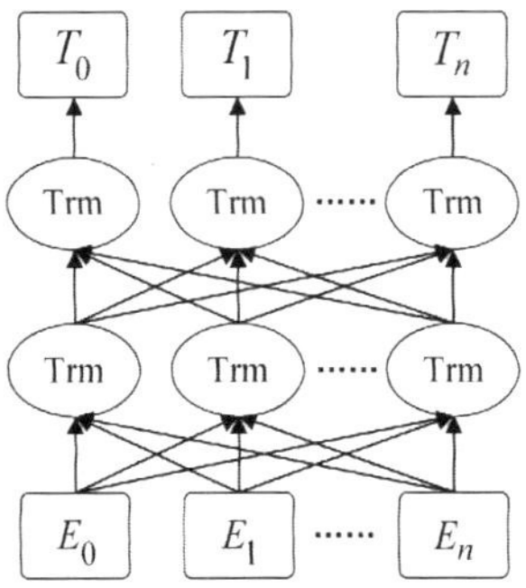

Figure 2. BERT structure diagram.

The BERT-base-Chinese model released by Google uses a single Chinese character as the granularity to segment text sentences and lacks Chinese lexical features. Therefore, this paper adopts the BERT-wwm model released by the HIT•iFLYTEK Language Cognitive Computing Lab [14]. The model adopts the Whole Word Masking (wwm) strategy. In the wwm strategy, when a single word is covered by the mask, other words that form vocabulary with this word will also be covered by the mask, so it can extract richer vocabulary features.

The input of BERT is the representation corresponding to each word, which consists of three parts: token embeddings, segmentation embeddings and position embeddings. A token embedding represents a one-dimensional vector of word conversions in Chinese sentences. The segment embedding, which is automatically learned during the training process and fused with the single-word semantic information, represents the global semantic information of the sentence. . Since different positions in a sentence often endow words with different semantic information, a position embedding is added to distinguish the positions of words. The addition of token embeddings, segment embeddings and position embeddings constitute the input of the BERT model.

3.3. BiGRU

GRU (Gated Recurrent Unit) [15] is a kind of Recurrent Neural Network (RNN). GRU and Long Short-Term Memory (LSTM) network [16] are both proposed to solve the long-term memory and back-propagation gradient problem of RNN. GRU can achieve similar effects as LSTM, but the structure of GRU is simpler, therefore, GRU is easier to train. The internal structure of GRU is shown in figure 3.

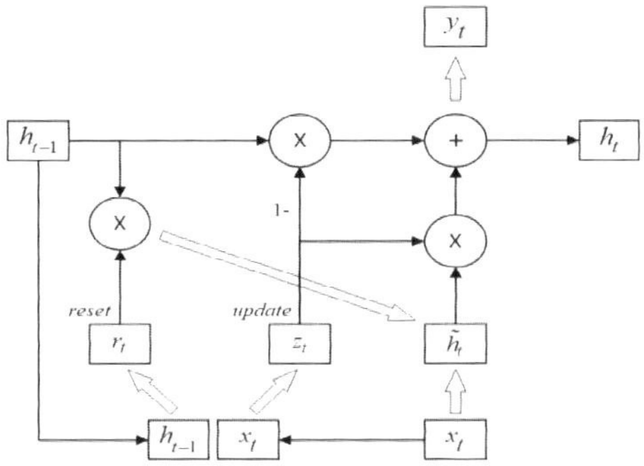

Figure 3. Internal structure of GRU.

The input and output of GRU are the same as the general RNN, where x_t represents the input of the current time, h_{t-1} indicates the state of the hidden layer at the previous time, y_t represents the output at the current time, h_t indicates the current state of the hidden layer. When x_t and h_{t-1} when inputting GRU, first obtain two gating States, r_t and z_t:

$$r_t = \sigma(W_r \cdot [h_{t-1}, x_t]) \quad (1)$$

$$z_t = \sigma(W_z \cdot [h_{t-1}, x_t]) \quad (2)$$

After obtaining the gating signal, first use the reset gate to reset the data and calculate the candidate hidden layer state $\tilde{h_t}$:

$$\tilde{h_t} = \tanh(W_{\tilde{h}} \cdot [r_t * h_{t-1}, x_t]) \quad (3)$$

The reset gate mainly controls how much historical information at the previous time is written in $\tilde{h_t}$. When r_t approaches 0, the historical information of the previous moment will be forgotten, on the contrary, when r_t approaches 1, the historical information of the previous time will be added to the candidate hidden layer state $\tilde{h_t}$. When $\tilde{h_t}$ is calculated, the update gate updates the memory information, and the update expression is shown in (4):

$$h_t = (1 - z_t) * h_{t-1} + z_t * \tilde{h_t} \quad (4)$$

It can be found that the update gate has completed the forgetting of h_{t-1} and the selection of $\tilde{h_t}$ at the same time, while in LSTM, multiple gates is required.

3.4. Attention Mechanism

Adding an attention mechanism to the neural network model can help the model learn and calculate the contribution of the input data to the output data, and help reduce the influence of noise information, thereby highlighting the key information of interest. At present, attention mechanism has been widely used in NLP.

Due to the complex description of the spatial relation between geographic entities in the network text, the extracted high-dimensional features often contain noise information. Therefore, this paper integrates the self-attention mechanism [17] in the model, allowing the sentence to calculate the vocabulary by comparing it with itself. It helps the model to extract the semantic-grammatical connections and dependencies between words, to a certain extent, overcomes the feature extraction problem caused by the diverse structure of network text in the geographic domain, and helps to improve the efficiency of relation extraction.

4. Experimental Results and Analysis

4.1. Experimental Data and Evaluation Criteria

In recent years, the research on entity relation extraction in the Chinese field has gradually developed, but the related research in the geographical field is still in its infancy, lacking professional large-scale public datasets, and the application of deep learning methods in the geographical field is also limited.

This paper uses web crawler technology to obtain web texts related to geographic fields from websites such as China National Geographic Network, Baidu Baike, Sina Microblog, etc. as raw data, and performs preprocessing operations such as data cleaning and removal of stop words. The preprocessed data is manually annotated with Stanza to construct a corpus of spatial relations of geographic entities. Since the description of metaphorical spatial relation is not easy to understand, such spatial relations are not considered in this paper [18]. Each piece of data in the corpus contains 2 geographic entities, 1 geographic entity spatial relationship descriptor and 1 complete sentence. Finally, a corpus of 4036 geographic texts was obtained, and the training set and the test set were allocated according to the ratio of 8:2, including 3229 pieces of data in the training set and 807 pieces of data in the training set. The spatial relation between entities in the corpus is divided into 3 categories and 14 sub-categories. The specific relations are shown in table 1:

Table 1. Classification of spatial relations of geographical entities.

Entity relation category	Entity relation subclass
Unknown	Unknown
Topological relation	Inclusion
	Connection
	Disjoint
	Overlay
	Equivalence
Directional relation	East
	South
	West
	North
	Southeast
	Northeast
	Southwest
	Northwest

In the experiment, the three indicators of precision (P), recall (R) and F1 value are used as the evaluation indicators of the model effect. If the positive instance is correctly classified as a positive instance, it is expressed as TP (true positive), otherwise denoted as FN. The calculation formulas of the three evaluation indicators are as follows:

$$P = \frac{TP}{TP+FP} \tag{5}$$

$$R = \frac{TP}{TP+FN} \tag{6}$$

$$F_1 = \frac{2 \cdot P \cdot R}{P+R} \tag{7}$$

4.2. Experimental Results and Analysis

This experiment of spatial relation extraction of geographic entities is based on the environment of python3.7, Tensorflow1.13.1, and keras2.3.1. The maximum length of the input text sentence is set to 128, the batch_size of the model training is 128, the epochs is 30, the Adam optimizer in keras is used, the default value of the learning rate is 0.001, and the droupout is 0.2. The BERT-wwm-ext, released by the HIT•iFLYTEK Language Cognitive Computing Lab is used, and the parameters of the model are not modified during the experiment.

The BERT-BiGRU-Attention model is trained and tested on the self-built dataset according to the above parameters. The change curves of loss and accuracy during the training process are shown in figure 4. The final loss and accuracy on the test set are 0.709 and 0.852.

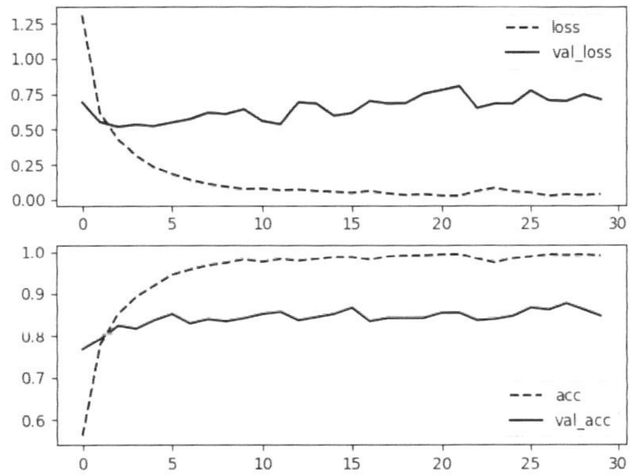

Figure 4. Change of loss and accuracy during training.

The results of spatial relation extraction on the test dataset of the trained BERT-BiGRU-Attention model are shown in table 2:

Table 2. Extraction results of relation of each category.

Spatial relation category	Precession (%)	Recall (%)	F1 (%)
Unknown	78.8	79.4	79.1
Topological relation—inclusion	91.3	88.4	89.8
Topological relation—connection	84.4	83.6	84.0
Topological relation—disjoint	71.9	64.1	67.8
Topological relation—overlay	83.3	92.5	87.7
Topological relation—equivalence	95.7	86.3	90.8
Directional relation—east	79.6	77.5	78.5

Spatial relation category	Precession (%)	Recall (%)	F1 (%)
Directional relation—south	84.1	85.2	84.6
Directional relation—west	95.1	95.7	95.4
Directional relation—north	91.9	94.3	93.1
Directional relation—southeast	79.2	93.1	85.6
Directional relation—northeast	85.8	83.6	84.7
Directional relation—southwest	88.6	84.2	86.3
Directional relation—northwest	97.8	80.5	88.3

In order to verify the reliability of the experiment, the BERT-BiGRU model without attention mechanism and two sets of BiGRU and BiGRU-Attention models that use word2vec to vectorize the input is added in the experiment. A total of three sets of experiments are compared with the BERT-BiGRU-Attention model, and the final results are shown in table 3.

Table 3. Comparison of experimental results of different models.

Model	Precession (%)	Recall (%)	F1(%)
word2vec-BiGRU	70.1	71.9	71.0
word2vec-BiGRU-attention	75.3	76.9	76.1
BERT-BiGRU	81.7	82.2	81.9
BERT-BiGRU-attention	85.1	85.3	85.2

Observing the experimental results, it can be found that using BERT to vectorize the input text can significantly improve the accuracy of relation extraction. When the model does not add the attention mechanism, the F1 value is increased by 10.9% using BERT for vectorization; when the model is added with the attention mechanism, the F1 value of the BERT model is increased by 9.1%. At the same time, whether word2vec or BERT is used to vectorize the input, adding an attention mechanism to the model can improve the efficiency of relation extraction, and the F1 value is increased by 5.1% and 3.3%, respectively.

In this experiment the corpus is relatively small in scale, however, the F1 value of the BERT-BiGRU-Attention model still reaches 85.2%. By using BERT to vectorize the input sentences from network text and integrating the attention mechanism, the F1 value of BERT-BiGRU-Attention model is improved by 14.2% compared with the word2vec-BiGRU model. The results show that compared with other baseline models, the performance of BERT-BiGRU-Attention model is better in the task of spatial relationship extraction in geographic entities.

5. Conclusion

In this paper, a classification system is designed for the spatial relation between geographic entities, and a Chinese annotated corpus corresponding to the system is constructed using crawler technology. Aiming at the problem that traditional methods cannot fully express the semantic feature information of texts in geographic field, a new relation extraction method based on BERT-BiGRU-Attention model is proposed. This method uses BERT to construct sentence embeddings according to the semantic features contained in the text, and extracts high-dimensional feature information in the text with BiGRU, then captures the dependencies between words in the sentence using attention mechanism, and finally classifies the output of the previous layers with softmax classifier and completes the spatial relationship extraction task of geographic entities.

The results of experiment show that in the task of spatial relation extraction of geographic entities, the model proposed in this paper has better performance, and compared with other baseline models the accuracy, recall value and F1 value are improved. The future work will integrate more text data resources, expand the corpus of spatial relations of geographic entities, and further improve efficiency of spatial relation extraction for geographic entities.

Acknowledgments
This research is partially supported by the achievement (project No.2022C61540).

References

[1] Wang H B, Wang J B, Huang Z C, Niu Y Y and Wu S 2018 *Geom. Sci. Technol.* **35** 633-36+42.
[2] Yu L, Lu F and Zhang H C 2015 *Geo-info. Sci.* **17** 127-34.
[3] Wu W Y, Chen Y F, Xu J A and Zhang Y J 2018 *Comp. Mod.* **8** 21-7.
[4] Xie B H, Li Y and Zhao H Y 2022 *Chi. Info. Proc.* **36** 49-58.
[5] Wang B, Guo J Y, Xian Y T, Wang H B and Yu Z T 2019 *Patt. Recog. Artif. Intell.* **32** 133-43.
[6] Xiang Y, Feng J, Xia P P and Lu J M 2020 *Comp. Sci.* **47** 131-8.
[7] Tan Y B, Hou M F, Zhang Z J, Li X L, Cheng P G and Zhang Z Z 2021 *Geo. Geo-info. Sci.* **37** 16-22.
[8] Liu C Y, Sun W B, Chao W H and Che W X 2013 *Inter. Conf. on Adv. Data Mining and Appl.* vol **8347** 231-42.
[9] Zhou P, Shi W, Qi Z Y, Li B C, Hao H W and Xu B 2016 *Conf. Proc. of the 54th Annual Meeting of the Association for Computational Linguistics*.
[10] Devlin J, Chang M W, Lee K and Toutanova K 2019 *CoRR* abs/1810.04805.
[11] Chen X L, Tang L Y, Hu Y, Jiang F, Peng W and Feng X C 2021 *Geo-info. Sci.* **23** 1208-20.
[12] Mikolov T, Chen K, Corrado G and Dean J 2013 *arXiv preprint arXiv:1301.3781*.
[13] Peters M, Neumann M, Iyyer M, Gardner M, Clark C, Lee K and Zettlemoyer L 2018 *CoRR* abs/1802.05365.
[14] Cui Y M, Che W X, Liu T, Qin B and Yang Z Q 2019 *CoRR* abs/1906.08101.
[15] Chung J Y, Gulcehre C, Cho K H and Bengio Y 2014 *CoRR* abs/1412.3555.
[16] Hochreiter S and Schmidhuber J 1997 *Neural Computation* **9** 1735-80.
[17] Vaswani A, Shazeer N, Parmar N, Uszkoreit J, Jones L, Gomez A, Kaiser L and Polosukhin I 2017 *CoRR* abs/1706.03762.
[18] Zhang X Y, Zhang C J and Zhu S N 2012 *Acta. Geod. Cartogr. Sin.* **41** 468-74.

AUTHOR INDEX